Revision
points for passing

1 Check your syllabus

Make sure that your teacher has not missed anything out from the syllabus and that your notes are not incomplete because you were away and have not copied them up. For effective revision you need a complete set of notes. Set about checking your notes against a recent copy of the syllabus which may be obtained from the addresses in the acknowledgements. Use a friend's notes or a textbook to fill in any of the gaps.

2 Use the book and your notes together

Use this book as a tool with your notes. The book's aim is to help you pass the exam by giving you practice in the way questions should be answered. Although notes are included in each chapter, these are only brief and are intended only to jog your memory about the topic. You will need to use this book in conjunction with your notes and the textbook you use.

3 Revision tips

Pace yourself when revising. Most people make the mistake of leaving their revision to the last couple of weeks before the exam. It is a good idea to prepare yourself a realistic revision timetable and try to stick to it. The important thing is to develop a routine.

4 Short bursts

Short sessions of revision are much more effective than a long stretch; try to do your revision in a quiet room away from any distractions.

5 Ask yourself questions

Be active in your revision. Ask yourself questions all the time – for example, 'Do I know how to perform the experiment to verify Hooke's Law for a spring?' You can do this in any spare moment. You could also ask friends or relatives to help you.

6 Exam question practice

Study the worked examples, since these have been designed to help you build up your knowledge and give advice about how questions should be answered.
Use this book by trying the examination questions at your particular tier and then checking your answer against the supplied one. Don't worry that your answer is different sometimes, because some questions can have a wide range of possible answers and the one supplied is only one of them.

7 Understanding better

People remember more if they write things down. You can do this, but try to avoid copying out your notes again.

How
to tackle the questions

Topic guide

1   How to tackle different types of examination questions
    [Calculations; Multiple choice questions; Short answer
    or structured questions; Free response type questions;
    Practical coursework]
2   A few important points
3   Some maths help
    [Using mathematical equations; Graphs; Significant
    figures; Powers of ten; Conversion of units; Solidus
    and negative index notation]

1   How to tackle different types of examination questions

Calculations

You will not obtain full marks for a calculation unless you start by stating the physical principles involved and show the steps by which you arrive at the answer. You will lose marks in an examination if your answers to calculations consist of numbers without any indication of the reasoning by which you arrive at them. Remember it is a physics examination and not an arithmetic examination! The physics of the question must be clearly stated. Unless the answer is a ratio it will have a unit, so you *must always remember to state the unit*.

You may find it helpful before starting a calculation to ask yourself:

1   What am I being asked to calculate?
2   Which principle, law or formula needs to be used?
3   What units am I going to use?

A further hint worth remembering is that an apparently complex problem involving many items of data may often be clarified using a simple diagram.

A numerical answer must not have more significant figures than any number used in the calculation (see section 3 (page viii): 'Significant figures').

Example of a calculation

A uniform plank of wood 3 m long is pivoted at its mid-point and used as a see-saw. Jean, who weighs 400 N, sits on one end. Where must John, who weighs 600 N, sit if the see-saw is to balance?

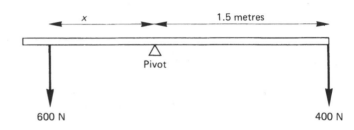

To answer this question first draw a diagram showing the plank and the forces acting on it.

All the relevant information is shown on the diagram. We need to calculate the distance x metres so that the see-saw is balanced. The physical principle must first be stated, i.e. for a body in equilibrium:

Anticlockwise moment about a point = clockwise moment about the same point.

We now use this equation. Taking moments about the pivot

$$(600 \text{ N} \times x) = (400 \text{ N} \times 1.5 \text{ m})$$

$$x = \frac{400 \text{ N} \times 1.5 \text{ m}}{600 \text{ N}}$$

$$x = 1 \text{ m}$$

John must sit 1 m from the mid-point on the opposite side to Jean.

Always check your answer at the end to ensure that it is physically reasonable. Had you got an answer to the above question of 3 m, then you should at once realise that this is unreasonable (John would be off the end of the see-saw!), and look back to discover where you have made a mistake. *And don't forget to put a unit after the answer.*

Multiple choice questions

Multiple choice questions are no longer part of most GCSE physics examinations. Check with a recent copy of the examination syllabuses to see if they form part of your examination. If all the questions have to be attempted, do not waste valuable time reading through the paper before you start. Start at question 1 and work steadily through the paper. If you come to a difficult question which you can't answer, miss it out and come back to it at the end. If you spend a lot of time thinking about questions which you don't find particularly easy, you may find yourself short of time to do some easier questions which are at the end of the paper.

When possible it is helpful to try to answer the question without first looking at the responses. This can reduce the chance of 'jumping to the wrong conclusions'. When you have found the response you think is correct, if time allows, check the other responses to see why they are incorrect.

When you have worked through the paper in this way, return to the questions you found difficult the first time through, but leave until the very end any questions about which you have little or no idea. In the closing minutes of the examination, if you still have not answered all the questions, sprint to the finish by guessing, if necessary.

As examination boards do not deduct marks for wrong answers it is essential that you answer every question. If you are not sure which of the choices is the correct answer try to eliminate one or two of the alternative answers, and, if necessary, guess which of the remaining alternatives is correct.

If you have to alter an answer, ensure that the previous one has been completely rubbed out. Only one answer for each question must appear on the answer sheet.

Example of a multiple choice question

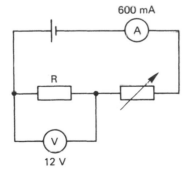

600 mA

R

V

12 V

In the circuit shown in the diagram the high-resistance voltmeter reads 12.0 V, and the ammeter reads 600 mA. The value of the resistance marked R is

A 7200 Ohm **B** 72 Ohm **C** 20 Ohm **D** 7.2 Ohm
E 0.02 Ohm

Solution

[We use the equation in chapter 12, section 1, which defines resistance, namely

$$R = \frac{\text{potential difference across object (volts)}}{\text{current through object (amperes)}}$$

The ammeter reading is 600 mA = 0.6 A

$$R = \frac{12\,\text{V}}{0.6\,\text{A}} = 20\,\text{Ohm.}]$$

Answer **C**

Short answer or structured questions

In this type of question most examining boards leave space for the answer to be written on the question paper. The amount of space left will be a guide to the length of answer that is required, as will the number of marks indicated by the question. It doesn't follow that if you don't fill all the space up you haven't answered the question correctly. But if five lines are left for the answer, and you have only written on one line, or if you can't possibly get your answer on five lines, then you certainly ought to have another think about the answer.

Example of a short answer or structured question

A system of pulleys is used to raise a load of 9 N through 2 m. The effort of 2 N needed to do this moves through 12 m. What is

(i) the potential energy gained by the load
(ii) the work done by the effort
(iii) the efficiency of the system?

Solution

(i) Work done on load = force × distance = 9 N × 2 m = 18 J
Potential energy gained by load = 18 J
(ii) Work done by effort = force × distance = 2 N × 12 m = 24 J
(iii) Efficiency $= \dfrac{\text{work out}}{\text{work in}} = \dfrac{18\,\text{J}}{24\,\text{J}} = \dfrac{3}{4}$
$= 0.75$ or 75%

(If you have problems understanding the answer, refer to chapter 3.)

Free response type questions

These are likely to be met in the optional papers taken by those seeking the highest grades. Make sure you attempt the full number of questions you have to answer.

If you are asked to describe an experiment you must normally draw a diagram of the apparatus or the relevant circuit diagram. A labelled diagram saves time, as information shown on the diagram need not be repeated in words. You should state clearly exactly what readings are taken, remembering that readings should be repeated as a check whenever possible. Finally mention any precautions necessary to obtain an accurate result.

Avoid saying vaguely at the end 'the result is calculated from the readings'. You must state exactly how it is calculated.

If the answer involves drawing a graph, be sure to label the axes and choose a scale so that the graph covers most of the paper.

Use the indicated mark scheme as a guide to the length of answer required.

Example of a free response type question

(a) Describe an experiment you would perform in order to measure the average power a girl can develop over a period of a few seconds.   (8 marks)

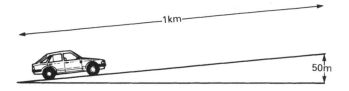

(b) A car of weight 6000 N climbs a hill 1 km long which raises the car a vertical distance of 50 m. The driver maintains a constant speed of 25 m/s while he travels up the hill.
  (i) How long does it take him to reach the top of the hill?   (2 marks)
  (ii) What is the gain in gravitational potential energy of the car when it reaches the top of the hill?   (4 marks)
  (iii) Neglecting frictional forces, what is the power developed by the car as it climbs the hill?   (2 marks)

(c) Outline the main energy changes which take place as the car climbs the hill.   (4 marks)

Solution

(a) Leg muscle power may be measured by running up a flight of steps and measuring the time it takes with a stopwatch. Start the watch as the girl starts to climb the stairs and stop it when she reaches the top. Suppose that the average of three runs is 6 s. Measure the total vertical height of the stairs by using a ruler to measure the average height of each step and multiply by the number of steps to get the total vertical height. If this height is 5 m and the weight of the person, found by using bathroom scales, is 720 N, then

Work done in 6 s = 720 N × 5 m = 3600 J

$$\text{Power} = \frac{\text{work done}}{\text{time taken}} \quad \text{[see chapter 4, section 4]}$$

$$\frac{3600\,\text{J}}{6\,\text{s}} = 600\,\text{W}$$

[The important thing in answering this question is to state clearly what must be measured: (1) the weight of the person, using bathroom scales; (2) the vertical height of the staircase, using a metre rule; (3) the time to climb the stairs, using a clock, taking the average of a number of runs. You must also show how to calculate her power.]

(b)   (i) The car travels 1 km (1000 m) at 25 m/s; hence

$$\text{Time taken} = \frac{1000\,\text{m}}{25\,\text{m/s}} = 40\,\text{s}$$

  (ii) Gravitational potential energy gained
        = weight × vertical height raised
        = 6000 N × 50 m = 300 000 J

  (iii) Power $= \dfrac{\text{work done}}{\text{time taken}}$

          $= \dfrac{300\,000\,\text{J}}{40\,\text{s}} = 7500\,\text{W}$

(c) The chemical energy of the fuel becomes heat energy in the engine cylinders, which eventually becomes gravitational potential energy of the care together with heat energy in the atmosphere as a result of frictional forces.

Practical coursework

Some points to bear in mind are:

1   Follow any instructions carefully step by step. Read each instruction twice to ensure that you have understood it.

2   Take great care in reading an instrument. Record each reading accurately, carefully, systematically and clearly. When reading a horizontal scale with a pointer above it, make sure your eye is vertically above the pointer as you take the reading. Don't forget to put the **unit**. Whenever you read a scale, make sure you have carefully thought through the value of each of the small divisions on the scale. Remember to correct for zero errors (for example, make sure a spring balance reads zero before you hang a weight from it). When reading multiscale instruments, be careful to read the correct scale.

3   Present your readings clearly. Tabulate them neatly (don't forget the **unit**) and present them graphically whenever possible. Choose a suitable scale for the graph (i.e. 1 large square = 1 unit) and don't forget to label the axes. Take great care over plotting the points. Don't join them with short straight lines but draw the best smooth curve or straight line through them.

4   Repeat all the readings and take the average. Be particularly careful to check any unexpected reading.

5   If you have to take the temperature of a liquid which is being heated, you must stir the liquid well before taking the temperature.

6   Be aware of the accuracy of any readings. For example, when reading a ruler marked in centimetres, you can only read it to the nearest 0.1 cm.

7   It is important to select the most appropriate apparatus. For example, if you need to measure a current of about 0.009 A (9 mA) and you have the choice of two ammeters, one reading 0–10 mA and the other 0–1 A, you should choose the 0–10 mA one.

8   You will need to know how to use, among other things, a measuring cylinder, a balance for measuring mass, a spring balance, a stopwatch, a thermometer, a voltmeter, an ammeter and a CRO.

9   You will be required to:
   (a) devise an investigation, select and set up suitable apparatus;
   (b) take precautions necessary for obtaining an accurate result;
   (c) show your skills at observation and measurement;
   (d) analyse and process data;
   (e) draw conclusions from your observations and readings.

10   Be prepared to criticise your own work. State your conclusion clearly, and whenever possible state a concise relationship between variables.

2   A few important points

1   Read the question carefully before you write anything. Make sure you know exactly what the question is asking.

2   Answer the question precisely as asked. Note carefully the phrase used in the question. For example, 'explain', 'define', 'state', 'derive' and 'describe' all mean different things and are meant to be taken literally.

3   Try to pace yourself during the exam. Your aim should be to complete the paper with enough time to check through at the end.

   So keep an eye on the clock and make sure you do not spend too much time on any one question. On the other hand, do make sure that you don't miss out the last parts of an easy question, because you are anxious to press on and finish the paper.

4   Set your work out neatly and clearly. Examiners are human and if they have done many hours of marking, they are likely to be far more sympathetic if your work is well set out and easy to follow.

5   Make sure you are familiar with the style of question set by your particular examining board and the length of time allowed for each question (syllabuses and past

papers may be obtained from the addresses given in the Acknowledgements).

6   Don't spend a lot of time on a multiple choice question or a short answer question with which you are having difficulty. Leave that question and come back to it later if you have time.

3   Some maths help

Using mathematical equations

Equations help us to relate certain quantities. The usefulness of the equation is extended if it can be rearranged. There are three helpful rules for rearranging an equation.

The plus/minus rule

A symbol or number may be moved from one side of an equation to the other provided the sign in front of the symbol or number is changed. That is, a 'plus' item on one side becomes a 'minus' item on the other side and vice versa. For example:

$X = Y + 30$

$X - Y = 30$ or $X - 30 = Y$

The diagonal rule

An item may be moved diagonally across an equals sign. For example, if

$$\frac{A}{B} \diagdown = \diagup \frac{C}{D}$$

then the arrows show possible moves, so that

$$\frac{A}{B \times C} = \frac{1}{D} \text{ or } \frac{A}{C} = \frac{B}{D} \text{ or } \frac{D}{B} = \frac{C}{A}$$

The 'do unto others' rule

Whatever is done on one side of the equation must be done to the other side. For example:

if $A = B$ then $A^2 = B^2$ (both sides have been squared)

if $C = \dfrac{1}{D}$ then $\dfrac{1}{C} = D$ (both sides have been inverted)

But if $\dfrac{1}{R} = \dfrac{1}{R_1} + \dfrac{1}{R_2}$ then $R \neq R_1 + R_2$

($\neq$ means 'does not equal')
The **whole** of each side must be inverted, i.e.

$$R = \frac{1}{\dfrac{1}{R_1} + \dfrac{1}{R_2}} = \frac{1}{\dfrac{R_2 + R_1}{R_1 R_2}} = \frac{R_1 R_2}{R_1 + R_2}$$

Graphs

Choose easy scales, such as one large square to represent 1, 2 or 5 units (or multiples of 10 of these numbers). Avoid scales where one large square is 3, 4, 6, 7, 8 or 9 units. If each large square is, say, 3 units, then a small square is 0.3 units and this makes the plotting of the graph much more difficult. Usually the spread of readings along the two axes should be about the same and they should cover most of the page. For example, on the graph shown below on the horizontal scale 2 small squares represent 1 second, and on the vertical scale 1 small square represents 2 m/s.

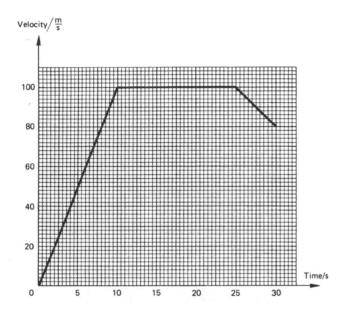

It is important to remember to label the axes and to put a title at the top of the page. In physics most relationships are either straight lines or smooth curves, so it is not correct to join adjacent points together by short straight lines. You must decide whether the relationship is a straight line or a curve, and then draw either the best straight line through the points or a smooth curve.

Significant figures

A useful rule to remember is that you must never give more significant figures in the answer than the number of significant figures given in the least precise piece of data. For example, if a cube of side 2.0 cm has a mass of 71.213 g, then we may calculate the density by using the equation

$$\text{Density} = \frac{\text{mass}}{\text{volume}} \quad [\text{see chapter 1 section 3}]$$

$$= \frac{71.213 \text{ g}}{8.0 \text{ cm}^3} = 8.901\ 625 \text{ g/cm}^3$$

according to my calculator! But the side was only given to two significant figures and we may not give the answer to more than two significant figures, i.e. 8.9 g/cm^3 (we certainly do not know the density to the nearest millionth of a g/cm^3!). The answer should be written 8.9 g/cm^3.

A problem arises when the examiners use whole numbers without a decimal point and expect you to take them as exact. In the above calculation, if the side were given as 2 cm we may only strictly give the answer to one significant figure, i.e. 9 g/cm^3. If you are in doubt in such a calculation, work the calculation to, say, three significant figures and then give the answer to one significant figure.

Powers of ten

100 may be written as 10^2 $1000 = 10^3$

$\dfrac{1}{10}$ may be written as 10^{-1} $\dfrac{1}{100} = 10^{-2}$ $\dfrac{1}{1000} = 10^{-3}$

$\dfrac{1}{100\ 000} = 10^{-5}$

Conversion of units

$1 \text{ m}^2 = 100 \text{ cm} \times 100 \text{ cm} = 10^4 \text{ cm}^2$
so $5 \text{ N/cm}^2 = 5 \times 10^4 \text{ N/m}^2 = 50 \text{ kN/m}^2$
$1 \text{ m}^3 = 100 \text{ cm} \times 100 \text{ cm} \times 100 \text{ cm} = 10^6 \text{ cm}^3$
so $1 \text{ g/cm}^3 = 10^6 \text{ g/m}^3 = 1000 \text{ kg/m}^3$
Therefore to convert g/cm^3 to kg/m^3 we multiply by 1000.

Solidus and negative index notation

Make sure you are familiar with the notation used by your examination board. Most boards use the solidus for GCSE examinations, i.e. m/s and kg/m^3. But m/s may be written ms^{-1} and kg/m^3 may be written kgm^{-3}.

Exam Board Addresses

For syllabuses and past papers contact the Publications office at the following addresses:

Northern Examinations and Assessment Board (NEAB)
12 Harter Street
MANCHESTER M1 6HL
Tel 0161 952 1170
Fax 0161 273 7572

Welsh Joint Education Committee (WJEC)
245 Western Avenue
Llandaff
CARDIFF CF5 2YX
Tel 01222 265112
Fax 01222 575987
(There is a shop at the above address and the hours of opening are 9.30am to 5.30pm, Monday to Friday.)

Midland Examining Group (MEG)
1 Hills Road
CAMBRIDGE CB1 2EU
Tel 0223 61111
Fax 0223 460278

University of London Examinations and Assessment Council (ULEAC)
ULEAC Publications
River Park
Billet Lane
BERKHAMSTED
Herts HP4 1EL
Tel 01442 876701
Fax 01442 876809

Southern Examining Group (SEG)
Publications Department
Stag Hill House
GUILDFORD
Surrey GU2 5XJ
Tel 01483 302302
Fax 01483 300152

Northern Ireland Council for the Curriculum, Examinations and Assessment (NICCEA)
29 Clarendon Road
BELFAST BT1 3BG
Tel 01232 261200
Fax 01232 261234

Scottish Examinations Board (SEB) [Use this address for the ordering of syllabuses]
Ironmills Road
DALKEITH
Midlothian EH22 1LE
Tel 0131 636 6601

or recent papers from the SEB's agent
Robert Gibson and Sons Ltd
17 Fitzroy Place
GLASGOW G3 75F
Tel 0141 248 5674

Remember to check your syllabus number with your teacher!

Contents

Topic

1 Density, pressure and Hooke's Law — 1

2 Motion, scalars and vectors — 13

3 Energy, work and power — 29

4 Moments and machines — 40

5 Kinetic theory and gas laws — 48

6 Heat and change of state — 59

7 Transfer of heat — 66

8 Sources of energy — 75

9 Optics — 89

10 Wave motion — 104

11 Electrostatics — 124

12 Electrical circuits — 132

13 Electronics — 147

14 Magnetism and electromagnetism — 163

15 Radioactivity — 178

16 Spectra and photons — 194

17 Communications — 201

18 The Earth, atmosphere and weather — 210

19 Space physics — 222

20 Fluids — 233

Index — 244

John Keighley and Stephen Doyle

Physics GCSE

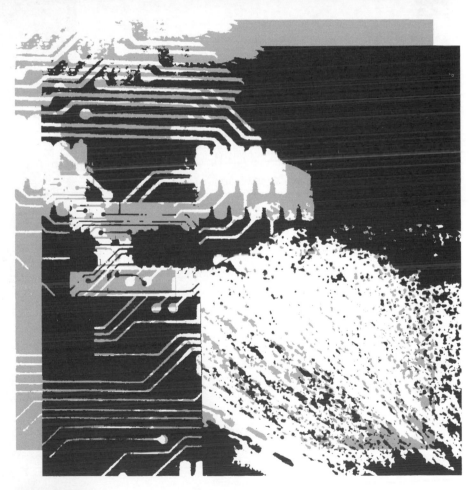

MACMILLAN

Acknowledgements

The authors and publishers thank the following for permission to reproduce copyright material:
Edexcel Foundation, London Examinations, Midland Examining Group, Northern Examinations and Assessment Board, Northern Ireland Council for the Curriculum Examinations and Assessment, Southern Examining Group, and the Welsh Joint Education Committee for past examination questions; University of Bath, for Figure 17.11, from J. Allen, *Telecommunications, University of Bath Macmillan Science 16–19 Project (1990)*.

Every effort has been made to trace the copyright holders but if any have been inadvertently overlooked the publishers will be pleased to make the necessary arrangement at the first opportunity.

First edition 1986 (reprinted once)
Second edition 1987
Third edition 1990 (reprinted six times)
Fourth edition 1998

Published by
MACMILLAN PRESS LTD
Houndmills, Basingstoke, Hampshire RG21 6XS
and London
Companies and representatives
throughout the world

ISBN 0–333–68032–4

A catalogue record for this book is available from the British Library.

This book is printed on paper suitable for recycling and made from fully managed and sustained forest sources.

10   9   8   7   6   5   4   3   2   1
07 06 05 04 03 02 01 00 99 98

Typeset by Wearset, Boldon, Tyne and Wear

Printed in Malaysia

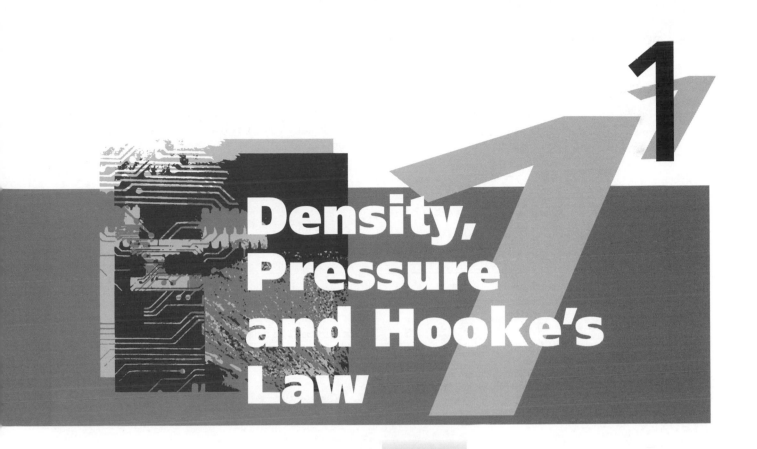

Density, Pressure and Hooke's Law

1

WJEC & NEAB	ULEAC Syll A	ULEAC Syll B	MEG	Topic	MEG Salters'	MEG Nuffield	SEG	NICCEA
✓	✓	✓	✓	**SI units**	✓	✓	✓	✓
✓	✓	✓	✓	**Weight and mass**	✓	✓	✓	✓
✓			✓	**Density**	✓	✓	✓	✓
✓	✓	✓	✓	**Pressure**	✓	✓	✓	✓
✓	✓	✓	✓	**Hooke's Law**		✓	✓	✓
✓	✓			**Floating and sinking**				✓

1 SI units

These are the basic units in physics from which the other units you come across may be defined. A table of them is shown in table 1.1.

Table 1.1

Physical quantity	Name of unit	Symbol of unit
Length	metre	m
Mass	kilogram	kg
Time	second	s
Current	ampere	A
Temperature	kelvin	K

Prefixes

Some of the more commonly used prefixes are given in table 1.2.

Table 1.2

Prefix	Sub-multiple	Symbol
centi-	10^{-2}	c
milli-	10^{-3}	m
micro-	10^{-6}	μ
nano-	10^{-9}	n
pico-	10^{-12}	p
kilo-	10^{3}	k
mega-	10^{6}	M
giga-	10^{9}	G

2 Weight and mass

The mass of a body (measured in kg) is constant wherever the body is situated in the universe. The weight of a body (measured in N) is the pull of the force of gravity on the body and this does depend on where the body is situated in the Universe.

The Earth's gravitational field strength is 10 N/kg, so the weight of a mass of 1 kg is 10 N. The pull of the Earth (weight) of an apple of average size is about 1 N. In outer space it is possible for the gravitational field strength to be zero and this gives rise to weightlessness. Weight may be calculated from the mass using the following formula:

Weight (N) = Mass (kg) × gravitational field strength (N/kg)

A beam balance compares masses, a spring balance measures weight.

3 Density

Density is obtained using the formula:

$$\text{Density} = \frac{\text{mass}}{\text{volume}}$$

If the mass is in kg and the volume in m^3, then the density is in kg/m^3.

Density may also be measured in the units of g/cm^3.

Density may be determined by measuring the mass of a measured volume.

4 Pressure

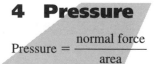

$$\text{Pressure} = \frac{\text{normal force}}{\text{area}}$$

If the force is measured in newtons and the area in $(\text{metre})^2$, then the pressure is in Pascals (Pa). $1\ Pa = 1\ N/m^2$. Pressure may also be measured in N/cm^2.

Pressure in fluids

The pressure due to a column of liquid (a) acts equally in all directions, (b) depends on the depth and the density of the liquid. It may be calculated using the equation

Pressure (Pa) = 10 (N/kg) × depth (m) × density (kg/m^3)

which may also be written as

Pressure (PA) = $\rho g h$

where ρ is the density in kg/m^3, g the Earth's gravitational field strength (10 N/kg) and h the depth in metres.

Pressure may be measured using a U-tube manometer or a Bourdon gauge. When any part of a confined liquid is subject to a pressure, the pressure is transmitted equally to all parts of the vessel containing the liquid. This principle, and the fact that liquids are virtually incompressible, are made use of in hydraulic machines. Such machines are useful force multipliers. Referring to the first illustration, the pressure on the

small piston is $\dfrac{20\ N}{10\ cm^2}$ or 2 N/cm^2. This pressure is

transmitted to the large piston and the force on it is

$$2\ \frac{N}{cm^2} \times 100\ cm^2 = 200\ N$$

The atmosphere above us exerts a pressure known as the atmospheric pressure. If the air is withdrawn from a metal can, the force due to atmospheric pressure acting on the outside of the can will collapse the can. It is because of the

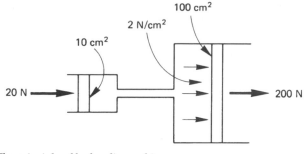

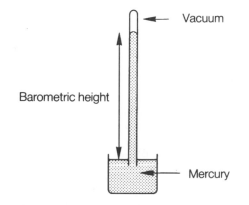

The principle of hydraulic machines.

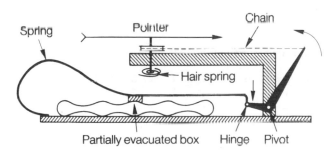

A mercury barometer. The atmosphere exerts a force on the surface of the mercury in the trough and this pushes the mercury up the tube. The atmospheric pressure balances the pressure due to the weight of the mercury column.

An aneroid barometer. When the atmospheric pressure increases, the centre of the partially evacuated box moves inwards and this small movement is magnified by a system of levers. The chain attached to the end lever moves the pointer. The large spring prevents the box from collapsing.

decrease in atmospheric pressure with height that aircraft cabins have to be pressurised. The atmospheric pressure may be measured using a mercury barometer or an aneroid barometer.

The pressure of a gas on a surface is caused by moving molecules colliding with the surface.

5 Hooke's Law

Hooke's Law states that provided loads are not used which would cause a spring or wire to exceed its *elastic limit*, the extension is proportional to the applied load.

Until the elastic limit is reached a spring (or wire) returns to its original length if the load is removed. (See example 13.)

6 Floating and sinking

When an object floats, there are two forces acting; the weight acting vertically down and an upward force called the upthrust provided by the liquid. The size of the upthrust depends on the amount of the object in the water. If an object floats it means that the upthrust and the weight balance. If a floating object is pushed down, an upward force is felt because the upthrust is now bigger than the weight. If an object is just pushed below the surface of the water and the upthrust is not big enough to equal the weight, then the object will sink.

Objects sink because their weight is greater than the upthrust which can be produced by the water.

Metal sinks and boats are made of metal and still float. The reason for this is that the metal is shaped so that it contains a lot of air. This means that they are able to displace (push aside) more water so their upthrust is greater and can now balance the weight.

If the object is solid, then if it has a density greater than that of water, it will sink and if it has a density less than water it will float.

The same situation applies to gases. Air expands when heated and rises through the cooler, denser air and the movement of hot air is called a convection current.

Worked examples

Example 1

A mass of 1 kg is secured to the hook of a spring balance calibrated on the Earth. The spring balance reading is observed when it is freely suspended at rest just above the Earth's surface, secondly inside a spaceship orbiting round the Earth, and finally at rest on the Moon's surface.

If the acceleration due to free fall on the Earth is 10 m/s^2 and acceleration due to free fall on the Moon is 1.6 m/s^2, the spring balance readings, in N, would be (table 1.3):

Table 1.3

	Point above Earth's surface	Inside a spaceship	On the Moon
A	1.0	0	0.16
B	1.0	0.84	0.16
C	10.0	0	1.6
D	10.0	0.84	0.16
E	10.0	11.6	1.6

Solution

[Weight on Earth = 1 kg × 10 N/kg = 10 N
Weight on Moon = 1 kg × 1.6 N/kg = 1.6 N
Inside the spaceship the weight is zero.]

Answer **C**

Example 2

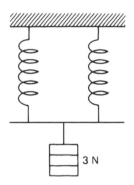

3 N

A student applies a force of 6 N to a helical spring and it extends by 12 cm. He then hangs the spring in parallel with an identical spring and attaches a load of 3 N as shown. The resulting extension of the system, in cm, will be:

A 3 **B** 4 **C** 6 **D** 12 **E** 24   (AEB)

Solution

[Force acting on end of each spring = 1.5 N (the total upward force must equal the total downward force)
6 N extends the spring by 12 cm. Assuming Hooke's Law applies (section 5),

1.5 N extends the spring by $\left(\dfrac{12}{6} \times 1.5\right)$ cm = 3 cm.]

Answer **A**

Example 3

A U-tube containing mercury is used as a manometer to measure the pressure of gas in a container. When the manometer has been connected, and the tap opened, the mercury in the U-tube settles as shown in the diagram.

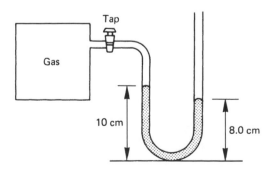

The pressure of the atmosphere is equal to that exerted by a column of mercury of length 76 cm. The pressure of the gas in the container is equal to that exerted by a column of mercury of length

A 2.0 cm **B** 58 cm **C** 74 cm **D** 78 cm **E** 94 cm

Solution

[The atmospheric pressure exerted on the open limb of the U-tube is greater than the gas pressure by 2 cm of mercury. The gas pressure is therefore 76 cm − 2 cm = 74 cm.]

Answer **C**

Example 4

The diagrams show three steps in an experiment to measure the density of a metal rod.

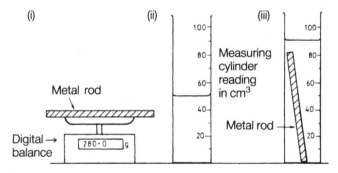

What is the density of the metal?

A 0.3 g/cm^3 **B** 3.1 g/cm^3 **C** 5.6 g/cm^3 **D** 7.0 g/cm^3
(LEAG)

Solution

[Mass of rod = 280 g
Volume of rod = 90 cm^3 − 50 cm^3 = 40 cm^3

$$\text{Density} = \frac{\text{mass}}{\text{volume}} = \frac{280 \text{ g}}{40 \text{ cm}^3} = 7.0 \text{ g/cm}^3.]$$

Answer **D**

Example 5

The extension of a piece of copper wire was measured for various loads placed on the end of the wire. A graph was plotted of extension against load. Which graph shows that the wire was loaded beyond its elastic limit?

(NEAB, Intermediate Tier)

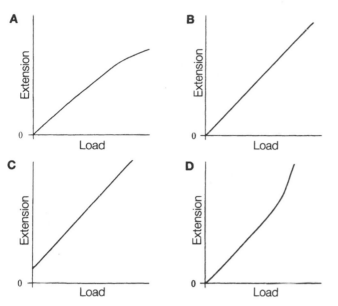

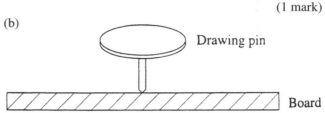

(iii) Explain how the results of this experiment affect the design of the wall of a dam.

(1 mark)

(b)

Drawing pin

Board

A flat-head drawing pin has an area of 1.2 cm^2 and the point of the pin has an area of 0.0002 cm^2. A force of 10 N is needed to press the pin into the board.

(i) Calculate the pressure produced by the pin on the board in N/cm^2. (2 marks)

(ii) Explain why the pressure on the flat-head of the pin is very much less than the pressure calculated in (b)(i). (1 mark)

(WJEC, Jun 95, Intermediate Tier, Q9)

Solution

[Before the elastic limit is reached, the graph is a straight line through the origin. When the elastic limit is passed, the extension increases more rapidly with an increase in the load. See graph on page 9 but note that in that graph the load axis is vertical.]

Answer **D**

Example 6

(a)

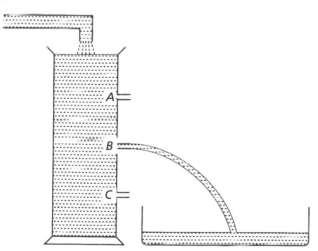

The diagram shows a tall cylinder which is kept filled with water. The path of the water escaping through hole *B* is also shown.

(i) Add to the diagram the path followed by the water escaping through holes *A* and *C*.

(1 mark)

(ii) What do the results of this experiment tell you about the pressure in a liquid?

(1 mark)

Solution

(a)   (i)

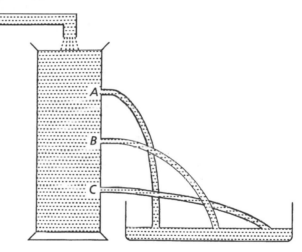

(ii) Pressure in a liquid increases with increasing depth.

(iii) Since the water pressure at the bottom of the dam is greater, it needs to be thicker at the bottom than at the top.

(b)   (i) Pressure (Pa) $= \dfrac{\text{Force (N)}}{\text{Area (cm}^2)}$

$= \dfrac{10 \text{ N}}{0.0002 \text{ cm}^2} = 50\,000 \text{ N/cm}^2$

(ii) The flat-head has a much larger area so although the force exerted is the same, the pressure is much less on the head than on the point.

Example 7

The diagram shows Jo using a hydraulic lift to raise her car.

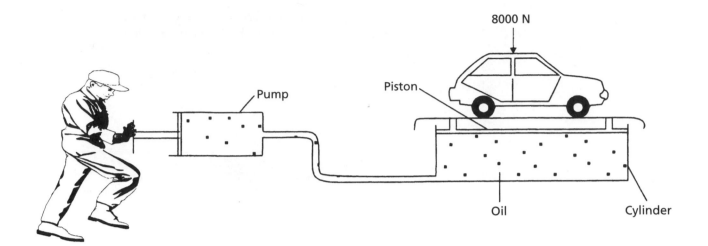

As she pumps oil into the cylinder, the piston moves up with the car.

(a) The car has a weight of 8000 N. When the piston exerts an upward force of 8400 N, the car moves up at a steady speed.
   (i) How large is the friction force between the piston and the cylinder? (1 mark)
   (ii) Draw an arrow on the diagram to show the direction of the friction force. (1 mark)

(b) The area of the piston is 210 cm².
   (i) State the formula linking pressure, force and area. (1 mark)
   (ii) Calculate the oil needed to raise the car at steady speed. (2 marks)

(c) Jo cannot raise the car without the hydraulic lift. Explain how the lift helps (2 marks)
   (MEG Salters', Jun 95, Intermediate Tier, Q3)

Solution

(a) (i) Friction force = 8400 N − 8000 N = 400 N

   (ii) *A* on the diagram, an arrow should be drawn pointing in the opposite direction to the push exerted by Jo.

(b) (i) $\text{Pressure (N/cm}^2) = \dfrac{\text{Force (N)}}{\text{Area (cm}^2)}$

   (ii) $\text{Pressure} = \dfrac{8400 \text{ N}}{210 \text{ cm}^2} = 40 \text{ N/cm}^2$

(c) A small force on a small area is able to exert a large force on a large area. Hence the person can effectively lift the car.

Example 8

A fitness enthusiast is using chest-expanders. The graph shows how the extension of the chest-expanders changes with the force pulling on them.

(a) What force would be required to stretch the chest expanders by 0.2 m? (2 marks)

(b) There are four springs in the chest-expanders. What force would be required to stretch just one of the

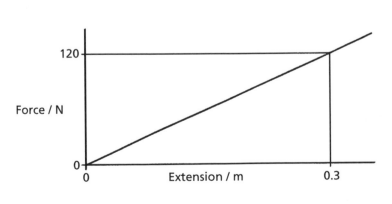

springs by 0.3 m? (1 mark)

(c) Use the graph or another method to calculate the work the enthusiast does stretching the expanders by 0.3 m (3 marks)

(d) An energy transfer takes place when the enthusiast stretches the expanders.
   (i) Where does the energy come from? (1 mark)
   (ii) Where does the energy go to? (1 mark)

(e) His burly friend stretches the expanders by 0.6 m.
   (i) What force is required to do this? (1 mark)
   (ii) Explain why he does four times as much work. (1 mark)
      (MEG, Jun 95, Nuffield Intermediate Tier, Q10)

Solution

(a) 120 N extends expanders by 0.3 m so 40 N would extend expanders by 0.1 m. Hence, 80 N would extend expanders by 0.2 m.

(b) One spring would need one quarter of the force. Force required = 20 N

(c) Work done is the area under the force-extension graph (i.e. the area of a triangle)
   $= 0.5 \times 120 \text{ N} \times 0.3 \text{ m} = 18 \text{ J}$

(d) (i) Chemical energy in muscles.
   (ii) Elastic potential energy stored in the springs.

(e) (i) 240 N.
   (ii) Work done $= 0.5 \times 240 \text{ N} \times 0.6 \text{ m} = 72 \text{ J}$ (which is four times the answer for part (c)).

Example 9

Describe, briefly, an experiment to find the density of a solid object, stating clearly:

(a) What measurements you would make; (1 mark)
(b) how you would make them; (2 marks)
(c) how you would use them to calculate the density. (1 mark)
   (WJEC, Jun 95, Intermediate Tier, Q18)

Solution

(a) The volume in cm^3 and the mass in grams.

(b) The mass is found by placing the object on a top-pan balance.
   The volume of the object is found by displacement of water. A measuring cylinder is partly filled with water and the level is noted. The object is then immersed in the water and the new reading noted. The difference between the reading will equal the volume of the object.

(c) The density is found by placing the numbers into the formula:

$$\text{Density} = \frac{\text{mass (g)}}{\text{volume (cm}^3)}$$

This gives the density in the units of g/cm^3.

Example 10

The drawing shows a foot-pump which may be used to pump air into car tyres. The diagram shows the internal details of the pump.

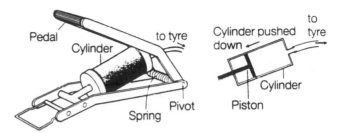

As the pedal is pressed down the cylinder is pushed and moves down the outside of the piston. (This is equivalent to pushing the piston into the cylinder.) This causes air to be forced along the connecting tube to the tyre.

(a) If the area of the piston is 25 cm^2 and the force applied to the cylinder is 800 N, show that the pressure exerted on the air by the piston is 320 kPa (that is, 3.2×10^5 Pa). (3 marks)

(b) With the type of pump shown the force exerted by the foot is not the same as the force exerted on the cylinder. Why is this? (2 marks)

(c) It is important that the area of the piston (and cylinder) is not too large or too small.
   (i) What problem would arise if the area of the piston were too large? (1 mark)
   (ii) What problem would arise if the area of the piston were too small? (1 mark)

(d) A car rests on four wheels, and each tyre is in contact with the ground over an area of 0.0075 m^2. When the pressure gauge (which reads the pressure difference between the inside and outside of the tyre) is connected to each tyre in turn, it reads 200 kPa. Calculate the mass of the car. ($g = 10$ N/kg) (4 marks)

(e) The manufacturer's handbook for this car says that the tyre pressures should be increased when it is heavily loaded. The owner thinks that this is unnecessary because the extra weight will automatically increase the pressure. What is wrong with his reasoning? (3 marks)
   (SEG, Intermediate Tier)

Solution

(a) $\text{Pressure} = \dfrac{\text{force}}{\text{area}} = \dfrac{800 \text{ N}}{25 \times 10^{-4} \text{ m}^2}$
   $= 320\,000 \text{ Pa} = 320 \text{ kPa}$

(b) The force is not applied directly to the cylinder. The force is applied via a lever.

(c) (i) The force required to operate the system would be very large, and thus difficult to apply by foot pressure.
   (ii) The amount of air going into the tyre at each depression of the foot pedal would be very small.

(d)   Pressure $= \dfrac{\text{force}}{\text{area}}$

Force = pressure × area
Force = 200 000 Pa × 0.0075 m^2 = 1500 N
Total force on 4 tyres = 4 × 1500 N = 6000 N
The Earth's gravitational field is 10 N/kg [see section 2], so mass of car is

$$\dfrac{6000 \text{ N}}{10 \text{ N/kg}} = 600 \text{ kg}$$

(e)   The owner has forgotten to take into account the fact that when the car is heavily loaded the area of tyre in contact with the ground increases. The pressure will not change much and since force = pressure × area, the increased area of contact means that the upward force has increased to balance the increased load. To ensure that the same area of tyre is in contact with the road, more air must be pumped into the tyre, thus increasing the pressure as recommended by the manufacturer.

Example 11

(a)   A block of stone measures 2 m × 2 m × 1 m. It has a mass of 8000 kg.
    (i) What is its density?   (3 marks)
    (ii) When it is standing on a bench what is the maximum pressure it can exert on the bench?   (3 marks)

(b)   A bath has some water in it and the depth of the water at the shallow end is 0.2 m. At the plug hole end it is 0.3 m. What pressure does the water exert on the plug?
(The Earth's gravitational field is 10 N/kg and the density of water is 1000 kg/m^3.)   (4 marks)

Solution

(a)   (i) Density $= \dfrac{\text{mass}}{\text{volume}}$ [section 3]

$$= \dfrac{8000 \text{ kg}}{4 \text{ m}^3} = 2000 \text{ kg/m}^3$$

    (ii) Maximum pressure occurs when the area of contact is a minimum.
Minimum area in contact with bench = 2 m^2.

$$\text{Pressure} = \dfrac{\text{force}}{\text{area}} \text{ [section 4]}$$

$$= \dfrac{80\,000 \text{ N}}{2 \text{ m}^2}$$

$$= 40\,000 \text{ Pa} = 40 \text{ kPa}$$

(b)   Pressure $= 10 \dfrac{\text{N}}{\text{kg}} \times \text{depth} \times \text{density}$ [section 4]

$$= 10 \text{ N/kg} \times 0.3 \text{ m} \times 1000 \text{ kg/m}^3$$
$$= 3000 \text{ Pa}$$

[The pressure on the plug depends on the vertical height of the water above the plug.]

Example 12

A vehicle designed for carrying heavy loads across mud has four wide low-pressure tyres, each of which is 120 cm wide. When the vehicle and its load have a combined mass of 12 000 kg each tyre flattens so that 50 cm of tyre is in contact with the mud as shown in the diagram.

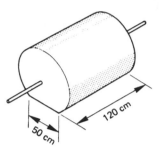

(a)   Calculate
    (i) the total area of contact of the vehicle tyres with the mud
    (ii) the pressure exerted on the mud.
(b)   A car of mass 1000 kg is unable to travel across the mud although it is much lighter than the loadcarrying vehicle. Why is this?   (6 marks)
                                                                              (L)

Solution

(a)   (i) Area = 4 × (120 × 50) cm^2 = 24 000 cm^2

    (ii) Pressure $= \dfrac{\text{force}}{\text{area}} = \dfrac{(12\,000 \times 10) \text{ N}}{24\,000 \text{ cm}^2}$

$$= 5 \text{ N/cm}^2$$

(b)   The area of car tyre in contact with the road is much less than that of the vehicle. The pressure exerted by the car tyre on the mud is therefore greater than that exerted by the vehicle, and the tyre sinks into the mud.

Example 13

(a)   Describe how you would obtain, as accurately as possible, a series of readings for the load and corresponding extension of a spiral spring.
                                                                              (6 marks)
(b)   A student obtained the following readings:

Load/N	0	1	2	3	4	5	6
Length of spring/cm	10.0	11.5	13.0	14.5	16.0	18.5	24.0

Using these results, plot a graph of load against extension and estimate the load beyond which Hooke's Law is no longer obeyed.   (7 marks)

(c) The spring is at rest with a mass of 0.2 kg on its lower end. It is then further extended by a finger exerting a vertical force of 0.5 N. Draw a diagram showing the forces acting on the mass in this position, giving the values of the forces.

(3 marks)

(d) Describe the motion of the mass when the finger is removed. Make your description as precise as possible, by giving distances. State the position where the kinetic energy of the mass will be greatest.

(4 marks)

Solution

(a) Clamp the top of the spring firmly to a support, making sure that the support is also firm and cannot move. Clamp a ruler alongside the spring and attach a horizontal pointer to the bottom of the spring in such a way that the pointer is close to the surface of the ruler (this will help to avoid a parallax error when taking the readings). Record the pointer reading. Hang a known load on the end of the spring and again record the pointer reading. Increase the load and record the new pointer reading. Continue in this way, thus obtaining a series of readings. The extension is calculated for each load by subtracting the unloaded pointer reading from the loaded pointer reading. A check may be made by again recording the pointer readings as the loads are removed one at a time (this is also a means of checking that the elastic limit has not been reached).

(b) [Graph below. Remember to label axes, choose suitable scales, and when the line is no longer straight, draw a smooth curve. You are asked to plot extension against load, so the original length of the spring, 10 cm, must be subtracted from each reading.] Hooke's Law is obeyed for loads up to 4 N but very soon after this the graph begins to curve and Hooke's Law is no longer obeyed.

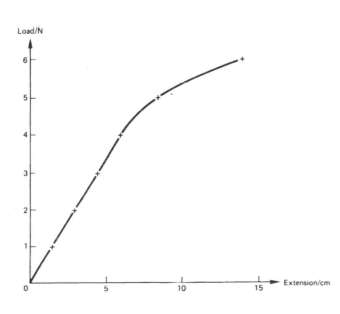

(c)

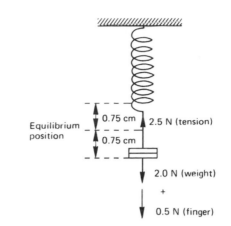

(d) The 0.2 kg mass (force 2 N) extends the spring 3 cm to a position of equilibrium. A further force of 0.5 N extends the spring 0.75 cm beyond this position. When the 0.5 N force is removed the spring oscillates about the original position of equilibrium. The oscillations will gradually decrease from an amplitude of 0.75 cm to zero. The spring will then be at rest with an extension of 3 cm. The maximum kinetic energy is when the spring passes through the equilibrium position (extension 3 cm).

Example 14

The diagram shows a beam resting on two supports and supporting a load.

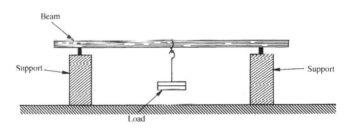

(a) Name three factors which will affect how a beam will bend when it is loaded. (3 marks)

(b) The following is an exaggerated diagram showing the deflection of a beam which is supporting a load.

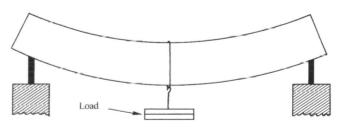

Add information to the diagram to show clearly which part of the beam is being subjected to
   (i) **tension** forces;
   (ii) **compression** forces. (4 marks)

(c) Explain each of the following:

(i) Steel is used to make the girders which form the framework of many buildings.

(ii) Steel rods under tension reinforce the concrete used to make many motorway bridges.

(6 marks)

(NEAB, Jun 95, Higher Tier, Q4)

Solution

(a) Any three from the following:
  • Distance between the supports;
  • Position of the load;
  • Size of the load;
  • Material from which the beam is made;
  • Shape of the cross-section of the beam.

(b)

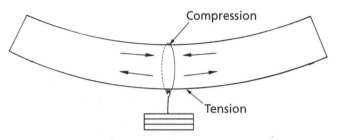

(c) (i) Steel is strong in both tension and compression and since load-bearing beams have both of these types of forces, this makes steel an ideal choice. Steel is also cheap compared with other materials of similar strength and is widely available.

(ii) Concrete is strong in compression but weak in tension. If used as a beam it needs to be reinforced by inserting a stretched steel bar in the region where the tension forces are experienced. The bar will try to contract and therefore exert a compressive force part of the beam which counteracts the tensile force.

Example 15

Crumple zones are parts of modern cars which are designed to collapse in a collision. A student sets up the following experiment to investigate crumple zones. Different test materials are attached to the front of the trolley. For each material the trolley is released from rest from the same position each time.

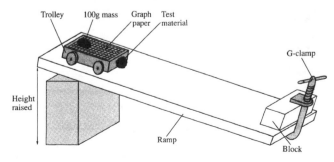

(a) Describe the motion of the trolley as it moves down the ramp. (1 mark)

(b) Explain why the 100 g mass continues to move when the trolley strikes the block. (2 marks)

(c) Various test results are used in the experiment (table 1.4). The mass moves the following distances when the trolley is stopped by the block.

Table 1.4

Material	Appearance	Movement of 100 g mass (cm)
Crumpled newspaper		10
Bubble pack		12
Plasticine (U-shape)		6
Solid plasticine		14
No absorber		16

How can these results be best explained?

(4 marks)

(ULEAC, Syll A, Jun 95, Higher Tier)

Solution

(a) The trolley accelerates down the ramp.

(b) According to Newton's first law, the 100 g mass will continue to move unless it is acted upon by an external force which in this case is only the small force of friction between the mass and the trolley. The mass therefore moves along the graph paper.

(c) The longer the time to stop, the less force is needed as registered by the smaller movement of the mass along the graph paper. Easily compressible or deformable materials like the U-shaped plasticine, crumpled newspaper and the bubble pack, spread out the time over which the force acts and this means the force is much smaller and the better the material is at absorbing the impact.

Examination questions

(Numerical answers and hints on solutions will be found at the end of the chapter.)

Question 1

Your teacher gives you a piece of pure metal. This has an irregular shape and size as shown below.

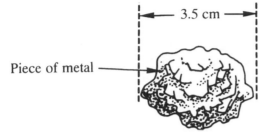

Piece of metal ⟶ ⟵ 3.5 cm ⟶

Your teacher now asks you to try to find out from which metallic element the piece of metal is made. You decide to find out the density of the metal.

(a)   What measuring device would you use to find the **mass** of the piece of metal?   (1 mark)

(b)   Next you need to find the volume of the piece of metal. Carefully describe how you would do this. (You may find it helpful to draw a labelled diagram of the apparatus you would use.)   (4 marks)

(c)   You obtain the following results:

Mass of metal = 87.0 g
Volume of metal = 32.0 cm^3

Calculate the density of the metal.   (3 marks)

(d)   The density of some elements are listed on page 6 of the Data Book in the 'Properties of Elements' table. Which element do you think the piece of metal is made from? Give a reason for your answer.   (2 marks)

(NEAB, Jun 95, Tier Q)

Question 2

(a)   Explain the following in terms of the pressure of the atmosphere.

(i)   When a rubber sucker is pushed onto a tiled wall it sticks to it.

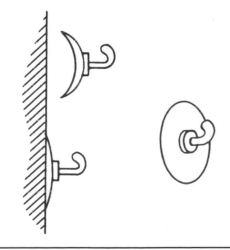

(ii)   You are able to drink lemonade through a straw.

(iii)   Aircraft which fly at a height of 10 000 m must have pressurised cabins.   (9 marks)

(b)   The illustration shows the variation of pressure on a diver with his depth in the sea.

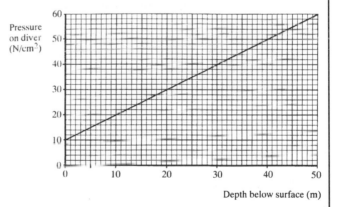

A deep sea-diver must have air in his suit and helmet at the same pressure as the sea water around him. He is lowered to a depth of 30 m.

(i)   Use the graph to find the total pressure exerted on the diver.

(ii)   Calculate the area of the rectangular glass window in the diver's helmet.

(iii)   Calculate the force exerted on the outside of the diver's window at a depth of 30 m.

(iv)   Explain why there is a pressure of 10 N/cm^2 at the surface of the sea.   (7 marks)

(NEAB, Jun 94, Intermediate Tier)

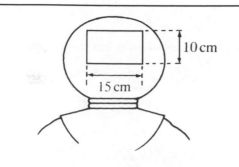

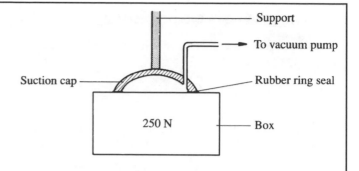

Question 3

A suction cap is used to pick up boxes. The diagram below shows how this is arranged.

(a)   Calculate the pressure difference which would be needed between the suction cap and surrounding air in order to just support the box. The area covered by the suction cap is 0.05 m² and the weight of the box 250 N. (4 marks)

(b)   If the atmospheric pressure is 100 000 Pa, calculate the pressure of the air between the box and suction cap. (2 marks)

(c)   What would need to be changed to lift heavier boxes? Explain your answer. (2 marks)

(ULEAC, Syll A, Jun 95, Intermediate Tier, Q2)

Answers to examination questions

1. (a) A top-pan balance.
   (b) A displacement can is placed on a level surface and filled with water till it overflowed. After allowing the excess water to drain away, the water will be level with the spout. Thin cotton is tied to the object and a measuring cylinder is placed under the spout. The object is lowered into the can until it is completely immersed and the water collected in the measuring cylinder. The volume of the water as measured in the cylinder will be the same as the volume of the object.

   (c) Density (g/cm³) = $\dfrac{\text{Mass (g)}}{\text{Volume (cm}^3)} = \dfrac{87.0 \text{ g}}{32.0 \text{ cm}^3}$

   $= 2.7 \text{ g/cm}^3$

   (d) By examination of a table of densities which are supplied as a data booklet by this examination board for use in the exams.
   Metal is probably aluminium since it has a density of 2.7 g/cm³

2. (a) (i)   When the sucker is pushed against the wall, some of the air is driven out of the inside. A partial vacuum is produced on the inside. The outside atmospheric pressure is much larger than the inside pressure and this forces the sucker against the tile.
      (ii) When you suck through the straw, the pressure inside the straw is lowered and the atmospheric pressure acting on the surface of the drink pushes down and forces the liquid up the straw.

   (iii) The air is less dense at this height and therefore only exerts a small atmospheric pressure. The human body is pressurised to balance the normal atmospheric pressure at the surface of the Earth. The low pressure would therefore make the passengers ill so the pressure inside the cabin is increased by pumping air in.

   (b) (i)   40 N/cm²
      (ii) Area = 15 cm × 10 cm = 150 cm²

      (iii) Pressure $= \dfrac{\text{Force}}{\text{Area}}$

      So, Force = Pressure × Area
      = 40 N/cm² × 150 cm²
      = 6000 N

      (iv) This is because at the surface, the atmospheric pressure which is 10 N/cm² is acting on the surface.

3. (a) Extra pressure needed $= \dfrac{\text{Force}}{\text{Area}} = \dfrac{250 \text{ N}}{0.05 \text{ m}^2}$

   $= 5000 \text{ Pa}$

   (b) Pressure = Atmospheric pressure − pressure difference

   = 100 000 Pa − 5000 Pa
   = 95 000 Pa

   (c) You could increase the area of the suction cap in contact with the box. This keeps the pressure difference the same since a larger force would be divided by a larger area.

2

Motion, Scalars and Vectors

Topic

WJEC & NEAB	ULEAC Syll A	ULEAC Syll B	MEG		MEG Salters'	MEG Nuffield	SEG	NICCEA
✓	✓	✓	✓	**Velocity and acceleration**	✓	✓	✓	✓
✓	✓	✓	✓	**Newton's laws of motion**	✓	✓	✓	✓
✓	✓	✓	✓	**Momentum**	✓	✓	✓	✓
	✓	✓	✓	**Scalars and vectors**		✓	✓	
✓	✓	✓	✓	**Graphs of motion**	✓	✓	✓	✓
✓	✓	✓	✓	**Friction**	✓	✓	✓	✓
✓				**Equations of motion**		✓	✓	✓
	✓	✓	✓	**Circular motion**			✓	✓
	✓	✓		**Projectiles**	✓	✓	✓	

1 Velocity and acceleration

$$\text{Velocity} = \frac{\text{displacement}}{\text{time taken}} \quad (\text{unit: m/s})$$

(Remember that displacement is distance in a specific direction)

$$\text{Speed} = \frac{\text{distance travelled}}{\text{time taken}} \quad (\text{unit: m/s})$$

$$\text{Acceleration} = \frac{\text{change in velocity}}{\text{time taken for change}} \quad (\text{unit: m/s}^2)$$

2 Newton's laws of motion

1 If a body is at rest, it will remain at rest; and if it is moving, it will continue to move in a straight line with constant velocity unless it is acted upon by an external force.
2 The acceleration of a body is directly proportional to the resultant force acting on it and inversely proportional to the mass of the body.
3 If a body A exerts a force on a body B, then B exerts an equal and opposite force on body A.

Newton's second law may be verified using a trolley and ticker timer as described in example 17. The second law may be summarised by the equation

$$F = ma$$

where F is the force in newtons, m the mass in kilograms and a the acceleration in metres per second per second.

3 Momentum

The greater the mass of an object and the greater its speed in a particular direction (i.e. the greater its velocity) then the more momentum the object has in that direction.

Momentum, mass and velocity are related by the following equation:

momentum (kg m/s) = mass (kg) × velocity (m/s)

There are two units for momentum which are equivalent: kg m/s and Ns. Momentum is a vector quantity and momentums to the right are usually taken to be positive and those to the left are taken as negative.

Law of conservation of momentum

In any collision/explosion, the momentum after the collision/explosion in a particular direction is the same as the momentum in that direction before the collision/explosion.

Elastic collisions

These are those collisions where kinetic energy is conserved so the total kinetic energy before the collisions is equal to the total kinetic energy after the collision. This can also apply to explosions. It is rare to get completely elastic collisions occurring.

Inelastic collisions

Here some of the kinetic energy is lost during the collision as heat or sound so the kinetic energy is less after than before. This equally applies to explosions.

If a body of mass m starts from rest and reaches a velocity v in t seconds as a result of a force F acting on it, then the acceleration is $\dfrac{v}{t}$ and

$$F = ma = \frac{mv}{t}$$

where $\dfrac{mv}{t}$ is the rate of change of momentum.

4 Scalars and vectors

Scalar quantities have magnitude (size) only whereas vector quantities have magnitude and direction (table 2.1).

Table 2.1

Scalar (quantities)	Vector quantities
Speed	Velocity
Distance	Displacement
Force	Mass
Energy	Momentum

Notice that speed is a scalar whereas velocity is a vector. This means that an object moving with a constant speed of 2 m/s in a circular path would have a changing velocity (i.e. it will be accelerating) because its direction is changing.

Vector quantities must be added by the rule for vector addition. Forces of 3 N and 4 N acting at right angles have a resultant force of 5 N.

5 Uniformly accelerated motion
Graphs of motion

• The gradient of a distance–time graph represents the speed.
• The gradient of a velocity–time graph represents the acceleration.
• The area under a velocity–time graph between two times represents the distance travelled during the time period.

Friction

Friction is a force which opposes motion and it acts when:

- an object moves through a fluid (a fluid is a liquid or a gas);
- solid surfaces slide across each other.

Friction may be reduced by

- making the object travelling through the fluid more streamlined;
- using a lubricant such as oil between two solid surfaces.

Friction always acts in the direction which will oppose the motion.

Friction causes objects to heat up and frequently causes kinetic energy (movement energy) to be lost as heat. The faster an object moves through a fluid, the greater will be the force of friction which acts on it. This is because the faster the object travels, the greater the volume of the fluid it has to push out of the way.

Terminal velocity

When an object such as a steel ball bearing is released in a fluid such as oil, the ball bearing is accelerated downward owing to its weight but as it starts to speed up, the upward frictional force starts to act. As the ball bearing moves faster the upward frictional force increases until a point is reached where the frictional force has grown to a size where it is equal to the weight of the ball bearing. When this happens, the two forces balance so no overall force acts and the ball bearing starts to move with a constant velocity called the *terminal velocity*.

The equations of motion

These are a set of equations which may be applied when a body moves with uniform acceleration.
The equations of motion are:

$$v = u + at$$

$$v^2 = u^2 + 2as$$

$$s = ut + \tfrac{1}{2}at^2$$

$$s = \frac{(u + v)t}{2}$$

Where the letters represent the following
u = initial velocity in m/s
v = final velocity in m/s
a = acceleration in m/s^2
s = distance travelled in m
t = time taken in s

Advice on using the four equations of motion.

1   Write down the quantities you are given in the question next to the letters used to represent them making sure that you also include the units that they are measured in.
2   You may need to convert some of the numbers to make sure that they are in the correct units for placing in the equations (e.g. a velocity in km/h would need to be converted into m/s).
3   Select the equation of motion. You should know three of the four quantities, the fourth quantity being the one you are trying to find.

Circular motion

Even if an object travels in a circle with constant speed its direction is constantly changing which means that its velocity will also be changing. The changing velocity means that there is an acceleration and this acceleration is directed towards the centre of the circle. If there is an acceleration there must also be a force acting and this is in the same direction as the acceleration (i.e. towards the centre of the circle). This force is called the centripetal force and its size depends on the following:

- mass: the greater the mass, the greater the centripetal force;
- velocity: the greater the velocity, the greater the centripetal force;
- radius: the smaller the radius of the circle, the greater the centripetal force.

6   Projectiles

Projectiles are objects which are given an initial velocity and are then acted upon by the force of gravity. They have no propelling force of their own.

When dealing with projectiles you consider the vertical components and horizontal components of the motion separately. The fact that a ball has a horizontal velocity has nothing to do with its downward motion. This means that a ball projected horizontally from a height will take the same time to fall as one just dropped vertically from the same height.

When considering two such balls, remember the following:

1   The downward acceleration of both balls will be the same (i.e. equal to the acceleration of free fall which is 10 m/s^2).
2   Because the acceleration only acts in the vertical direction, the horizontal velocity remains unaltered.

The illustration overleaf shows the main points.

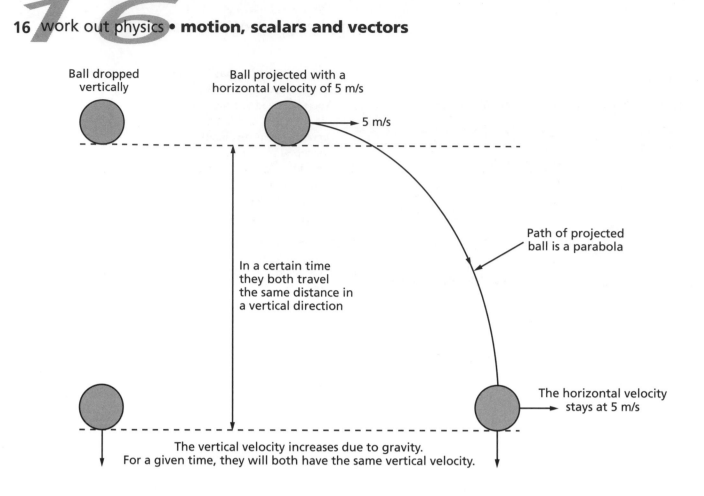

Ball dropped
vertically

Ball projected with a
horizontal velocity of 5 m/s

5 m/s

In a certain time
they both travel
the same distance in
a vertical direction

Path of projected
ball is a parabola

The horizontal velocity
stays at 5 m/s

The vertical velocity increases due to gravity.
For a given time, they will both have the same vertical velocity.

Worked examples

Example 1

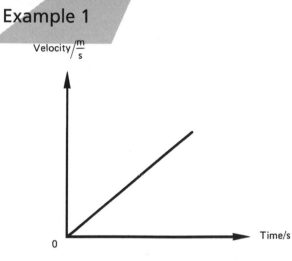

Velocity/$\frac{m}{s}$

0

Time/s

The graph shows how the velocity of an object varies with
time. The object is

A   moving with decreasing acceleration
B   moving with a constant acceleration
C   moving with increasing acceleration
D   moving with constant velocity

Solution

[The increase in the velocity in a given time is always the
same, i.e. the gradient of the graph is constant, so the
acceleration is constant.]

Answer **B**

Example 2

A ticker timer makes 50 dots every second.
- The tape shown above (drawn in actual size) is pulled
through this timer. What speed does the tape show?

A   10 cm/s
B   50 cm/s
C   100 cm/s
D   250 cm/s

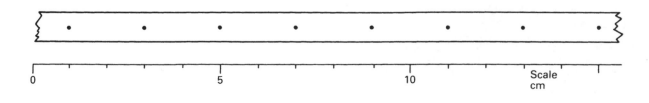

0   5   10   Scale
                                    cm

Solution

[The distance between successive dots is 2 cm. It travels the distance in 0.02 s.

$$\text{Speed} = \frac{\text{distance travelled}}{\text{time taken}} = \frac{2\text{ cm}}{0.02\text{ s}} = 100\text{ cm/s.}]$$

Answer **C**

Example 3

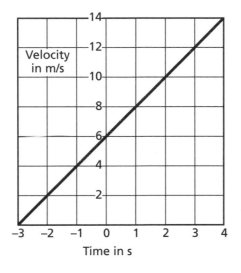

The acceleration, in m/s², of the body whose motion is represented by the graph shown is

A 1.5 **B** 2.0 **C** 3.5 **D** 6.0 **E** 14.0

Solution

$$[\text{Acceleration} = \frac{\text{change in velocity}}{\text{time taken for change}}$$

$$= \text{gradient of graph}$$

$$= \frac{4\text{ m/s}}{2\text{ s}} = 2\text{ m/s}^2]$$

Answer **B**

Example 4

The diagram shows a parachutist who is falling with a constant velocity. Which line shows possible values for both the weight of the parachutist and the air resistance acting on the parachute?

	Weight of parachutist	Air resistance
A	70 N	60 N
B	700 N	600 N
C	700 N	700 N
D	7000 N	6000 N

Solution

[A body falls with a constant velocity when the resultant force acting on it is zero. When the weight and the air resistance are equal and opposite, the resultant force is zero. See also section 5.]

Answer **C**

Example 5

The diagram shows the directions of three forces on a moving lorry.

The lorry is travelling at constant speed.
Which of these is correct?

	Air resistance	Friction	Forward thrust
A	2000 N	1000 N	3000 N
B	2000 N	2000 N	5000 N
C	3000 N	3000 N	3000 N
D	3000 N	4000 N	6000 N

Solution

[When a body is travelling at constant speed the net (resultant) force on it is zero. If a resultant force is acting on it then the body accelerates. In this question, (air resistance) + (friction) = (forward thrust).]

Answer **A**

Examples 6 to 8

You are the officer in charge of a spaceship travelling from our Galaxy to the Andromeda Galaxy. You are in deep space where the gravitational field due to surrounding galaxies is zero.

6. If your drive motors are off then you must be
   A stationary
   B decreasing in speed
   C increasing in speed
   D maintaining your present speed
7. To slow the ship you must fire the main rocket motor. You are looking ahead, towards the Andromeda Galaxy, then the direction you must fire the motor is
   A forwards, towards Andromeda
   B backwards, towards the Milky Way
   C to the right of your path
   D to the left of your path
8. Your spaceship has a mass of 10^6 kg. If the rocket produces a thrust of 10^5 N, what is your acceleration?
   A 0.1 m/s^2
   B 1 m/s^2
   C 10 m/s^2
   D 11 m/s^2

Solutions

6. [A body continues to move with constant velocity unless it is acted upon by a force. You are in deep space where there is no force on your spaceship.]

   Answer **D**
7. [When the motors are fired forwards a jet of burning gases will escape towards the Andromeda. This will result in a force acting on the rocket away from Andromeda and the ship will slow down. This is an example of Newton's third law of motion (section 2). The spaceship exerts a force on the burning gases and the gases exert an equal and opposite force on the spaceship.]

   Answer **A**
8. [Use Force = mass × acceleration
   10^5 N = $10^6 \times a$
   Hence $a = 0.1$ m/s^2.]

   Answer **A**

Example 9

Three trolleys are pulled by the forces shown in the diagrams. Friction and air resistance are negligible.

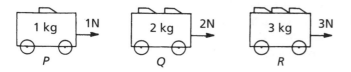

Which statement about the ACCELERATION of the trolleys is correct?
A P, Q and R all accelerate at the same rate
B P has the biggest acceleration
C Q has the biggest acceleration
D R has the biggest acceleration

Solution

Use Force = mass × acceleration or

$$\text{acceleration} = \frac{\text{force}}{\text{mass}}$$

Applying this equation to each trolley we have

$$a = \frac{1}{1} = \frac{2}{2} = \frac{3}{3} = 1 \text{ m/s}^2$$

[To produce the same acceleration, twice the mass needs twice the force. This is the argument you need to give if you are asked to explain why all bodies fall vertically with the same acceleration. The force of gravity on a 2 kg mass is twice that on a 1 kg mass.]

Answer **A**

Example 10

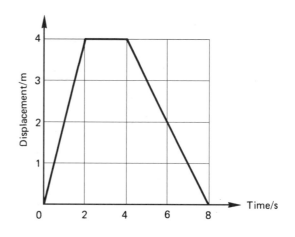

The graph shows how the displacement of a body varies with time.

(a) Describe the motion of the body (4 marks)
(b) What is the velocity of the body during the first 2 s? (3 marks)
(c) How far does the body travel in the first 4 s? (2 marks)

Solution

(a) The displacement at time $t = 0$ s is zero. The body then moves with a constant velocity until it has moved 4 m, and this takes 2 s. From 2 s to 4 s the body is stationary. It then returns to its original position at a constant velocity, and this velocity is half the original velocity.

(b) Velocity = $\dfrac{\text{distance travelled}}{\text{time taken}}$ = $\dfrac{4 \text{ m}}{2 \text{ s}}$ = 2 m/s

(c) The body travels 4 m in the first 2 s and is then stationary for 2 s. Distance travelled is 4 m.

Example 11

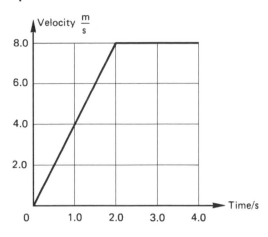

The graph is a velocity–time graph for a trolley.
(a) What is the acceleration of the trolley during the first 2 s?(3 marks)
(b) What is the acceleration of the trolley between 2 s and 4 s?(1 mark)
(c) How far does the trolley travel in 4 s? (4 marks)

Solution

(a) Acceleration $= \dfrac{\text{change in velocity}}{\text{time taken for change}}$

[see section 1]

$= \dfrac{8.0 \text{ m/s}}{2.0 \text{ s}} = 4.0 \text{ m/s}^2$

(b) Zero. [The velocity is constant.]
(c) Distance travelled = area under graph

[see section 1]

= (area of triangle) + (area of rectangle)
= ($\frac{1}{2}$ × 2.0 s × 80 m/s) + (8.0 m/s × 2.0 s)
= 8.0 m + 16.0 m = 24.0 m

Example 12

A free-fall parachutist jumps from an aircraft on a calm day. The figure shows a graph of his speed plotted against his time of fall.

(a) (i) Name the force pulling the parachutist to the ground.
(ii) Name another force acting on the parachutist.
(iii) What happens to this other force as his speed increases?
(b) Between 20 s and 40 s the speed is steady.
(i) How can you tell from the graph that the speed is steady?
(ii) Suggest why the speed is steady.
(iii) Calculate how far the parachutist falls between the time of 20 s and the time of 40 s.

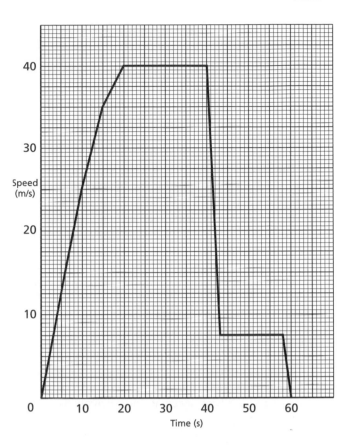

(c) (i) What is happening to his speed between 40 s and 43 s?
(ii) Explain why the speed is changing between 40 s and 43 s.
(iii) At what speed does the parachutist first touch the ground?

Solution

(a) (i) Force of gravity.
(ii) Air resistance.
(iii) The force increases.
(b) (i) The graph is horizontal.
(ii) the resultant force on the parachutist is zero (the force due to air resistance is equal to the force due to gravity).
(iii) He travels at 40 m/s for 20 s.
Distance = 40 m/s × 20 s = 800 m.
(c) (i) It decreases from 40 m/s to 7.5 m/s.
(ii) He opens his parachute and the force due to air resistance increases. Therefore he decelerates.
(iii) 7.5 m/s [in the 2 s after he touches the ground his speed drops to zero].

Example 13

This question is about a road accident involving a car and a van in a head-on collision.

The drawing shows the situation before the vehicle crash. The car and the van were travelling in opposite directions along a straight road where the speed limit is 60 miles per hour (26 m/s). They are involved in a head-on crash which locks the vehicles together and brings them to rest on the spot. The drivers were wearing seat belts and no-one was seriously hurt.

The police have the job of working out what happened.

They know that the van (of mass 2000 kg) was travelling at a speed of 15 m/s because this vehicle was fitted with a tachometer. But they will have to do some calculations to find the speed of the car (of mass 1000 kg).

(a) Calculate the momentum of the van (in kg m/s) before the collision. (2 marks)

(b) Explain how you can use momentum to show that the car must have been speeding and calculate the speed of the car. (4 marks)

(c) Tachometers have to be accurate to within 10%. Allowing for this, could the police prosecute either driver for speeding? Explain your reasoning. (3 marks)

(d) In the collision, the van comes to rest in 0.5 seconds.
   (i) Calculate the deceleration of the van. (3 marks)
   (ii) Calculate the force on the van while it is stopping. (3 marks)

Solution

(a) Momentum = mv [see section 3]
$$= 2000 \text{ kg} \times 15 \text{ m/s}$$
$$= 30\,000 \text{ kg m/s}$$

(b) Total momentum before collision = total momentum after collision.
$(1000 \times v) - 30\,000 = 0$ (where v is the velocity of the car).
[Remember that momentum is a vector quantity. This means that the momentum of the van must be given a negative sign because the van is travelling in the opposite direction to the car. The final momentum is zero because the vehicles are stationary.]
$v = 30$ m/s.
[This is about 67 mph and it seems extremely unlikely that the drivers would not be seriously hurt.]

(c) [The error is 10%, i.e. $\frac{1}{10}$ of 30 m/s.]
The speed of the car was between $(30 - 3)$ m/s and $(30 + 3)$ m/s, i.e. 27 m/s, i.e. 27 m/s and 33 m/s. It is travelling faster than the speed limit. The van was travelling below the speed limit. The police could prosecute the driver of the car but as they could only

prove he was travelling 1 m/s above the speed limit they would be unlikely to do so.

(d) (i) Deceleration = $\dfrac{\text{change in velocity}}{\text{time taken}}$

[see section 1]

$$= \frac{15 \text{ m/s}}{0.5 \text{ s}} = 30 \text{ m/s}^2$$

(ii)
$$F = ma \text{ [see section 2]}$$
$$F = 2000 \text{ kg} \times 30 \text{ m/s}^2$$
$$= 60\,000 \text{ N}$$
$$\text{Force on van} = 60\,000 \text{ N}$$

Example 14

The diagram shows a raindrop of mass 0.0001 kg falling freely in air.

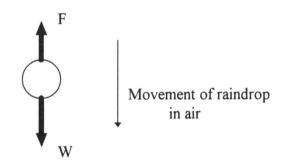

(a) What is force W called?

(b) If the acceleration of free fall is 10 m/s², calculate the value of the downward force.

(c) F is an upward force whose value varies according to the speed of the raindrop. What is the maximum value that this force may have?

(d) Initially, the raindrops' velocity will increase to a constant value. What is this velocity called?

(e) What do we know about the upward and downward forces when this velocity is reached?

Solution

(a) W is the weight of the drop.

(b) Weight = mass × acceleration of free fall = 0.0001 kg × 10 m/s² = 0.001 N.

(c) F has its maximum value when it is equal to the weight, i.e. 0.001 N.

(d) It's terminal velocity.

(e) They are both equal in size but opposite in direction.

Example 15

The following diagram shows a hovercraft, which is moving from left to right.

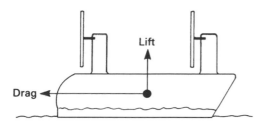

Solution

(a)

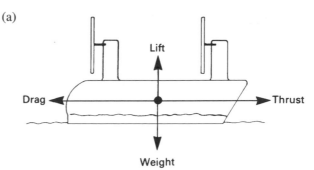

Four forces acting on a hovercraft are: DRAG, LIFT, THRUST, WEIGHT.

(a) Two of the forces have been marked in for you. Draw and label **two** arrows to show the other **two** forces. (4 marks)

(b) The mass of the hovercraft is 500 kg. Calculate its weight. (2 marks)

(c) State the size of the lift required to get the hovercraft to just 'hover'. (2 marks)

(d) The hovercraft is travelling with a constant velocity and at a constant height. Will the drag be more than, equal to, or less than the thrust? (2 marks)

(e) Ignoring the drag, calculate the force needed to accelerate the hovercraft forward at 3 m/s². (2 marks)

(f) The drag on the hovercraft is actually 200 N. What is the thrust needed to give an acceleration of 3 m/s²? (2 marks)

(b) Weight = 500 kg × 10 N/kg = 5000 N
[The Earth's gravitational field strength is 10 N/kg.]

(c) 5000 N
[The lift must just balance the weight.]

(d) The drag and the thrust will be equal.
[When a body is travelling at a constant velocity the resultant force on it is zero.]

(e) Force = mass × acceleration
= 500 kg × 3 m/s²
= 1500 N

(f) Thrust = 1700 N
[The resultant force must be 1500 N.]

Example 16

A war plane is travelling horizontally at constant speed. The diagram shows the position of the plane after equal time intervals. It releases a bomb at position **A**.

(a) In the space provided, sketch the position of the bomb when the plane is at position B, C, D and E. (Neglect the effects of air resistance on the bomb.) (4 marks)

(b) Explain why the bomb follows the path that you have shown above. (4 marks)

(WIEC, Jun 95, Higher Tier Q8)

A   B   C   D   E

Bomb ⟶ o

Solution

(a) On page 22.

(b) Gravity acts in a vertical direction and will not therefore alter the horizontal motion. This means that the horizontal component of the velocity remains constant so the bomb will travel equal distances horizontally in the time intervals. The vertical component will accelerate due to gravity and will travel increasing distances during each time interval. This is the reason for the parabolic shape shown.

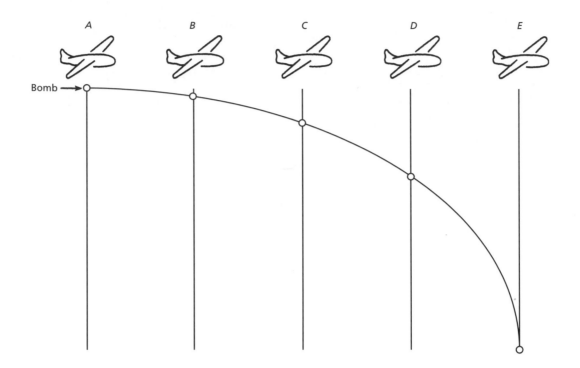

Example 17

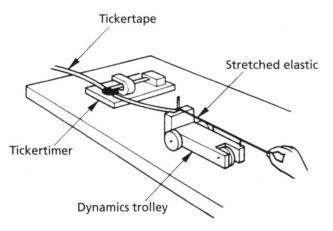

(a) In order to investigate the relationship between force and acceleration, an experiment was carried out using a ticker tape attached to a trolley as shown in the diagram. The tape was fed through a ticker timer which made 50 dots on the tape every second. The trolley was placed on a gently sloping inclined plane. A force was applied by pulling on an elastic band attached to the trolley.

   (i) Explain why an inclined plane was used and what experiment would have been conducted in order to get the correct inclination of the plane.   (3 marks)

  (ii) How would the force applied to the trolley have been kept constant?   (1 mark)

 (iii) How would a force which was twice the magnitude of the original force have been applied?   (1 mark)

(b) The diagram below shows two sections of one tape obtained from such an experiment.

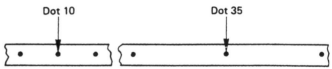

The first section shows dots 9, 10 and 11; the second section dots 34, 35 and 36.

   (i) Measure the distance between dots 9 and 11 and calculate the average velocity of the trolley between dots 9 and 11.   (3 marks)

  (ii) Calculate the average velocity of the trolley between dots 34 and 36.   (2 marks)

 (iii) What time interval elapsed between dot 10 and dot 35?   (1 mark)

 (iv) Calculate the acceleration of the trolley.   (3 marks)

(c) The acceleration of the trolley was calculated for five different forces.

   (i) Sketch a graph which represents the results of the above experiment.   (2 marks)

  (ii) State the relationship between force and acceleration.   (2 marks)

Solution

(a)   (i) The inclined plane is to compensate for friction. The plane is tilted until the component of the gravitational force accelerating the trolley is equal to the frictional force. This is done by tilting the plane until the trolley moves with a constant velocity when given a push. When the correct tilt is obtained, a ticker tape

attached to the trolley will have dots on it
equally spaced.

(ii) The rubber band must be kept stretched by the
same amount throughout the run.

(iii) Two identical rubber bands each stretched the
same amount as the original rubber band.

(b) (i) Distance between dot 9 and dot 11 = 2 cm

Time between dot 9 and dot 11

$$= \frac{2\ s}{50} = 0.04\ s$$

$$\text{Velocity} = \frac{\text{distance gone}}{\text{time taken}}$$

$$= \frac{2\ cm}{0.04\ s} = 50\ cm/s$$

(ii) Distance between dot 34 and dot 36 = 5 cm

$$\text{Velocity} = \frac{5\ cm}{0.04\ s} = 125\ cm/s$$

(iii) 0.5 s. [This is the time for 25 dots.]

(iv) $\text{Acceleration} = \dfrac{\text{change in velocity}}{\text{time taken for change}}$

$$= \frac{125 - 50\ cm/s}{0.5\ s}$$

$$= 150\ cm/s^2.$$

(c) (i)

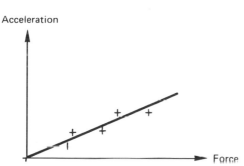

(ii) The acceleration is proportional to the resultant
force, provided the mass is kept constant.
[Or resultant force = mass × acceleration.]
[The relationship between mass and acceleration may
be investigated by keeping the force constant (band at
constant stretch) and varying the mass (add masses to
the trolley). The acceleration due to gravity, g, may be
measured if the ticker timer is arranged so that the
tape (with a mass on its end) falls vertically.]

Example 18

(a) What do Newton's laws tell us about the effect of a
force on a body? (3 marks)

(b) A body of mass 5 kg is at rest when a horizontal force
is applied to it. The force varies with time as shown

on the graph below. Use the figures on the graph to
calculate how the velocity varies with time and plot a
graph of velocity against time for the first 30 s of its
motion. (11 marks)

(c) How far does it travel in the first 10 s? (3 marks)

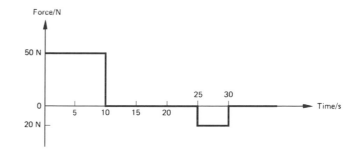

Solution

(a) When a resultant force acts on a body, the body
accelerates – that is, changes its velocity. The
acceleration is proportional to the force and inversely
proportional to the mass.

(b) $F = ma$. During the first 10 s, 50 N = 5 kg × a;
hence, $a = 10\ m/s^2$. For the next 15 s no force acts
and the body continues with constant velocity. From
25 s to 30 s the body decelerates, and
$-20\ N = 5\ kg \times a$; hence, $a = -4\ m/s^2$. Since
$v = at$ [see section 5], after 10 s
velocity = 10 m/s^2 × 10 s = 100 m/s. From 10 s to
25 s the body continues at 100 m/s. Between 25 s and
30 s the change in velocity is 4 m/s^2 × 5 s = 20 m/s.
So, after 30 s the velocity is
(100 − 20) m/s = 80 m/s.

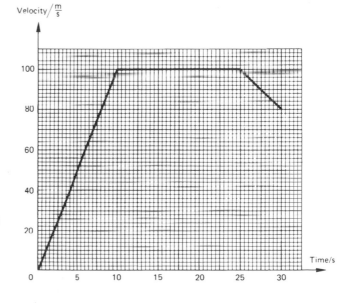

(c) Distance travelled = average
velocity × time = 50 × 10 = 500 m.
[This is also the area under the graph for the first
10 s.]

Examination questions

(Numerical answers and hints on solutions will be found at the end of the chapter.)

Question 1

A racing driver is driving his car along a straight and level road as shown below.

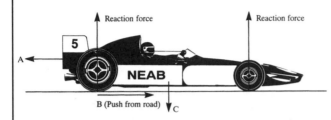

(a) The arrows on the diagram show forces acting on the car. Write down the names of the forces shown by the arrows *A* and *C*. (2 marks)

(b) (i) At first, force *B* is larger than force *A*. What will happen to the speed of the car?

   (ii) After a time, forces *A* and *B* are balanced. What will happen to the speed of the car?

   (iii) Explain why forces *A* and *B* eventually balance. (4 marks)

(c) Each lap of the circuit is 4500 m. The racing car takes 90 seconds to complete one lap. Calculate the average speed of the car. (4 marks)

(d) During the race the amount of fuel in the car gets less. Explain what effect this will have on the time taken to complete one lap. (2 marks)

(e) The racing car has a mass of 1250 kg. When the brake pedal is pushed down, a constant braking force of 10 000 N is exerted on the car. Calculate the acceleration of the car. (5 marks)

(NEAB, Jun 95, Intermediate Tier)

Question 2

Table 2.2 shows information about the stopping of a car travelling at different speeds.

The thinking distance is the distance the car travels whilst the driver reacts and before the brakes begin to act. The braking distance is the distance the car travels after the brakes begin to act.

(a) (i) Describe the way in which the thinking distance changes as the speed of the car increases. (2 marks)

   (ii) Why does it change in this way? (2 marks)

(b) The car is travelling at 48 km/h. The driver brakes

Table 2.2

Speed of car (km/h)	Speed of car (m/s)	Thinking distance (m)	Braking distance (m)
32	8.9	6	6
48	13.3	9	14
64	17.8	12	24
80	22.2	15	38
96	26.7	18	55
112	31.1	21	75

constantly until the car stops. The car and driver have a mass of 900 kg.

   (i) Calculate the change in kinetic energy (KE) of the car. (4 marks)

   (ii) Use your answer to (b) (i) and the information from table 2 to calculate the braking force which acts on the car. (3 marks)

(MEG, Jun 95, Higher Tier)

Question 3

(a) Define the newton. (4 marks)

(b) Describe, in detail, and with the help of a diagram, an experiment to investigate how the mass of an object is related to the acceleration of the object, under the action of a constant force.

In your answer you should clearly state how the accelerating force is produced, how frictional forces are allowed for, what measurements are taken and how they are used to show the relationship between mass and acceleration.

(12 marks)

(c) A hot air balloon is travelling vertically upwards at a constant speed.

   (i) Draw a circle to represent the balloon and mark on the diagram the **three** forces acting on the balloon, naming each and showing its direction. (3 marks)

   (ii) Write down an equation to show how these forces are related. (2 marks)

(d) A car is moving along a level road with an acceleration of 0.5 m/s² when the engine is providing a forward driving force of 750 N.

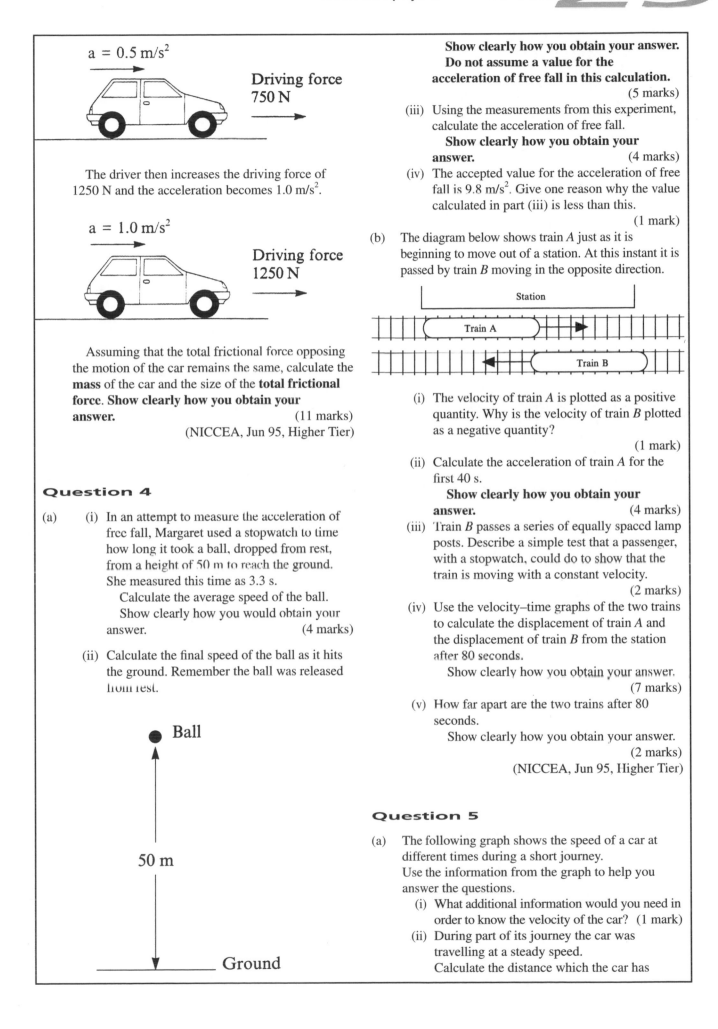

a = 0.5 m/s²

Driving force
750 N

The driver then increases the driving force of
1250 N and the acceleration becomes 1.0 m/s².

a = 1.0 m/s²

Driving force
1250 N

Assuming that the total frictional force opposing
the motion of the car remains the same, calculate the
mass of the car and the size of the **total frictional
force. Show clearly how you obtain your
answer.**   (11 marks)
(NICCEA, Jun 95, Higher Tier)

Question 4

(a)   (i)   In an attempt to measure the acceleration of
free fall, Margaret used a stopwatch to time
how long it took a ball, dropped from rest,
from a height of 50 m to reach the ground.
She measured this time as 3.3 s.
Calculate the average speed of the ball.
Show clearly how you would obtain your
answer.   (4 marks)

(ii)   Calculate the final speed of the ball as it hits
the ground. Remember the ball was released
from rest.

● Ball

50 m

Ground

**Show clearly how you obtain your answer.
Do not assume a value for the
acceleration of free fall in this calculation.**
(5 marks)

(iii) Using the measurements from this experiment,
calculate the acceleration of free fall.
**Show clearly how you obtain your
answer.**   (4 marks)

(iv)   The accepted value for the acceleration of free
fall is 9.8 m/s². Give one reason why the value
calculated in part (iii) is less than this.
(1 mark)

(b)   The diagram below shows train A just as it is
beginning to move out of a station. At this instant it is
passed by train B moving in the opposite direction.

Station

Train A

Train B

(i)   The velocity of train A is plotted as a positive
quantity. Why is the velocity of train B plotted
as a negative quantity?
(1 mark)

(ii)   Calculate the acceleration of train A for the
first 40 s.
**Show clearly how you obtain your
answer.**   (4 marks)

(iii) Train B passes a series of equally spaced lamp
posts. Describe a simple test that a passenger,
with a stopwatch, could do to show that the
train is moving with a constant velocity.
(2 marks)

(iv)   Use the velocity–time graphs of the two trains
to calculate the displacement of train A and
the displacement of train B from the station
after 80 seconds.
Show clearly how you obtain your answer.
(7 marks)

(v)   How far apart are the two trains after 80
seconds.
Show clearly how you obtain your answer.
(2 marks)
(NICCEA, Jun 95, Higher Tier)

Question 5

(a)   The following graph shows the speed of a car at
different times during a short journey.
Use the information from the graph to help you
answer the questions.
(i)   What additional information would you need in
order to know the velocity of the car? (1 mark)
(ii)   During part of its journey the car was
travelling at a steady speed.
Calculate the distance which the car has

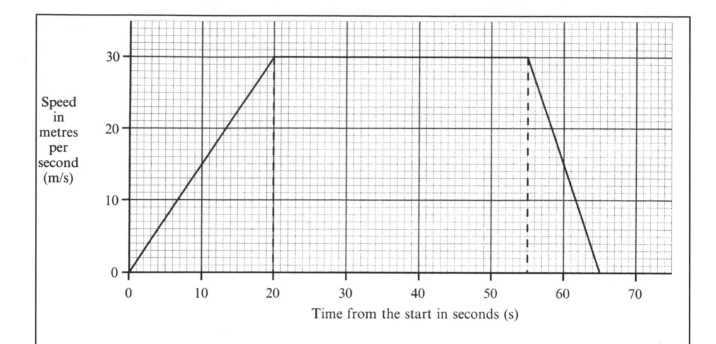

travelled during that part of its journey. Include in your answer the equation you are going to use. Show clearly how you would get to your final answer and give the unit. (4 marks)

(iii) During part of its journey the car was slowing down.

Calculate the deceleration of the car during that part of its journey. Include in your answer the equation you are going to use. Show clearly how you get to your final answer and give the unit. (4 marks)

(b) Later the car accelerates at 2.5 metres per second per second (m/s²).

The mass of the car is 800 kilograms (kg).

Calculate the force needed to accelerate the car. Include in your answer the equation you are going to use. Show clearly how you get your final answer and give the unit. (3 marks)

(c)

The car driver and passengers wear seatbelts. Explain how a seatbelt reduces the risk of injury if a car stops suddenly. You should include ideas about acceleration and force in your explanation.

(4 marks)

(SEG, Jun 95, Science Double Award, Intermediate Tier)

Question 6

(a) An astronaut is on the Moon. He drops a hammer from a height of 3.2 m and it takes 2.0 s to hit the lunar landscape.

What is the acceleration due to gravity on the Moon? (4 marks)

(b) He now throws the hammer horizontally at a speed of 10 m/s from the same height.

(i) How long does it take to hit the ground? (1 mark)

(ii) How far will it travel horizontally before it hits the surface? (1 mark)

(MEG Nuffield, Jun 95, Higher Tier, Q11)

Answers to examination questions

1. (a) A = Drag (or total frictional forces) B = Weight.
   (b)   (i) The speed will increase. That is, the car will accelerate.
       (ii) The speed will stay constant since there is no acceleration.
       (iii) Force B is determined by the force delivered by the engine. Force B depends on the speed of the car and this force increases with increasing speed. As the car goes faster, eventually force B equals force A and there is no overall force acting on the car, so the cars move at constant speed.

   (c) Average speed = $\dfrac{\text{distance travelled}}{\text{time taken}}$

   $$= \dfrac{4500 \text{ m}}{90 \text{ s}} = 50 \text{ m/s}$$

   (d) The force provided by the engine and the brakes is constant for a particular car. As the race proceeds, fuel is used up making the mass less. This means that the acceleration and deceleration will increase and this will make the time to complete one lap less.

   (e) Acceleration = $\dfrac{\text{Force}}{\text{Mass}} = \dfrac{10\,000 \text{ N}}{1250 \text{ kg}} = 8 \text{ m/s}^2$

2. (a)   (i) The thinking distance is proportional to the speed since a doubling in speed results in a doubling of the thinking distance.
       (ii) The reaction time will be the same at all speeds since this is the time the brain registers that the brakes need to be applied and the time they are actually applied. The distance travelled during this 'thinking time' will increase with increasing speed.

   (b) Change in KE = $\frac{1}{2} \times$ mass $\times$ velocity^2
                 = $\frac{1}{2} \times 900$ kg $\times (13.3$ m/s$)^2$
                 = $80\,000$ J
       (ii)   Change in KE = work done by braking force
         $80\,000$ J = Braking force $\times$ distance travelled whilst braking
         $80\,000$ J = Braking force $\times$ 14 m
     Giving braking force = 5714 N

3. (a) One newton is the force required to give a mass of 1 kg an acceleration of 1 m/s^2.
   (b) See the experiment in example 17.
   (c)   (i) A circle should be drawn with three arrows. One arrow points upward and should be labelled as the lift and the other two arrows point down and should be labelled the weight and drag (or total frictional force).
       (ii) Lift = Weight + Drag.
   (d) Accelerating force = Driving force − Frictional force.
   For the first situation we have
   Mass $\times$ 0.5 m/s^2 = 750 − Frictional force
   For the second situation we have
   Mass $\times$ 1.0 m/s^2 = 1250 − Frictional force
   Solving these two equations together (i.e. simultaneously) gives
   Mass = 1000 kg and Frictional force = 250 N

4. (a)   (i) Average speed = $\dfrac{\text{Distance travelled}}{\text{Time taken}}$

   $$= \dfrac{50 \text{ m}}{3.3 \text{ s}} = 15.2 \text{ m/s}$$

       (ii) Using the equation of motion
   $s = \frac{1}{2}(u + v)t$ with $u = 0$ m/s
   $50 = \frac{1}{2} \times v \times 3.3$
   Giving $v = 30.4$ m/s
       (iii) Using the equation of motion
   $s = ut + \frac{1}{2}at^2$ with $u = 0$ m/s
   $50 = \frac{1}{2}g\, 3.3^2$
   Giving $g = 9.2$ m/s^2
       (iv) Air resistance slows the ball down.
   (b)   (i) Velocity is a vector quantity and since it is moving backwards its velocity must be negative.
       (ii) Acceleration = $\dfrac{15 \text{ m/s}}{40 \text{ s}} = 0.38$ m/s^2
       (iii) He could time the intervals between each lamp post. If the train is travelling at constant velocity, these time intervals will be the same.
       (iv) Displacement of train A = Area under the graph between $t = 0$ s and $t = 80$ s
         = $\frac{1}{2} \times 40 \times 15 + 15 \times 40 = 900$ m
   Displacement of train B = Area under the graph between $t = 0$ s and $t = 80$ s
   = $-(-20) \times 80 = -1600$ m
       (v) Distance apart = 900 + 1600 m = 2500 m
   [We add these because the displacements are in opposite directions.]

5. (a)   (i) The direction. [Since velocity is a vector quantity it has both size and direction.]
       (ii) Distance whilst travelling at steady speed = area under the graph between $t = 20$ s and $t = 55$ s.
   Area = 30 m/s $\times$ 35 s = 1050 m

(iii) Deceleration = gradient of the graph
between $t = 55$ s and $t = 65$ s

$$= \frac{30 \text{ m/s}}{10 \text{ s}} = 3 \text{ m/s}^2$$

(b) Force (N) = Mass (kg) $\times$ Acceleration (m/s^2)
= 800 kg $\times$ 2.5 m/s^2
= 2000 N

(c) When the car stops suddenly the passengers
keep moving forward when they are not wearing
a seatbelt. The only force restraining them is
the small force of friction between them and the
seat. Using the seatbelts provides a large
restraining force which prevents them moving
forward and injuring themselves when they hit the
windscreen.

6. (a) Using the equation of motion $s = ut + \frac{1}{2}at^2$

with initial velocity, $u = 0$ m/s, s = 3.2 m and
$t = 2$ s we have
$3.2 = \frac{1}{2}a \times 2^2$
Giving acceleration, $a = 1.6$ m/s^2

(b) (i) 2 s

(ii) [Speed stays the same horizontally because
it is unaffected by gravity which acts only in
the vertical direction.]
Distance = velocity $\times$ time
= 10 m/s $\times$ 2 s = 20 m

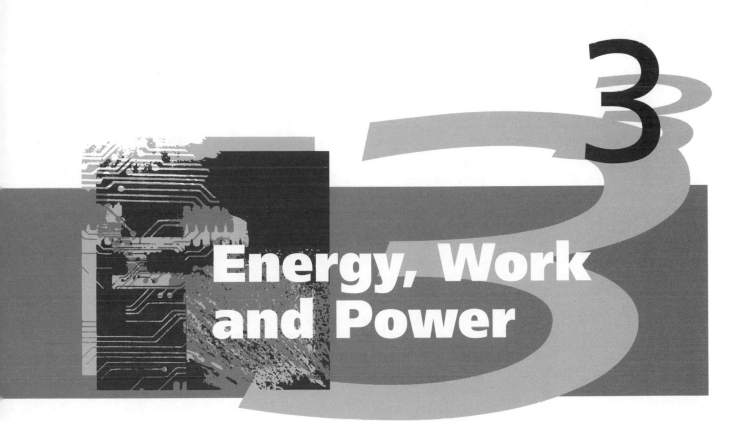

Energy, Work and Power

3

WJEC & NEAB	ULEAC Syll A	ULEAC Syll B	MEG	Topic	MEG Salters'	MEG Nuffield	SEG	NICCEA
✓	✓	✓	✓	**The various forms of energy**	✓	✓	✓	✓
✓	✓	✓	✓	**Energy changes**	✓	✓	✓	✓
✓	✓	✓	✓	**Work**	✓	✓	✓	✓
✓	✓	✓	✓	**Power**	✓	✓	✓	✓
✓	✓	✓	✓	**Efficiency**	✓	✓	✓	✓

1 The various forms of energy

Energy

Energy is the capacity to do work. When work is done, one form of energy is changed into another. Energy and work are measured in the same units of Joules (J).

The Principle of conservation of energy states that energy cannot be created or destroyed, although it can be changed from one form to another.

Forms of energy

Chemical

Chemical energy is the energy trapped in certain substances such as fuels (oil, gas and coal). Food is also a fuel which is used by our bodies.

Nuclear

Nuclear

Nuclear energy is the energy stored in the nuclei of certain radioactive atoms of elements such as uranium or plutonium. When the nuclei split up, this energy is released as a large amount of heat energy.

Heat

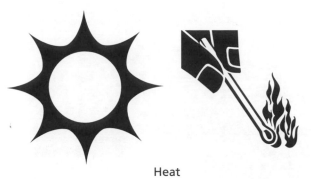

Heat

Heat energy is possessed by all objects to some extent. The amount of heat energy a body has depends on the amount of energy contained by its vibrating molecules.

Sound

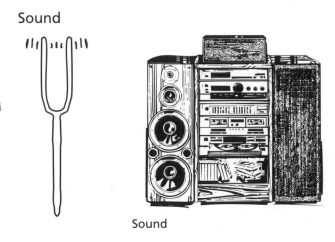

Sound

Sound energy is produced when a medium such as a solid, liquid or gas starts to vibrate.

Light/Electromagnetic energy

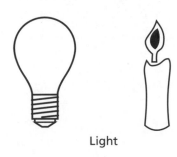

Light

All electromagnetic waves including light have energy associated with them.

Electrical energy

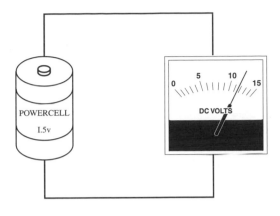

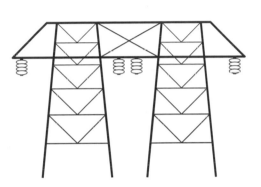

This is the energy produced by a battery or a generator and is the form of energy which is easiest to transfer into other forms of energy.

Potential energy

This is the energy possessed by a body by virtue of its position or the state that the body is in. If a body is lifted vertically, then it gains gravitational potential energy. The gain in gravitational potential energy may be calculated from the equation.

Gravitational potential energy
gained = (weight) × (vertical height raised)

Examples of bodies possessing gravitational potential energy (gpe) are shown below.

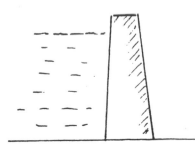

Other examples of bodies possessing potential energy are shown below

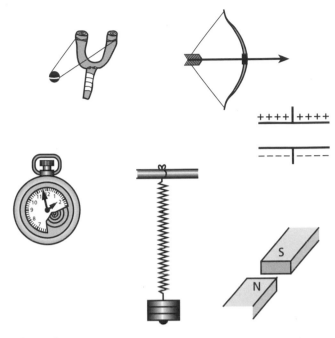

Kinetic energy

All moving objects have this form of energy. For a body whose mass is in kg, velocity in m/s, the kinetic energy in J is given by the equation:

Kinetic energy $= \frac{1}{2} \times$ mass $\times$ (velocity)^2

Often, there is an interchange between kinetic energy and gravitational potential energy. The following equation may be used provided that there is no energy lost from the system as heat (due to friction) or sound.

Kinetic energy = Gravitational potential energy

$\frac{1}{2} \times$ mass $\times$ (velocity)$^2 = mgh$

2 Energy changes

The law of conservation of energy states that energy cannot be created or destroyed, it may only be changed from one form to another.

Energy changes are therefore extremely important in physics. Take for example the following energy changes:

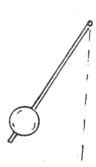

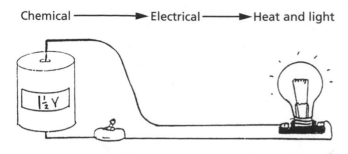

Chemical ———→ Electrical ———→ Heat and light

A battery changes chemical energy to electrical energy.

The bulb lights and converts the electrical to heat and light energy.

Light is the useful form of energy in this case.

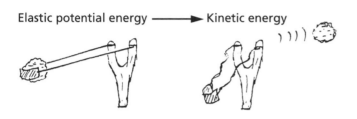

Elastic potential energy ———→ Kinetic energy

The elastic is stretched and stores elastic potential energy and when released, the energy is converted into the kinetic energy of the stone.

Chemical ————————————→ Kinetic and heat

Food

Kinetic

37°C

Heat (body temperature)

Food supplies the chemical energy which are body uses to convert into heat (to maintain body temperature) and kinetic energy (energy of movement).

Energy changes in a coal-fired power station

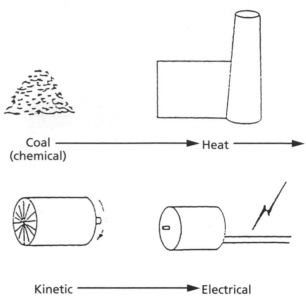

Coal ——————————→ Heat ——————→
(chemical)

Kinetic ————————————→ Electrical

1 The chemical energy in coal is released as heat.
2 Heat is used to turn the water into steam.
3 Steam is passed into the turbine which rotates at high speed.
4 The kinetic energy is then changed into electrical energy by the generator (dynamo).

Energy changes in a hydro-electric power station

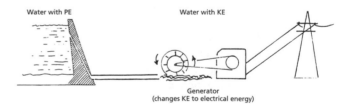

Water with PE   Water with KE

Generator
(changes KE to electrical energy)

1 Water at one side of the dam is higher and therefore possesses gravitational potential energy.
2 Water is allowed to flow down the pipe thus changing the potential energy into kinetic energy.
3 Turbine rotates and causes the generator to rotate and change the kinetic energy to electrical energy.

3 Work

Work is energy transfer, and may be calculated from the equation

Work = (force) × (distance moved in the direction of the force) (unit: joule (J))

1 *joule* of work is done when a force of 1 N moves its point of application through a distance of 1 m.

4 Power

Power is the rate at which work is done and is measured in watts.

Power may also be considered at the rate at which one form of energy is converted into another. For instance, a 100 W light bulb converts 100 J of electrical energy to heat and light energy every second.

Power may be calculated using the formula:

$$\text{Power} = \frac{\text{work done (J)}}{\text{time taken (s)}}$$ and this gives the power in

watts.

Electrical power may be calculated using the formula

Power (W) = Current (A) × Voltage (V)

Large amounts of power are measured in kilowatts (kW) with 1000 W = 1 kW or megawatts with 1000 kW = 1 MW (so, 1 000 000 W = 1 MW)

5 Efficiency

Motors are used to change one form of energy into kinetic energy. For example petrol engines convert the chemical energy in petrol into mainly kinetic energy and heat energy and electric motors converts electrical energy to kinetic and heat energy.

All the energy input to a motor is not changed into kinetic energy since a large amount of the energy is converted to unwanted heat energy by friction.

Petrol engines like all engines are not very efficient and for every 100 J input only about 25 J is available as kinetic energy.

To compare the energy input with that output we calculate a quantity called the efficiency using the formula:

$$\text{Efficiency} = \frac{\text{work output}}{\text{work input}}$$

Since the work out is always less than the work put in, the efficiency will always be a number less than one. Efficiency may also be expressed as a percentage and to do this the number for the efficiency is multiplied by one hundred.

Efficiency may also be expressed in terms of the powers using the following formula:

$$\text{Efficiency} = \frac{\text{Power output}}{\text{Power input}}$$

Worked examples

(When needed, take the Earth's gravitational field strength as 10 N/kg.)

Example 1

Energy may be defined as
A   producing power
B   causing motion
C   ability to do work
D   ability to exert a force

Solution

Answer **C** [see section 1]

Example 2

The diagram shows a hydro-electric generation system.

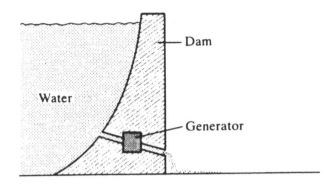

Which of the following gives the correct order of the energy changes that occur?
A   kinetic → electric → potential
B   kinetic → potential → electric
C   potential → electric → kinetic
D   potential → kinetic → electric

Solution

[The water behind the dam has potential energy. When it flows through the generator it has kinetic energy. Electrical energy is produced by the generator. See section 2.]

Answer **D**

Example 3

A coconut is falling from a palm tree.

Which of these is correct?

Velocity of coconut	Gravitational potential energy of coconut
A decreasing	decreasing
B increasing	decreasing
C increasing	increasing
D decreasing	increasing

Solution

[Potential energy increases as work is done in raising an object and decreases when the body falls.]

Answer **B**

Example 4

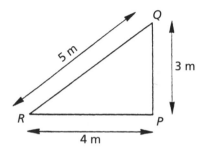

A weight of 5 N is moved up a frictionless inclined plane from *R* to *Q* as shown. What is the work done in joules?
A 15 **B** 20 **C** 35 **D** 60

Solution

[Work done = weight × vertical height raised (see section 1)
= 5 N × 3 m
= 15 J]

Answer **A**

Example 5

A person who has a mass of 50 kg runs up some stairs in 9 s. The stairs are 8 m high. His power output is

A $\dfrac{50 \times 8}{9}$ W **B** $\dfrac{50 \times 9}{8}$ W **C** $\dfrac{50 \times 10 \times 8}{9}$ W

D $\dfrac{50 \times 10 \times 9}{8}$ W

Solution

[The Earth's gravitational field strength is 10 N/kg and the weight of the man is therefore
= (50 kg)(10 N/kg) = (50 × 10) N
Work done = force × distance [see section 3]
= (50 × 10 N)(8 m) = (50 × 10 × 8 J)

$$\text{Power} = \frac{\text{work done}}{\text{time taken}} = \frac{(50 \times 10 \times 8)\,\text{J}}{9\,\text{s}}$$

$$= \frac{(50 \times 10 \times 8)}{9}\,\text{W}$$

Hence **C** is the correct answer.]

Answer **C**

Example 6

A man does 10 press-ups in 30 seconds. His weight is 600 N and each time he lifts his body a distance of 0.5 m.
(a) What gravitational potential energy does he gain with each lift?
(b) What amount of work does he do altogether?
(c) What is his power in watts?
(d) Working at the power in part (c), how long would it take for him to climb a flight of stairs 10 m high?

(Gravitational field strength = 10 N/kg.)

Solution

(a) PE = Mass (kg) × gravitational field strength
       (N/kg) × height (m)
   = Weight (N) × height (m)
   = 600 × 0.5
   = 300 J

(b) Work done (J) = 300 J × 10 = 3000 J

(c) Power (W) = $\dfrac{\text{Work done (J)}}{\text{Time taken (s)}} = \dfrac{300\,\text{J}}{30\,\text{s}} = 100\,\text{W}$

(d) Power (W) = $\dfrac{\text{Work done (J)}}{\text{Time taken (s)}}$

$$100\,\text{W} = \frac{(600\,\text{N}) \times (10\,\text{m})}{t}$$

$$t = 60\,\text{s}$$

Example 7

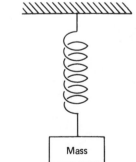

A spring is fixed to a support at one end and a mass is hung on the other end. The mass is raised until the spring just goes slack. It is then released. From the moment the mass is released until it reaches its lowest point describe the changes that take place in

    (i) the kinetic energy of the mass   (3 marks)
    (ii) the gravitational potential energy of the mass
                      (2 marks)
    (iii) the energy stored in the spring.   (2 marks)

Solution

    (i) The kinetic energy is initially zero. [Remember that kinetic energy is energy possessed by a moving body.] The kinetic energy increases until the spring passes through its equilibrium position, when the kinetic energy is a maximum, and then the kinetic energy decreases until it reaches zero when the spring is at its lowest position.
    (ii) The gravitational potential energy is a maximum and it gradually decreases until the spring reaches its lowest position.
    (iii) This is initially zero and increases until it is a maximum when the spring is at its greatest extension.

Example 8

(a) What power is produced by a machine which lifts a mass of 2 kg through a vertical height of 10 m in 2 s?
(b) A mass of 3 kg is thrown vertically upwards with a kinetic energy of 600 J. To what height will it rise?
                                    (6 marks)

Solution

(a) Force needed to lift 2 kg = 20 N

$$\text{Power} = \frac{\text{work done}}{\text{time taken}} = \frac{20 \times 10}{2} \text{ J/s} = 100 \text{ W}$$

(b) 600 J = mgh. [See section 1, 'Kinetic energy'.]
    ⇒ 600 J = (3 kg) (10 N/kg) h
    ⇒ h = 20 m
    Height = 20 m

Example 9

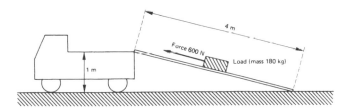

The diagram shows a ramp being used to get a load, which has a mass of 180 kg, onto a lorry. The ramp is 4 m long and the end of the lorry is 1 m above the ground. A force of 600 N is needed to pull the load up the ramp.

(a) Calculate
    (i) The gravitational potential energy gained by the load as it goes from the bottom to the top of the ramp;   (3 marks)
    (ii) the work done by the 600 N force in pulling the load up the ramp;   (2 marks)
    (iii) the efficiency of the system   (2 marks)
(b) As the lorry starts, the load topples off and falls to the ground. What is the kinetic energy of the load just before it hits the ground?   (2 marks)
(Take the Earth's gravitational field strength as 10 N/kg.)

Solution

(a)   (i) Gravitational potential energy gained = mgh [see section 1]
        (180 kg) (10 N/kg) (1 m) = 1800 J
        [This is equal to the force in Newtons multiplied by the vertical height raised, see section 1 potential energy.]
    (ii) Work done = force × distance [see section 3]
        = 600 N × 4 m = 2400 J

    (iii) Efficiency $= \dfrac{\text{work out}}{\text{work in}}$ [see section 5]

                $= \dfrac{1800 \text{ J}}{2400 \text{ J}} = \dfrac{3}{4} = 75\%$

    [The gravitational potential energy gained is equal to the work done against gravity, and a force of 1800 N is lifted vertically through 1 m, so 1800 J of work are done, and this is the work got out of the machine.]

(b) Just before the load hits the ground all the gravitational potential energy will have turned into kinetic energy. Kinetic energy = 1800 J.

Example 10

Pumped-storage power stations are used by electricity boards to produce electricity during periods of peak demand. Water is stored in one reservoir and allowed to flow through a pipe to another reservoir at a lower level. The rush of water is used to turn turbines which are connected to generators.

(a) Why are pumped-storage power stations used to produce electricity in short bursts and not for the continuous generation of electricity?   (1 mark)
(b) In one such power station 400 kg of water passes through the turbines every second after falling through a vertical height of 500 m. (g = 10 m/s^2) Assuming that no energy is wasted, calculate
    (i) the decrease in gravitational potential energy of 400 kg of water when it falls 500 m;   (3 marks)
    (ii) the power delivered to the turbines by the water;   (1 mark)

(iii) the power output of the generator; (1 mark)
(iv) the current produced by the generator if the output voltage is 20 000 V. (3 marks)
(c) At night the water is pumped back to the reservoir to be used again.
    (i) Why is this done at night? (1 mark)
    (ii) Why is more energy needed to pump the water back than is released when it falls?
(1 mark)

Solution

(a) There is only sufficient water stored in the reservoir to produce short bursts at peak times.

(b)   (i) Potential energy $= mgh$ [see section 1]
$$= 400 \text{ kg} \times 10 \text{ m/s}^2 \times 500 \text{ m}$$
$$= 2 \times 10^6 \text{ J}$$
$$= 2 \text{ MJ}$$

  (ii) 2 MJ per second = 2 MW
[1 J/s = 1 W. Note that the answer assumes that the water has no kinetic energy after it has passed through the turbines.]

  (iii) 2 MW. [No energy is wasted, so the output is the same at the input.]

  (iv) Power = voltage × current
$$2 \times 10^6 \text{ W} = 20\,000 \text{ volt} \times \text{current}$$

$$\text{Current} = \frac{2 \times 10^6}{20\,000} \text{ A}$$

$$= 100 \text{ A}$$

(c)   (i) Demand for electricity is low and electrical power can be used for pumping the water up to the reservoir.

  (ii) Heat energy is produced because of friction and other inefficiencies of the system.

Example 11

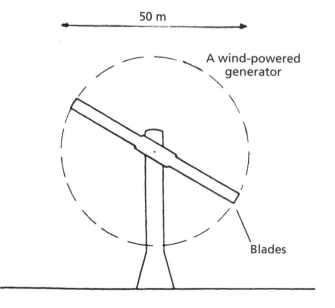

50 m

A wind-powered generator

Blades

A large wind-powered generator has blades which sweep out a circular area of diameter 50 metres. It faces head-on into a wind of average speed 10 m/s. Calculations show that in 1 second approximately 25 000 kg of air passes through the area swept out by the blades.

(a) Calculate the total kinetic energy of all the air that passes in 1 second through the area swept out by the blades when the wind speed is 10 m/s. (3 marks)

(b) The generator has an efficiency of 10% in converting the kinetic energy of the wind to electrical energy. Calculate the power output of the generator. (2 marks)

(c) If the wind speed doubles, what mass of air now passes in 1 second through the area swept out by the blades? (1 mark)

(d) If the wind speed doubles, how many times bigger will the power output of the generator be? (2 marks)

(e) The power output of a modern coal-fired power station may be 1000 MW (1×10^9 W). Use the figures in this question to discuss whether or not wind-powered generators are a practical alternative. (3 marks)

Solution

(a) Kinetic $= \frac{1}{2}mv^2$ [see section 1]
$$= (\tfrac{1}{2} \times 25\,000 \times 10^2 \text{ J})$$
$$= 1\,250\,000 \text{ J} = 1.25 \text{ MJ}$$

(b) $\dfrac{10}{100} \times 1.25$ MJ/s = 0.125 MW

(c) 50 000 kg

(d) Kinetic energy $= \frac{1}{2}mv^2$. m and v both double (doubling the velocity quadruples v^2). Therefore the kinetic energy is 8 times bigger and the output of the generator is 8 times bigger. [This part is difficult and only grade A candidates got it correct.]

(e) Power output of the wind-powered generator when wind velocity is 20 m/s is
(0.125×8) MW = 1 MW. 1000 wind-powered generators would be needed to produce the same power as one coal-fired power station. The space required would be very large and they would only work when the wind was blowing. The power available would be totally dependent on the wind. They could never replace coal-fired power stations but can be useful for generating small quantities of electrical power on windy days especially in remote locations.

Examination questions

(Numerical answers and hints on solutions will be found at the end of the chapter.)

Question 1

(a) In an oil-fired power station the burning of oil is used to heat water to produce steam. The steam drives a turbine which turns generators to produce electrical energy. The block diagram shows three parts of the system described above.

(i) Block *A* is the ....

(ii) Block *B* is the ....

(iii) The energy from *C* is.......................................

(2 marks)

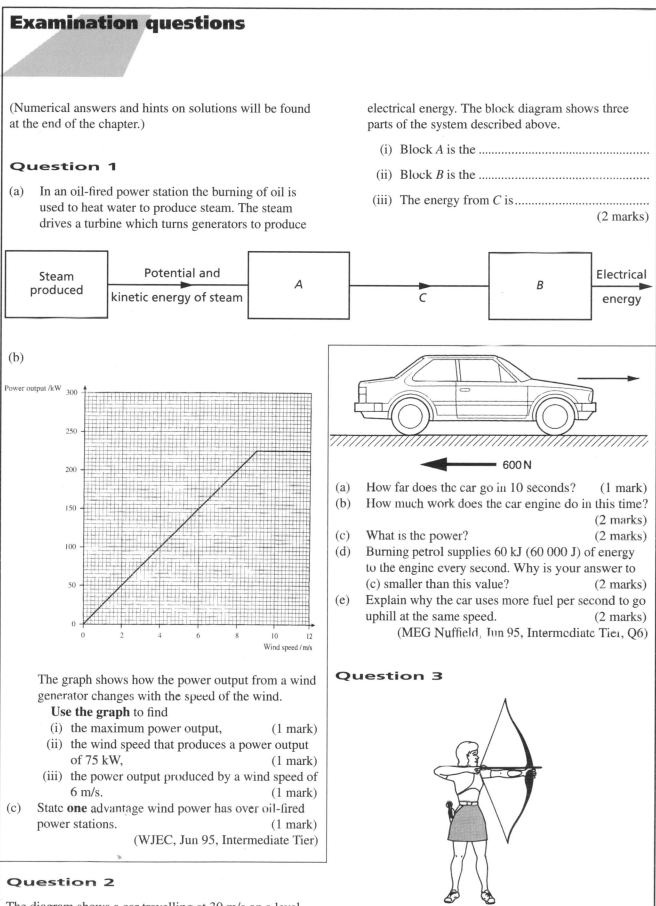

(b)

The graph shows how the power output from a wind generator changes with the speed of the wind.

Use the graph to find

(i) the maximum power output, (1 mark)

(ii) the wind speed that produces a power output of 75 kW, (1 mark)

(iii) the power output produced by a wind speed of 6 m/s. (1 mark)

(c) State **one** advantage wind power has over oil-fired power stations. (1 mark)

(WJEC, Jun 95, Intermediate Tier)

Question 2

The diagram shows a car travelling at 30 m/s on a level road. At this speed the car has to overcome a total force opposing motion of 600 N.

(a) How far does the car go in 10 seconds? (1 mark)

(b) How much work does the car engine do in this time? (2 marks)

(c) What is the power? (2 marks)

(d) Burning petrol supplies 60 kJ (60 000 J) of energy to the engine every second. Why is your answer to (c) smaller than this value? (2 marks)

(e) Explain why the car uses more fuel per second to go uphill at the same speed. (2 marks)

(MEG Nuffield, Jun 95, Intermediate Tier, Q6)

Question 3

(a) State the type of energy stored in the stretched bow. (1 mark)

(b) An arrow of mass 25 g is fired from the bow with a speed of 50 m/s.
   (i) Calculate the kinetic energy of the arrow.
   (ii) How much energy was stored in the stretched bow? Assume that the energy transfer to the arrow was 60% efficient.
   (5 marks)

(c) The arrow is fired at a block of wood of mass 175 g which is floating on water a few metres away. The arrow hits and attaches itself to the block of wood as shown below.

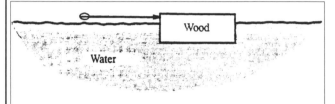

Calculate the velocity of the block and arrow just before impact. (5 marks)
(ULEAC, Syll B, Jun 95, Higher Tier, Q6)

Question 4

The diagram shows a toy car race set.

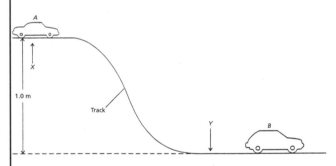

A toy car A was released at the top of the track. The car follows the track down and eventually collides with car B which is at rest.

(a) (i) State the formula linking the change in gravitational potential energy, mass, gravitational field strength and change of height. (1 mark)
   (ii) The mass of car A is 0.10 kg. Calculate the change in gravitational potential energy as it moves from X to Y. ($g = 10$ N/kg) (2 marks)

(b) All of the gravitational potential energy at X becomes kinetic energy at Y.
   (i) State the formula linking kinetic energy, mass and speed. (1 mark)
   (ii) Calculate the speed of the car A at point Y. (2 marks)

(c) Car A goes on to collide with car B. The mass of car B is 0.20 kg. The two cars stick together when they collide.
   (i) State the formula linking momentum, mass and speed. (1 mark)
   (ii) Calculate the momentum of car A at Y. (1 mark)
   (iii) Calculate the speed of both cars just after the collision. (2 marks)

(d) Sketch a graph on Fig. 3.26 to show how the speed of car A changes with time from the moment of release until after it collides with car B.

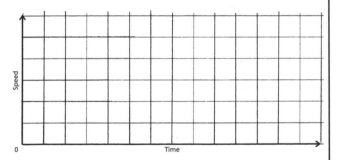

(MEG Salters', Jun 95, Higher Tier, Q9)

Answers to examination questions

1. (a) (i) Turbine
      (ii) Generator (dynamo)
      (iii) Kinetic energy.
   (b) (i) 225 kW
      (ii) 3 m/s
      (iii) 150 kW
   (c) No greenhouse or polluting gases are produced.

2. (a) 30 m/s × 10 s = 300 m
   (b) Work done (J) = Force (N) × Distance travelled in direction of force (m)
      = 600 N × 300 m
      = 180 000 J

(c) Power = $\dfrac{\text{Work done (J)}}{\text{Time taken (s)}}$

   = $\dfrac{180\ 000\ \text{J}}{10\ \text{s}}$ = 18 000 W

(d) Most of this energy is wasted as heat in the cooling system or in the exhaust gases. Some heat is also lost due to friction between the moving parts in the engine.

(e) When going up a hill, the car gains gravitational potential energy so the fuel will need to provide this additional energy as well as the energy to keep the speed at 30 m/s.

3. (a) Elastic potential energy.

(b) (i) KE $= \frac{1}{2} \times$ mass $\times$ velocity^2
$= \frac{1}{2} \times 0.025$ kg $\times (50$ m/s$)^2 = 31.25$ J

(ii) $\dfrac{\text{Energy output}}{\text{Energy input}} = \dfrac{60}{100}$

So, energy input $= \dfrac{31.25}{0.6} = 52.1$ J

(c) By conservation of momentum.
Momentum of arrow = Momentum of arrow + block after collision
[The block has no momentum before collision because it is not moving.]
Hence, $0.025 \times 50 = (0.025 \times 0.175) \times$ velocity
giving velocity = 6.25 m/s

4. (a) (i) Change in gpe = Mass (kg) $\times$ gravitational strength (N/kg) $\times$ change in height (m)

(ii) $= 0.10$ kg $\times 10$ N/kg $\times 1.0$ m
$= 1$ J

(b) (i) KE $= \frac{1}{2} \times$ mass (kg) $\times$ velocity (m/s)^2

(ii) [We assume that all the gpe has been converted into kinetic energy and that this is the only energy transfer.]
1 J $= \frac{1}{2} \times 0.10$ kg $\times v^2$
Giving velocity, $v = 4.5$ m/s

(c) (i) Momentum (kg m/s) = mass (kg) $\times$ velocity (m/s)

(ii) Momentum (kg m/s) $= 0.10$ kg $\times 4.5$ m/s $= 0.45$ kg m/s

(iii) Momentum before collision = momentum after collision
$0.45 = (0.1 + 0.2) \times v$
giving velocity of both cars after collision = 1.5 m/s

(d)

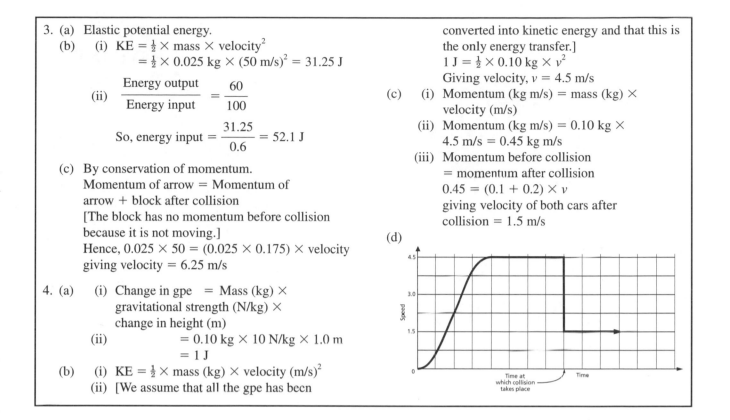

4

Moments and Machines

WJEC & NEAB	ULEAC Syll A	ULEAC Syll B	MEG	Topic	MEG Salters'	MEG Nuffield	SEG	NICCEA
✓		✓	✓	**Equilibrium and principle of moments**	✓	✓	✓	✓
✓			✓	**Bending of beams**			✓	
✓	✓	✓	✓	**Centre of gravity (or mass)**	✓	✓		✓
✓	✓	✓	✓	**Types of equilibrium**	✓	✓		
✓		✓	✓	**Efficiency of a machine**	✓	✓		✓
✓		✓	✓	**Pulleys**	✓			
✓	✓	✓	✓	**The inclined plane**	✓			✓
✓	✓	✓	✓	**Heat engines**	✓	✓	✓	✓

1 Equilibrium and the principle of moments

When a body is in equilibrium, the resultant force on its is zero and the sum of the clockwise moments about a point is equal to the sum of the anticlockwise moments about the same point. In the diagram a beam is balanced at its centre point. Using the measurements shown in the diagram,

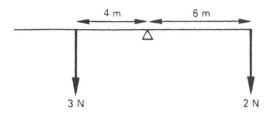

Clockwise moments = Anticlockwise moments = 12 Nm
Clockwise moment = 2 N × 6 m = 12 Nm
Anticlockwise moment = 3 N × 4 m = 12 Nm
The clockwise moment equals the anticlockwise moment and the beam is in equilibrium. In an experiment the forces arc usually applied by hanging weights on the beam.

Bending of beams

Beams bend when placed under stress. In the figure below the top of the beam is in compression and the underside is in tension. A force that causes stretching is called a *tensile force*. All parts of a stressed beam are in equilibrium under the action of the internal forces between neighbouring atoms. The greater the cross-sectional area of the beam, the greater the resistance to bending under stress.

Whenever a body changes shape by bending or stretching or compressing, it is said to be *elastic* if it returns to its original shape or length when the stress is removed. It is said to be *plastic* if it stays in its new shape or if it stays in its stretched position.

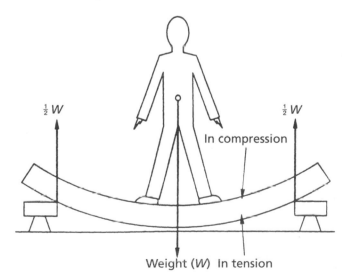

A man of weight W standing on the middle of a plank. The plank sags because the plank is not perfectly stiff. The top of it will shorten and is in compression. The underside will get longer and so is in tension.

2 Stability

Centre of gravity (centre of mass)

The centre of gravity of a body is the point through which its whole weight may be considered to act.

For a flat sheet (e.g. a protractor), its position may be determined by suspending the lamina from three points well spread around its edge and hanging a plumb-line from each point. The centre of gravity is the point on the sheet where the verticals from each point cross.

Types of equilibrium

A body is said to be in *stable equilibrium* if when given a small displacement and then released it returns to its original position.

A body is said to be in *unstable equilibrium* if when given a small displacement and then released it moves further from its original position.

A body is said to be in *neutral equilibrium* if when given a small displacement and then released it stays in its new position.

Referring to the illustration below, the equilibrium is stable in (a), unstable in (b) and neutral in (c).

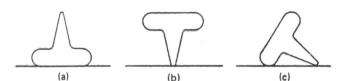

(a) Stable equilibrium (b) Unstable equilibrium (c) Neutral equilibrium

For a body in stable equilibrium, the stability is increased by having a large base area and/or a low centre of gravity. The object in (a) will return to its original position flat on the table when released, but in the situation shown in (b) the object will fall when released. The stance adopted in karate ensures a strong position: the position of the feet gives in effect a large base area and the bending of the knees lowers the centre of gravity.

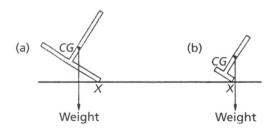

In (a) the moment about X of the weight tends to turn the body anticlockwise. In (b) the moment about X tends to turn the body clockwise and the body falls over.

3 Machines

Efficiency of a machine

$$\text{Efficiency} = \frac{\text{work got out of a machine}}{\text{work put into a machine}}$$

$$= \frac{\text{power output}}{\text{power input}}$$

Efficiency is a ratio and has no units. It is sometimes expressed as a percentage, but in the above equation it is always a fraction.

Pulleys

For the machine shown below, the work got out of the machine is equal to the potential energy gained by the load and the work put into the machine is the work done by the effort. If the load rises 1 m then each of the strings A, B and C must shorten by 1 m and the effort moves 3 m.

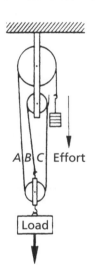

A simple pulley system.

The inclined plane

When the effort is lifted a vertical height h (see the next diagram) and the effort moves along the plane

$$\text{Efficiency} = \frac{\text{work got out}}{\text{work put in}}$$

$$= \frac{\text{energy gained by the load}}{\text{work done on the load}}$$

$$= \frac{\text{weight} \times \text{vertical height raised}}{\text{effort} \times \text{distance effort moves}}$$

$$= \frac{mgh}{\text{effort} \times d}$$

[Remember that the gravitational potential energy gained by the load is the weight (mg) multiplied by the vertical height raised.]

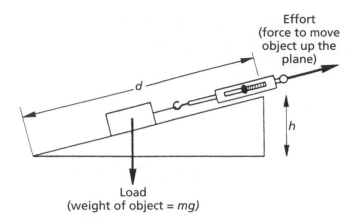

An inclined plane.

Heat engines

In a petrol engine the mixture of petrol and air is ignited and heat energy is produced. The force of the resulting explosion gives the piston kinetic energy. The burnt gases are discharged at a lower temperature. Thus, chemical energy produces heat energy which is turned into kinetic energy of the piston.

Worked examples

Example 1

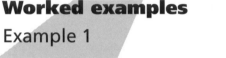

The diagram illustrates a uniform beam pivoted at its centre. The marks on the beam show equal distances each of 1 m in length. Loads of 6 N and 2 N are hung from the positions shown. Which of the following additional loads would keep the beam in equilibrium.

A   20 N at X
B   6 N at Y
C   7 N at Y
D   4 N at Z

Solution

[Total anticlockwise moment about
pivot = 6 N × 4 m = 24 Nm. Thus, total clockwise moment must be 24 Nm.
    The 2 N force provides a clockwise moment of 2 N × 2 m = 4 Nm, so the additional load must provide a moment of (24 − 4) Nm = 20 Nm.
    The additional moments are:
A   20 N × 1 m = 20 Nm
B   6 N × 3 m = 18 Nm
C   7 N × 3 m = 21 Nm
D   4 N × 4 m = 16 Nm]

Answer A

Example 2

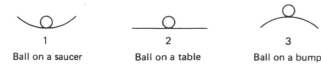

	1	2	3
	Ball on a saucer	Ball on a table	Ball on a bump

The diagrams show a ball resting on different-shaped surfaces. They represent three different types of equilibrium, which are

	1	2	3
A	Stable	Unstable	Neutral
B	Unstable	Neutral	Stable
C	Stable	Neutral	Unstable
D	Neutral	Stable	Unstable
E	Unstable	Stable	Neutral

Solution

[Use the summary in section 2, 'Types of equilibrium'.]

Answer **C**

Example 3

A wheelbarrow is shown in the diagram below. Five lifting positions are shown on the handle.

At which one is it easiest to lift the wheelbarrow?

(NICCEA, Jun 95, Intermediate Tier Q1)

Solution

[The wheel acts at the pivot and the weight acts at the centre of mass which will be situated at about the centre of the container. The moment of the weight about the wheel (pivot) will exert a clockwise moment. To lift the barrow you have to exert an anticlockwise moment. Since moment = force × distance, the force can be smaller provided that the distance is increased. Hence it will be easiest to lift the wheelbarrow at A.]

Answer **A**

Example 4

Leslie found that his 30 cm rule balanced at its mid-point. When he put one large coin on the right 12 cm from the pivot, and three small coins on the left 8 cm from the pivot, the rule balanced again.

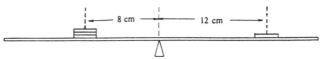

Leslie's experiment shows that the large coin weighs the same as

A   six small coins
B   four small coins
C   three small coins
D   two small coins

(ULEAC, Syll B, Jun 95, Intermediate Tier)

Solution

[Let the small coin weigh x grams and the large coin weigh y grams. Using the principle of moments we have
$24x = 12y$ (anticlockwise moments = clockwise moments)

So $y = \left(\dfrac{24}{12}\right)x = 2x$]

Answer **D**

Example 5

A painter stands on a horizontal platform which has a mass of 20 kg and is 5.0 m long. The platform is suspended by two vertical ropes, one attached to each end of the platform. The mass of the painter is 70 kg. If he is standing 2.0 m from the centre of the platform, calculate the tension in each of the ropes.

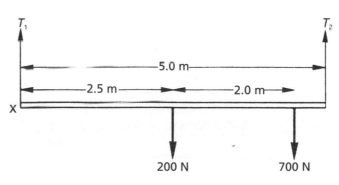

The diagram shows the forces acting on the platform. The platform is in equilibrium, so, taking moments about X,
Anticlockwise moments = clockwise moments
$T_2 \times 5\text{ m} = (200 \times 2.5)\text{ Nm} + (700 \times 4.5)\text{ Nm}$
$5\text{ m} \times T_2 = 500\text{ Nm} + 3150\text{ Nm}$

$$T_2 = \frac{3650\text{ Nm}}{5\text{ m}}$$

$T_2 = 730\text{ N}$

Since the platform is in equilibrium, the total force upwards must equal the total force downwards; hence,
200 N + 700 N = 730 N + T_1
T_1 = 170 N
The tensions are 730 N in the rope nearest the painter, and 170 N in the rope furthest from the painter.

[There are two unknowns in this problem, namely T_1 and T_2. To obtain an equation with only one unknown, choose a point about which to take moments so that the moment of one of the forces is zero. In this case the moment of T_1 about X is zero. T_2 could have been calculated by taking moments about the other end of the platform.]

Examination questions

(Numerical answers and hints on solutions will be found at the end of the chapter.)

Question 1

(a) Diagram 1 below shows a simple machine which uses pulleys to lift a load.
     The forces acting on the load can be represented by diagram 2. You can ignore the weight of the rope and pulley, and assume that the machine is frictionless.

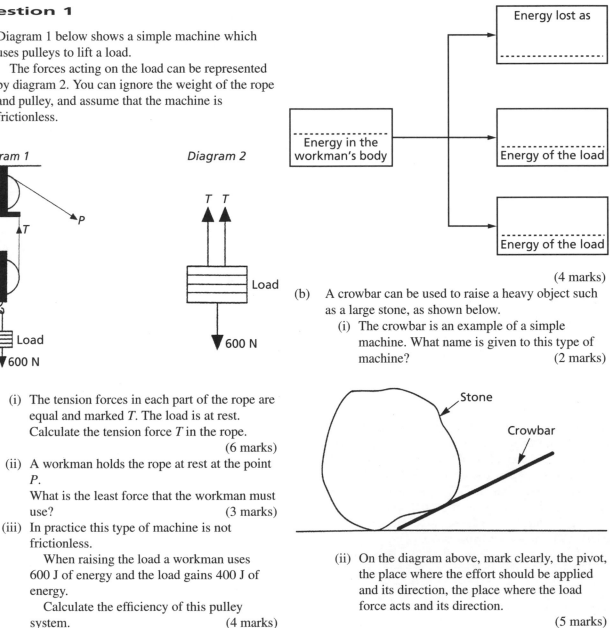

Diagram 1   *Diagram 2*

(i)   The tension forces in each part of the rope are equal and marked T. The load is at rest. Calculate the tension force T in the rope.
                                                              (6 marks)
(ii) A workman holds the rope at rest at the point P.
      What is the least force that the workman must use?   (3 marks)
(iii) In practice this type of machine is not frictionless.
      When raising the load a workman uses 600 J of energy and the load gains 400 J of energy.
      Calculate the efficiency of this pulley system.   (4 marks)

(iv) Complete the energy flow diagram below to show the energy changes that occur as the load is being raised.

Energy lost as ....

Energy in the workman's body

Energy of the load

Energy of the load
                                                              (4 marks)

(b) A crowbar can be used to raise a heavy object such as a large stone, as shown below.
     (i) The crowbar is an example of a simple machine. What name is given to this type of machine?   (2 marks)

Stone
Crowbar

     (ii) On the diagram above, mark clearly, the pivot, the place where the effort should be applied and its direction, the place where the load force acts and its direction.
                                                              (5 marks)

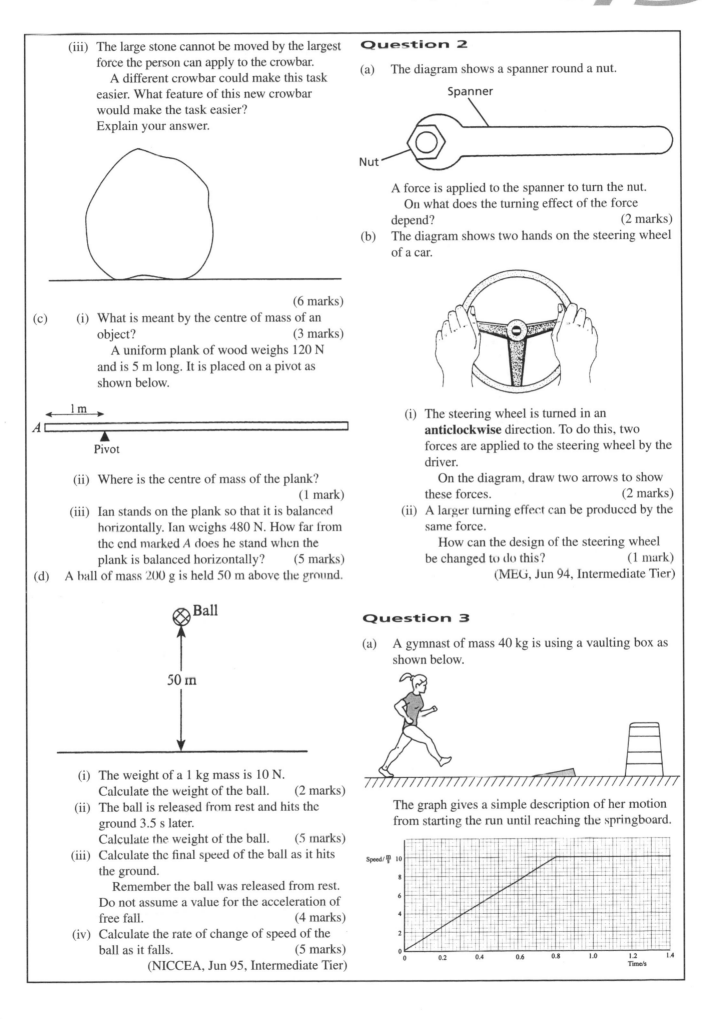

(iii) The large stone cannot be moved by the largest force the person can apply to the crowbar.

A different crowbar could make this task easier. What feature of this new crowbar would make the task easier?
Explain your answer.

(6 marks)

(c) (i) What is meant by the centre of mass of an object? (3 marks)

A uniform plank of wood weighs 120 N and is 5 m long. It is placed on a pivot as shown below.

(ii) Where is the centre of mass of the plank? (1 mark)

(iii) Ian stands on the plank so that it is balanced horizontally. Ian weighs 480 N. How far from the end marked A does he stand when the plank is balanced horizontally? (5 marks)

(d) A ball of mass 200 g is held 50 m above the ground.

(i) The weight of a 1 kg mass is 10 N.
Calculate the weight of the ball. (2 marks)

(ii) The ball is released from rest and hits the ground 3.5 s later.
Calculate the weight of the ball. (5 marks)

(iii) Calculate the final speed of the ball as it hits the ground.
Remember the ball was released from rest. Do not assume a value for the acceleration of free fall. (4 marks)

(iv) Calculate the rate of change of speed of the ball as it falls. (5 marks)

(NICCEA, Jun 95, Intermediate Tier)

Question 2

(a) The diagram shows a spanner round a nut.

A force is applied to the spanner to turn the nut.
On what does the turning effect of the force depend? (2 marks)

(b) The diagram shows two hands on the steering wheel of a car.

(i) The steering wheel is turned in an **anticlockwise** direction. To do this, two forces are applied to the steering wheel by the driver.
On the diagram, draw two arrows to show these forces. (2 marks)

(ii) A larger turning effect can be produced by the same force.
How can the design of the steering wheel be changed to do this? (1 mark)

(MEG, Jun 94, Intermediate Tier)

Question 3

(a) A gymnast of mass 40 kg is using a vaulting box as shown below.

The graph gives a simple description of her motion from starting the run until reaching the springboard.

(i) What was her initial acceleration?
(ii) Calculate the size of the force which produces this acceleration. (5 marks)

(b) In a second event, a uniform beam of length 6.0 m is supported 0.5 m from each end. The mass of the beam is 120 kg.

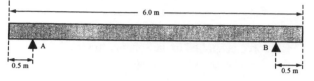

(i) Label the centre of mass of the beam with the letter *M*.
(ii) The gymnast weighs 400 N and stands on the beam 2.0 m from support *A*.
Calculate the moment of the weight of the gymnast about support *A*.
(iii) Calculate, showing all your working, the forces exerted by the supports on the beam when the gymnast is standing on it. Draw a diagram of the beam in the space below if you wish.
(Gravitational field strength = 10 N/kg.)
(11 marks)
(ULEAC, Syll B, Jun 95, Higher Tier, Q4)

Question 4

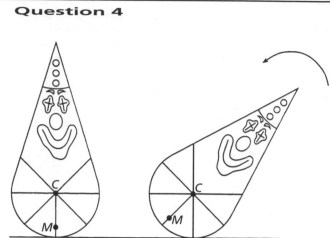

A child's toy clown can be pushed over but quickly returns to the upright position. Its base is circular with centre *C*. The centre of mass is at *M*. Explain why the clown returns to the vertical position. (2 marks)
(WJEC, Jun 95, Higher Tier, Q12)

Answers to examination questions

1. (a) (i) Considering the bottom pulley and the forces acting on it.

Forces up = Forces down
$$2T = 600 \text{ N}$$
$$T = 300 \text{ N}$$

(ii) 300 N

(iii) Efficiency = $\dfrac{\text{work out}}{\text{work in}} = \dfrac{400}{600} = 0.67$

(iv)

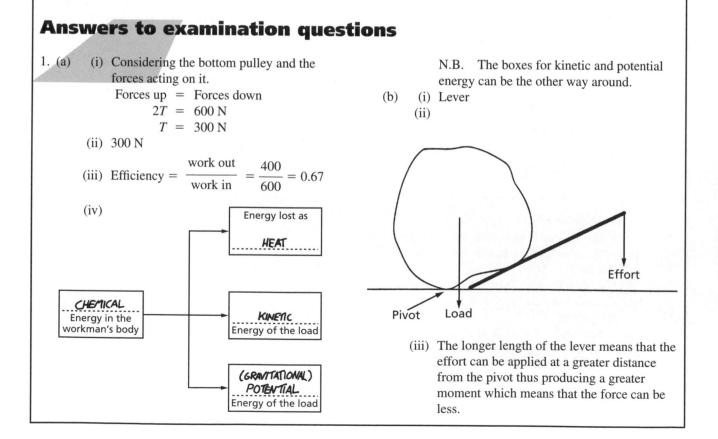

N.B. The boxes for kinetic and potential energy can be the other way around.

(b) (i) Lever
(ii)

(iii) The longer length of the lever means that the effort can be applied at a greater distance from the pivot thus producing a greater moment which means that the force can be less.

(c) (i) This is the point through which all the weight appears to act. It is also the point at which it would balance.

(ii) At the centre which is 2.5 m from either end.

(iii)

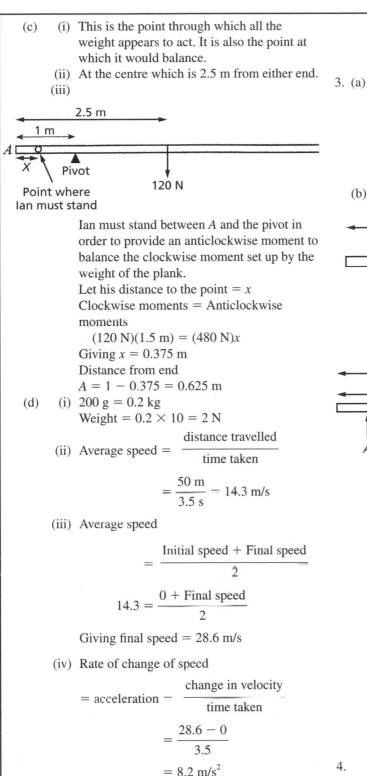

Ian must stand between A and the pivot in order to provide an anticlockwise moment to balance the clockwise moment set up by the weight of the plank.

Let his distance to the point $= x$

Clockwise moments = Anticlockwise moments

$$(120 \text{ N})(1.5 \text{ m}) = (480 \text{ N})x$$

Giving $x = 0.375$ m

Distance from end

$A = 1 - 0.375 = 0.625$ m

(d) (i) 200 g = 0.2 kg

Weight = $0.2 \times 10 = 2$ N

(ii) Average speed $= \dfrac{\text{distance travelled}}{\text{time taken}}$

$$= \frac{50 \text{ m}}{3.5 \text{ s}} = 14.3 \text{ m/s}$$

(iii) Average speed

$$= \frac{\text{Initial speed + Final speed}}{2}$$

$$14.3 = \frac{0 + \text{Final speed}}{2}$$

Giving final speed = 28.6 m/s

(iv) Rate of change of speed

$$= \text{acceleration} = \frac{\text{change in velocity}}{\text{time taken}}$$

$$= \frac{28.6 - 0}{3.5}$$

$$= 8.2 \text{ m/s}^2$$

2. (a) The size of the force and the distance between the force and the pivot (the pivot is the nut in this case).

(b) (i)

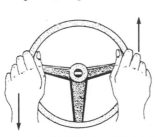

(ii) The steering wheel may be made with a larger diameter which makes the same forces exert a larger turning effect.

3. (a) (i) Initial acceleration = gradient

$$= \frac{10 \text{ m/s}}{0.8 \text{ s}} = 12.5 \text{ m/s}^2$$

(ii) Force (N) = mass (kg) × acceleration (m/s^2)

$$= 40 \text{ kg} \times 12.5 \text{ m/s}^2 = 500 \text{ N}$$

(b) (i)

6.0 m

M

(ii) Moment = force × distance

= 400 N × 2 m

= 800 Nm

(iii)

Taking moments about A [this eliminates the normal reaction acting at A because it has no moment about the point A owing to its zero distance to this point] gives:

Clockwise moments = Anticlockwise moments

$$400 \times 2 + 1200 \times 2.5 = R_2 \times 5$$

$$800 + 3000 = 5R_2$$

$$R_2 = 760 \text{ N}$$

Taking moments about B

Clockwise moments = Anticlockwise moments

$$5R_1 = 400 \times 3 + 1200 \times 2.5$$

$$5R_1 = 4200$$

$$R_1 = 840 \text{ N}$$

[Notice that the total upward forces, 1600 N, equals the total downward forces, 1600 N.]

4. The weight of the clown acts vertically down through the centre of mass. The point on the base of the clown in contact with the ground, acts as the pivot. When weight acts through the pivot there is no moment so the clown stays at rest. When displaced to one side, the pivot and the line of action of the weight are separated by a distance setting up a moment. This moves the clown back to its upright position.

5

Kinetic Theory and Gas Laws

Topic

WJEC & NEAB	ULEAC Syll A	ULEAC Syll B	MEG	Topic	MEG Salters'	MEG Nuffield	SEG	NICCEA
✓	✓	✓	✓	**Brownian motion**	✓	✓	✓	
✓	✓	✓	✓	**Diffusion**	✓	✓	✓	
✓	✓	✓	✓	**Kinetic theory**	✓	✓		✓
✓	✓	✓	✓	**The gas laws**	✓	✓	✓	

1 Evidence for the kinetic theory (Brownian motion and diffusion)

Brownian motion (see examples 1 and 9)

This provides evidence for the kinetic theory of matter. Brownian motion may be observed by using a microscope to view illuminated smoke particles in a smoke cell. The specks of light (the light reflected and scattered by the smoke particles) jostle about in a random irregular manner as they are bombarded by the invisible moving air molecules.

Diffusion (see example 8)

This also provides evidence for the motion of molecules.

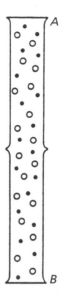

The glass disc has been removed and diffusion has occurred.

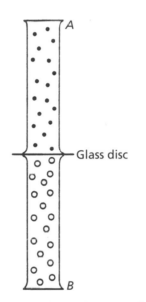

The denser gas in B is separated from the gas in A by a glass disc.

The diagram represents molecules of two different gases in two gas jars separated by a glass disc. The gas in jar *B* is denser than the gas in jar *A*. When the glass disc is removed, the gases gradually mix. The next diagram represents the situation some time later. The process is known as *diffusion*.

2 Atoms and molecules

All matter consists of tiny particles called atoms or molecules. A molecule consists of two or more atoms chemically bonded together.

Solids

The particles in a solid are close together and in fixed positions and are held together by strong forces of attraction which means that the solid has a fixed shape. The particles vibrate about a mean position but are unable to move from place to place.

Liquids

The particles in a liquid are still close together but now they are able to move independently since the forces of attraction are not as great as in a solid. Liquids do not have a fixed shape but they have a fixed volume.

Gases

These particles are widely spaced and have little or no force of attraction between their particles. Gases have no fixed shape or volume and will take up any space which is offered to them.

3 Using the kinetic theory to explain properties of matter (expansion, melting, evaporation and boiling)

The molecules of a solid vibrate about fixed positions. When heat is supplied to a solid, the molecules vibrate faster and through greater distances than before. The molecules are pushed further apart and the solid expands. If sufficient heat energy is supplied, the solid melts; the molecules are still close together and the resulting change in volume is small. In a liquid the molecules no longer vibrate about fixed positions but move among one another. The forces between the molecules are enough to give a definite volume but not a definite shape. As more heat energy is supplied, the molecules move faster and gain enough energy to break right away from one another and move independently. The liquid has started to boil and changes to a gas. In the gaseous state, the molecules are a long way apart and are in random motion, with varying velocities. They are continually colliding with one another and with the walls of the container, causing a pressure on the walls of the containing vessel. If the temperature is increased, the molecules move faster and hit the walls more often and harder, causing the pressure to increase. At the absolute zero of temperature (0 k or $-273\ ^\circ$C) the molecules have their lowest possible kinetic energy.

If a gas is compressed at constant temperature, the average distance between the molecules decreases. There are more collisions with the walls every second and the pressure increases.

Evaporation is the escape of the faster-moving molecules from the surface of a liquid, and this takes place at all temperatures. The average kinetic energy of the remaining molecules has decreased and so the temperature has fallen. The rate of evaporation may be increased by (i) increasing the surface area of the liquid, (ii) blowing air over the surface or (iii) increasing the temperature of the liquid. Boiling occurs when bubbles of vapour form in the body of the liquid, and rise and escape from the surface of the liquid. Boiling takes place at a particular temperature.

Evaporation of water from the skin reduces the body temperature.

If the body is exposed to cold windy weather, the risk of hypothermia (the cooling of the body to a low temperature) will be increased if moisture evaporates from the surface of the skin. The drop in body temperature can be reduced by wearing clothing made from materials which contain trapped air and are poor conductors of heat.

4 The behaviour of gases

For a fixed mass of gas at constant temperature,

$P_1V_1 = P_2V_2, p \propto \dfrac{1}{V}$ (Boyle's Law). When the

temperature, T, of a gas is raised, either the pressure, P, or the volume, V, or both increase. They are related by the equation

$$\frac{P_1V_1}{T_1} = \frac{P_2V_2}{T_2}$$

where the suffix 1 represents the initial state of the gas and the suffix 2 represents the final state. *Remember that* T *must be in kelvin – that is* ($^\circ$C + 273).

For a fixed mass of gas at constant volume, $P \propto T$ (the pressure law).

For a fixed mass of gas at constant pressure, $V \propto T$ (Charles' Law).

Following is a graph of pressure (P) against temperature (T), the volume being kept constant. The dotted part of the line is what we might expect to happen if we went on cooling the gas and it never liquefied. The absolute zero of

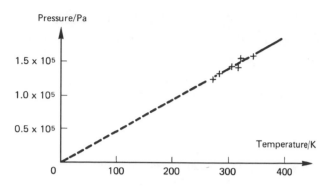

Graph of pressure against temperature for a fixed mass of gas at constant volume. A graph of volume against temperature (pressure constant) is the same shape and goes through 0 K.

temperature is the temperature at which the pressure would become zero.

Worked examples

Example 1

Brownian motion is observed by observing smoke particles under a microscope. Which of the following statements is correct?

A The smoke particles move about with uniform velocity
B The motion is caused by the air molecules colliding with the smoke particles
C If larger smoke particles were used, the Brownian motion would be more rapid
D The experiment would work just as well in a vacuum

Solution

[**A** is incorrect and **B** is correct. Larger smoke particles would show less rapid motion. They are more massive than smaller particles and are therefore accelerated less for a given force. In a vacuum there would be no air particles to bombard the smoke particles.]

<u>Answer **B**</u>

Example 2

Which one of the following graphs correctly represents the variation of pressure with absolute temperature for a fixed mass of an ideal gas at constant volume?

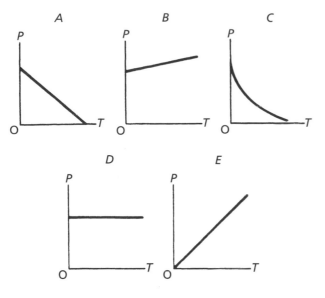

Solution

[The pressure is directly proportional to the kelvin (absolute) temperature. Direct proportion is represented by a straight line through the origin.]

Answer **E**

Example 3

When a kettle is boiling, the water in it is turning to steam, what is happening to the water *molecules*?
A They are losing energy
B They are beginning to vibrate
C They are expanding
D They are getting lighter
E They are moving further apart

Solution

[The molecules *gain* energy and move further apart. See section 3.]

Answer **E**

Example 4

(a) The diagrams below show a simple model of the arrangement of the molecules of a substance in its solid and liquid states.

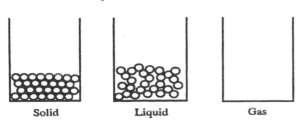

(i) Complete the diagram to show the arrangement of the molecules when the substance is in its gaseous state. (1 mark)

(ii) The boiling point of the substance is 56 °C and its melting point is 5 °C. In what state will the substance be at room temperature? Explain your answer. (2 marks)

(b) The diagram below shows a method of opening and closing a valve which depends on the temperature of the surrounding water.

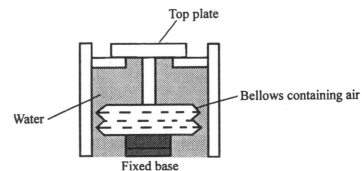

When the surrounding water is cold, the top plate closes the valve. When the temperature increases, the top plate moves up and opens the valve.
(i) Explain, in terms of molecules, how the air in the bellows is able to exert a pressure. (2 marks)
(ii) Explain why the top plate is pushed up when the temperature of the surrounding water is increased. (2 marks)
(iii) State TWO properties which the material of the bellows should have if the valve is to respond to small changes in temperature. Give a reason in each case. (4 marks)

(c) When the bellows are surrounded by water at 20 °C, the pressure of the air in them is 100 000 Pa. Calculate the new pressure when the temperature is increased to 50 °C. You can assume that the volume of air remains constant.
(ULEAC, Syll A, Jun 95, Intermediate Tier)

Solution

(a) (i) There should be few molecules drawn separated by a large distance.
(ii) Liquid because room temperature is between the melting point and the boiling point of the substance which means it will have melted but not boiled.

(b) (i) The molecules are moving randomly and are repeatedly colliding with the walls thus exerting a force on the wall during each collision. This force exerts a pressure on the container.
(ii) The temperature increase causes the molecules to move faster and therefore make more frequent and harder collisions with the walls. The pressure increases and causes the bellows to expand.
(iii) It must be thin so that the heat gets through quickly and it can respond rapidly to temperature changes.
It should be flexible so that it is easy to move.

(c) [Since the volume is constant, the volume terms may be left out of the equation.]

$$\frac{P_1}{T_1} = \frac{P_2}{T_2}$$

$$\frac{100\ 000\ \text{Pa}}{293\ \text{K}} = \frac{P_2}{323\ \text{K}}$$

Giving $P_2 = 110\ 239$ Pa

Example 5

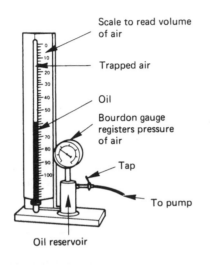

Scale to read volume of air

Trapped air

Oil

Bourdon gauge registers pressure of air

Tap

To pump

Oil reservoir

The diagram shows an apparatus for investigating Boyle's Law. Air is trapped in the vertical tube by the oil. The tap is opened and a car pump is used to pump the oil up the tube, decreasing the volume of the gas. The volume of air is read on the scale behind the vertical tube and the Bourdon gauge registers the pressure. By opening and closing the tap a series of readings of volume and pressure may be obtained.

  (i) Each time the oil is lowered, a short time must be allowed to elapse before the reading of the volume is taken. Why is this?   (1 mark)

 (ii) The Bourdon gauge registers the pressure. What are the units of pressure?   (1 mark)

(iii) Sketch the graph you would obtain by plotting the pressure of the air against l/volume.

  (2 marks)

Solution

  (i) To allow the oil on the sides of the tube to run down and for the air in the tube to return to room temperature.

 (ii) N/m^2

(iii)

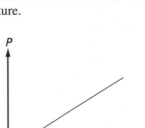

Example 6

Miriam and Myo used the apparatus below to investigate the changes in pressure in a gas as the temperature is varied.

To vary the temperature of the air in the flask the water bucket was used. The temperature inside the bucket could be raised by adding boiling water.

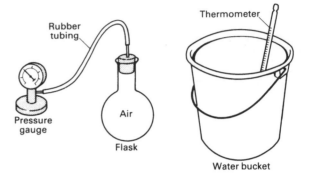

Thermometer

Rubber tubing

Pressure gauge

Air

Flask

Water bucket

They obtained the results in table 5.1:

Table 5.1

Temperature in °C	Pressure in kPa
0	95
15	100
30	105
45	111
60	116
75	122
90	127

(a) How do you think the girls obtained the starting temperature of 0 °C in the bucket?   (2 marks)

(b) What precautions would they have taken to ensure that the temperature of the air inside the flask is the same as that shown on the thermometer in the bucket?   (2 marks)

(c) Their results are plotted on the graph below.

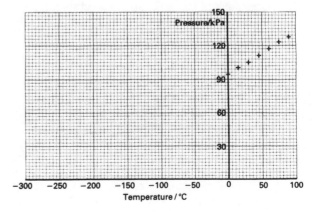

(i) On the graph draw the line of best fit.

(ii) Miriam suggests that their results showed that the pressure is proportional to temperature.
Myo did not agree, saying 'If that were true then doubling the temperature should double the pressure. Our results don't show that!'
Do you agree with Miriam or Myo? Discuss.
(2 marks)

(iii) Use your graph to predict the temperature of the air when the pressure drops to zero. (Assume that the air behaves as an ideal gas.)
(2 marks)

(d) The kinetic theory is based on the assumption that gases consist of millions of tiny molecules in continual random motion.
Use the theory to explain:

(i) the change in pressure as the temperature of the flask is increased. (4 marks)

(ii) the significance of the answer you got in (c)(iii). (2 marks)

Solution

(a) Putting ice in the bucket and stirring it until the temperature reached 0 °C.

(b) Stir the water in the bucket. Wait for some time before taking the temperature.

(c) (i)

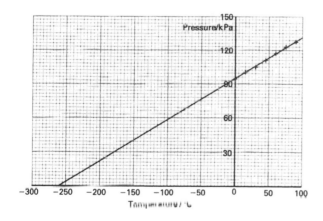

(ii) Myo is right. For direct proportion a straight line graph goes through (0,0). The pressure is proportional to the kelvin temperature (by this experiment −260 °C (taken as 0 K)).

(iii) −260 °C.

(d) (i) As the temperature of the flask rises the temperature of the air in it rises. The average velocity of the molecules increases. The collisions of the molecules with the sides of the container are harder and more frequent. The resultant force on each unit area of the sides has increased, i.e. the pressure has increased.

(ii) When the maximum possible energy has been extracted from the air, the kinetic energy of the molecules is zero and the pressure has dropped to zero.

Example 7

(a) When air in a bicycle pump is compressed by moving the piston, the temperature of the air in the pump increases. Explain this temperature rise in terms of the kinetic theory of gases. (2 marks)

(b) Describe what happens to the gas molecules as a gas at constant volume is cooled and indicate how this results in a reduction of pressure.
(2 marks)
(SEG, Higher Tier, Part Question)

Solution

(a) As the moving molecules strike the moving piston, they rebound with increased velocity. The average kinetic energy of the molecules has increased, and therefore the temperature has increased. [Or, a force has been applied to the piston in order to move it and so compress the gas (which due to its pressure exerts a force on the piston). The work done (force × distance) in compressing the gas increases the average kinetic energy of the molecules and hence increases the temperature of the gas.]

(b) As the gas is cooled the average kinetic energy of the gas molecules decreases, so the average velocity of the molecules falls. The impacts on the side of the vessel are not so hard and are less frequent, so the pressure falls.

Example 8

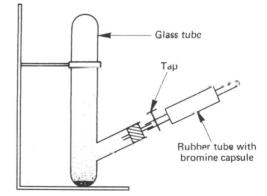

(a) The diagram shows a glass tube. The rubber tube on the side contains a bromine capsule. The end of the tubing is sealed with a glass rod. The bromine capsule is broken by squeezing the rubber tubing with a pair of pliers and the tap is opened.

(i) What will be observed in the vertical glass tube? (1 mark)

(ii) Explain this observation. (2 marks)

(iii) What would be observed if the air had been withdrawn from the vertical glass tube before the tap was opened? (1 mark)

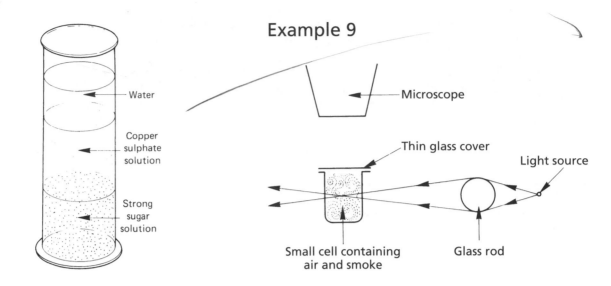

Example 9

Water

Copper sulphate solution

Strong sugar solution

Microscope

Thin glass cover

Light source

Small cell containing air and smoke

Glass rod

(b) The diagram shows the result of pouring a strong sugar solution into a gas jar. A strong copper sulphate solution was then poured very slowly and very carefully on top of the sugar solution. Finally, water was carefully poured on top of the copper sulphate solution. The result was a fairly sharp dividing line between the layers.

  (i) What will be observed in the gas jar after it has been left standing for some time?

(2 marks)

  (ii) Explain the observation. (2 marks)

Solution

(a)   (i) The dark colour of the bromine vapour will move (diffuse) slowly up the tube.

  (ii) The moving molecules of bromine continually collide with the air molecules in the tube and their progress up the tube is impeded.

  (iii) The dark brown colour of the bromine vapour would immediately fill the tube. [There are no air molecules to impede the movement of the bromine molecules. The bromine molecules are moving very fast and the moment the tap is opened the tube is filled with bromine molecules.]

(b)   (i) The blue copper sulphate will be observed to diffuse upwards into the water and downwards into the sugar solution.

  (ii) When two liquids are in contact with one another, the continuous movement of the molecules means that the molecules pass across the boundary between the liquids. Some will collide almost immediately with molecules of the other liquid, others will move a little further into the spaces between the molecules before colliding with another molecule. In this way the molecules of copper sulphate gradually move into the water and sugar solution.

The diagram shows one form of apparatus used to observe the Brownian motion of smoke particles in air. A student looking through the microscope sees tiny bright specks which he describes as 'dancing about'.
(a) What are the bright specks?
(b) Why are the specks 'dancing about'?
(c) What is the purpose of the thin glass cover?

(5 marks)

(L)

Solution

(a) The bright specks are light which is reflected and scattered by the smoke particles.
(b) The smoke particles are being bombarded by the air molecules which are in rapid random motion.
(c) To reduce air (convection) currents and to stop the smoke escaping.

Example 10

Use the kinetic theory to explain
(a) why the air in a car tyre which is kept at constant temperature and constant volume exerts a constant pressure; (4 marks)
(b) why heat energy must be supplied to turn water into steam; (4 marks)
(c) the existence of an absolute zero of temperature.

(4 marks)

Solution

(a) The molecules of air are in rapid motion with a large range of speeds. Large numbers strike the walls of the tyre every second. The average speed of the molecules is constant (because the temperature is constant), so the average force due to the bombardment is constant. The pressure is the force per unit area, so the pressure is constant.
(b) Intermolecular forces maintain the shape of the

liquid. When heat energy is supplied, the molecules gain kinetic energy and eventually break away from the attractive forces which exist between them when they are in the liquid state. They break free, and so the water becomes steam. Energy is also used to push back the air molecules above the liquid.

(c) As heat is withdrawn from a substance, the energy of the molecules becomes less and their average velocity, and hence the temperature, decreases. Eventually there comes a time when no more energy can be withdrawn and we have reached the absolute zero of temperature..

Examination questions

(Numerical answers and hints on solutions will be found at the end of the chapter.)

Question 1

In terms of the forces of attraction between the particles, the particle spacing and their motion.

(a) describe and explain the difference in density between liquids and gases (4 marks)
(b) explain why gases diffuse more quickly than liquids (4 marks)
(c) describe and explain the difference in compressibility between solids and gases (6 marks)
(d) describe and explain the change in volume that occurs on boiling. (5 marks)
(NEAB, Jun 94, Intermediate Tier)

Question 2

(a) The particles that make up solids, liquids and gases have different spacing and forces of attraction.

(i) Complete table 2:

Table 2

Forms of matter	Forces of attraction	Spacing
Solid		
Liquid	Strong	Close
Gas		

(ii) Using the idea of particles, explain why gases are easier to compress than liquids. (8 marks)

(b) A smoke cell is illuminated with a lamp and viewed using a microscope.

(i) Explain why you see bright specks of light.
(ii) Describe the motion of the specks.
(iii) This is called Brownian motion. The observed motion is caused by

convection of the air ☐

random impacts of the air molecules with the smoke particles ☐

smoke particles being pushed by the heat of the lamp. ☐

Tick **one** box only

(5 marks)

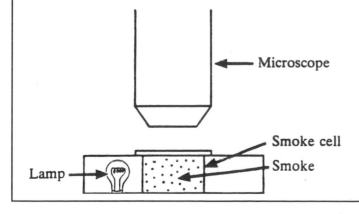

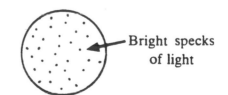

View through the microscope

(c) A soluble deep purple crystal is placed carefully at the bottom of a beaker of water as shown in the diagram.

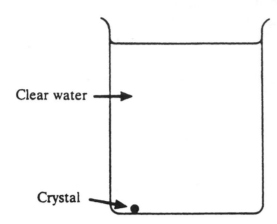

Clear water →

Crystal →

After a few days, the purple colour has spread through the water.

  (i) What is this process called?
 (ii) Using the ideas from part (b), explain why this happens.   (3 marks)
                (NEAB, Jun 92, Intermediate Tier)

Question 3

(a) Here are some statements each describing either a solid, liquid or gas. Read the statements then write the correct word, solid, liquid or gas, next to each one.
    (i) The particles are close together but slip against each other in all directions.
   (ii) The particles continually hit each other and their container.
  (iii) The particles are arranged in a regular pattern.
  (iv) The particles spread out to fill all the spaces available.   (4 marks)
(b) Explain in terms of molecules the following observations.
    (i) Bromine molecules travel at about 200 m/s in a vacuum yet it takes many minutes for the bromine to travel 50 cm inside a tube of air.
                                (3 marks)

 (ii) Before having an injection the doctor wipes some ethanol onto your arm. As the ethanol evaporates your arm feels cold.
                            (3 marks)
           (SEG, Spec, Intermediate Tier)

Question 4

(a) A mass of oxygen occupies a volume of 0.01 m^3 at a pressure of 100 000 Pa and a temperature of 0°C. If the pressure is increased to 5 000 000 Pa and the temperature is increased to 25°C, what volume will the gas now occupy?

(b) A student investigates the relationship between the temperature and the pressure of a fixed mass of gas using the apparatus shown in the figure below.

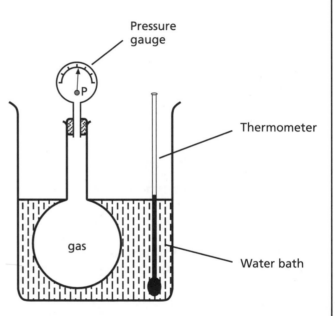

The water bath is heated continuously using a Bunsen burner and the temperature and the pressure are recorded every minute.
    (i) State two ways in which the experiment may be improved.
   (ii) When the results are obtained from the improved experiment, how could the relationship between the pressure and the temperature on the Kelvin scale be found?

Answers to examination questions

1. (a) The particles in a liquid are close together and are almost touching whereas gas particles are widely spaced. If the same volume of a liquid and a gas are taken, then there will be more particles in the liquid since they are closer together. This means that the mass in this volume is greater for a liquid and its density is greater.

   (b) Particles in liquids and gases are free to move individually; however in a gas there is virtually no force between particles whereas in a liquid there are still quite strong forces. There are large spaces between molecules in gases whereas liquids are closely packed and this enables gas particles to easily move between each other. The speed of the particles in a gas are therefore higher and the molecules diffuse more quickly.

   (c) Solids consist of particles almost touching each other whereas gas particles are widely spaced so gases may be compressed more easily. Another reason is the fact that there are large forces between particles in a solid so this also makes them harder to compress compared to the small (almost zero) forces between gas particles.

   (d) When boiling occurs the liquid particles gain enough energy to escape the attraction of the other particles to break through the surface and become a gas. On becoming a gas, the particles move apart and the force of attraction between particles decreases considerably. Both these factors mean that the volume increases considerably

2. (a) (i)

   Table 2

Forms of matter	Forces of attraction	Spacing
Solid	Strong	Close
Liquid	Strong	Close
Gas	Weak	Far apart

   (ii) Liquid particles are close together and there are strong forces between them. There is little room to push them closer and the force required to do this is large. Gases however have large distances between molecules and only very small forces of attraction between the molecules which makes it easier to push the molecules closer together.

   (b) (i) These bright dots are the light reflected off the smoke particles.

   (ii) The specks move randomly (with different speeds and directions).

   (iii) Caused by random impact of the air molecules with the smoke particles. [The other two options would mean that all the smoke particles would move in the same direction.]

   (c) (i) Diffusion

   (ii) The purple substance when dissolved has particles which move randomly and the water also has randomly moving particles. These particles collide with each other and sometimes manage to move past each other which means that they eventually mix and give a uniform purple colour throughout.

3. (a) (i) liquid (ii) gas (iii) solid (iv) gas

   (b) (i) There are collisions between the bromine and the air molecules.
   The bromine molecules change directions more often due to the collisions.
   The upward speed is therefore reduced.

   (ii) The alcohol will evaporate, which means the faster moving molecules will escape the surface and turn into a gas. Those molecules left behind will have their average speed lowered which means the temperature will fall.

4. (a) Using the equation $P_1 V_1 / T_1 = P_2 V_2 / T_2$ we have $100\,000 \times 0.01/273 = 5\,000\,000 \times V_2/298$ giving $V_2 = 0.00022 \text{ m}^3$

   (b) (i) Use a stirrer to ensure that all of the water (and hence the air in the flask) will be at the same temperature.

Not all of the air in the flask will be at the same temperature as the water since part of the flask rises above the water level so the flask needs to be moved down.

(ii) All the temperature measurements obtained from the thermometer in °C will need to have 273 added to them to convert them to temperatures in the Kelvin scale. A graph of pressure plotted on the y-axis is plotted against the Kelvin temperature on the x-axis and the graph should be a straight line which passes through the origin which proves that the pressure is proportional to the absolute temperature.

Heat and
Change of state

WJEC & NEAB	ULEAC Syll A	ULEAC Syll B	MEG	Topic	MEG Salters'	MEG Nuffield	SEG	NICCEA
✓	✓	✓	✓	**Heat and temperature**	✓	✓	✓	✓
✓	✓			**Effects of impurities and pressure on bpts and mpts**				
✓	✓	✓		**Refrigerators**				
✓	✓	✓	✓	**Specific heat capacity**	✓	✓	✓	
✓	✓	✓		**Specific latent heat**				

1 Heat and temperature

Heat is a form of energy and is measured in joules.

Temperature is a measure of the amount of kinetic energy (movement energy) that each molecule has whereas heat depends on the mass of the object.

Temperature is the degree of hotness and is measured in either °C or K. It is possible for an object to be at a very high temperature yet possess very little heat energy. Take for example a spark falling on your hand from a sparkler. The piece of metal may be white hot but since its mass is so small the heat it possesses is very small and therefore it does not burn your hand.

2 Effect of impurities and the pressure on boiling and melting points

If salt is added to water it raises the boiling point and lowers the freezing point. Salted water for cooking vegetables therefore boils above 100 °C and salt put on roads in cold weather lowers the freezing point of water, and any ice or snow on the roads melts. Increase in pressure raises the boiling point and lowers the freezing point. In pressure cookers the water boils at a temperature above 100 °C. Snowballs do not 'bind' on a very cold day because the increase in pressure on squeezing does not lower the freezing point below that of the snow, so the snow does not melt. Usually on squeezing, the snow melts and refreezes when the pressure is removed, so the snowball 'binds'.

3 Refrigerators

A compressor compresses a very volatile vapour (the refrigerant) and it liquefies. The refrigerant expands through a valve into tubes which are in the ice compartment. As it expands it again becomes a vapour. The energy needed to convert the liquid to a vapour is drawn from the ice compartment, which is therefore cooled. The vapour is compressed outside the refrigerator, where it gives out its latent heat (see example 7).

4 Specific heat capacity

The *specific heat capacity* of a substance is the heat energy required to raise 1 kg of it through 1 K or 1 °C. Its units are J/(kg K) or J/(kg °C).

It follows that if a mass m of a substance of specific heat capacity c is raised through a temperature T, then

Heat required, Q = mass × specific heat capacity × temperature change or, using symbols, $Q = mcT$.

To measure c, heat may be applied to a known mass, m, of substance by an immersion heater connected to the power supply via a joulemeter (which measures the energy supplied, Q). The rise in temperature, T, is measured, and the equation above is used to calculate c (see example 6). If

the substance is a solid, a hole is drilled in a block of it so that the immersion heater may be inserted. The thermometer is inserted into a second hole in the block.

Water has a high specific heat capacity compared with other liquids. Hence, for a given mass and a given quantity of heat, the temperature rise is comparatively small. Climate is therefore more temperate near large expanses of water, and fruit growing districts are often near large lakes.

Heating and cooling systems which use liquids work better the greater the mass of liquid used and the greater the specific heat capacity of the liquid.

5 Specific latent heat

The *specific latent heat* is the quantity of heat required to change the state of 1 kg of substance without a change in temperature. Its units are J/kg (you might meet kJ or MJ; 1 kJ = 10^3 J and 1 MJ = 10^6 J). If the change is from liquid to vapour, then it is the specific latent heat of vaporisation (note the spelling!), and if the change is from solid to liquid, then it is the specific latent heat of fusion. It follows that to change the state of a mass m of a substance of specific latent heat L

Heat required, Q = mass × specific latent heat capacity or in symbols, $Q = mL$

The specific latent heat of vaporisation of water is a large value (over 2 MJ), and much more heat is needed to boil away 1 kg of water than to raise its temperature from 0 °C to 100 °C.

Worked examples
Example 1

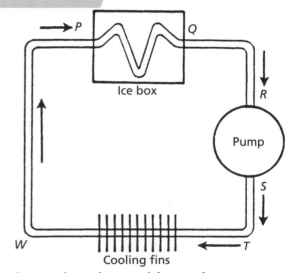

Ice box

Pump

Cooling fins

The diagram shows the essential parts of a compressor refrigerator which uses Freon as refrigerant.

In which region does the Freon evaporate?

A   *PQ*

B   *RS*

C   *TW*

D   *WP*

Solution

[Evaporation takes place in the ice box. For evaporation to take place heat energy must be supplied. This heat energy comes from the ice box and the temperature of the ice box falls.]

Answer **A**

Example 2

5632 joules of heat energy raise the temperature of 0.4 kg of aluminium from 20 °C to 36 °C. The specific heat capacity of aluminium, in J/(kg K), is given by

A $\quad 5632 \times 0.4 \times 16$

B $\quad \dfrac{5632}{0.4 \times 16}$

C $\quad \dfrac{5632 \times 16}{0.4}$

D $\quad \dfrac{5632 \times 0.4}{16}$

E $\quad \dfrac{0.4 \times 16}{5632}$

Solution

$[Q = mcT$ (see section 4)
$= 5632$ J
$= 0.4$ kg $\times c \times 16$ K
$c = \dfrac{5632}{0.4 \times 16}$ J/(kg K).]

Answer **B**

Example 3

The graph shows the change in temperature when heat is applied at 20 000 joule/minute to 1 kilogram of a substance.

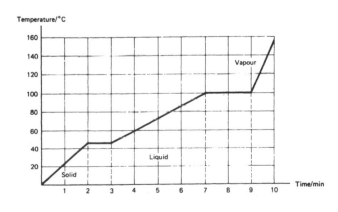

The specific latent heat of fusion of the substance in joule/kilogram is

A   2000
B   10 000
C   20 000
D   40 000
E   80 000

Solution

[It takes 1 min for the solid to melt and become a liquid. Heat supplied in 1 min = 20 000 J.]

Answer **C**

Example 4

If the specific latent heat of steam at 100 °C is 2.26×10^6 J/kg the heat, in J, required to evaporate 2 g of water at 100 °C is

A 2.00×10^2
B 1.13×10^3
C 4.52×10^3
D 1.13×10^6
E 4.52×10^6

Solution

[Heat supplied $= mL$ (see section 5)
$= (0.002$ kg$)(2.26 \times 10^6$ J/kg$)$
$= 4.52 \times 10^3$ J.]

Answer **C**

Example 5

A 50 W immersion heater is switched on and immersed in some water which is at 0 °C. The graph below shows the temperature rise plotted against the time.

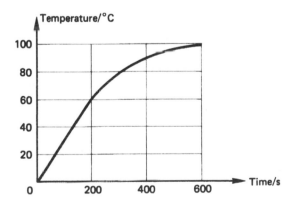

(i) How much energy is supplied by the heater in 400 s? (3 marks)

(ii) What is the rise in temperature during the first 200 s? (1 mark)

(iii) Complete table 6.1 (over page) showing the energy supplied by the heater and the rise in temperature.

Table 6.1

Time/s	0	200	400	600
Energy supplied by heater/kJ	0			
Temperature rise/K	0			

(5 marks)

(iv) Why is the temperature rise during the first 200 s greater than the rise in temperature during the last 200 s? (3 marks)

Solution

(i) [50 W is 50 J/s.]
Energy supplied = (50 J/s) (400 s) = 20 000 J

(ii) 60 °C

(iii)

Table 6.1

Time/s	0	200	400	600
Energy supplied by heater/kJ	0	10	20	30
Temperature rise/K	0	60	90	100

(iv) More heat is escaping into the atmosphere because the body is at a higher temperature, so more heat must be supplied by the heater to obtain the same temperature rise. Also, the higher the temperature, the faster the water is evaporating and more heat is supplied to change the state of the water.

Example 6

(a) (i) A 25 W immersion heater was placed in 100 g of a liquid and the temperature of the liquid rose by 20 °C in 4 min. Calculate the specific heat capacity of the liquid. (3 marks)

(ii) Assuming the figures given above are accurate, why is the result only an approximate one? (2 marks)

(iii) State two precautions necessary in doing this experiment in order to obtain an accurate result. (2 marks)

(b) In a domestic oil-fired boiler central heating system, 0.5 kg of water flows through the boiler every second. The water enters the boiler at a temperature of 30 °C and leaves at a temperature of 70 °C, re-entering the boiler after flowing round the radiators at 30 °C. 3.0×10^7 J of heat is given to the water by each kilogram of oil burnt. The specific heat capacity of water is 4200 J/(kg K).

Using this information to calculate
(i) the energy absorbed by the water every second as it passes through the boiler, (3 marks)

(ii) the mass of oil which would need to be burnt in order to provide this energy. (2 marks)

(c) If the owner of the above central heating system wished to reduce his fuel bill, suggest *three* ways he could do it. (3 marks)

Solution

(a) (i) Heat supplied by immersion heater in 4 min
$= (25 \text{ J/s}) (4 \times 60) \text{ s}$
$= 6000 \text{ J}$
Assume that all the heat supplied by the heater goes into the liquid
$Q = mcT$ [see section 4]
$6000 \text{ J} = 0.1 \text{ kg} \times c \times 20 \text{ K}$
[Don't forget that the mass must be in kg, and an interval of 1 °C is the same as an interval of 1 K.]

$$c = \frac{6000 \text{ J}}{0.1 \text{ kg} \times 20 \text{ K}} = 3000 \text{ J/(kg K)}$$

The specific heat capacity of the liquid is 3000 J/(kg K)

(ii) The liquid must be in a container, and we have neglected the heat taken up by the container and also the heat lost to the atmosphere.

(iii) Lag the container to reduce the heat lost to the atmosphere and stir the liquid before taking the temperature.

(b) (i) $Q = mcT$
$= (0.50 \text{ kg/s}) (4200 \text{ J/(kg K)}) (40 \text{ K})$
$= 84\,000 \text{ J/s}$

(ii) Mass $= \dfrac{84\,000 \text{ J}}{3.0 \times 10^7 \text{ J/kg}}$

$= 0.0028 \text{ kg} = 2.8 \text{ g}$

(c) He could insulate his loft with fibre glass, double glaze his windows and put draught-proof material at the bottom of his doors.
[Alternatively, you could mention filling the walls with insulating material.]

Example 7

The diagram shows the basic components of a refrigerator system which contains a volatile liquid known as the refrigerant.

(i) Explain how the action of the circulating refrigerant as it passes through the freezer unit reduces the temperature of its contents. (4 marks)

(ii) What is the purpose of the pump? (2 marks)

(iii) Explain why the external metal fins become

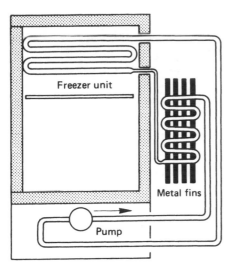

warm while the refrigerator is in operation. What is the source of this heat energy? (3 marks)

(iv) When a tray of water is placed in the freezer unit the temperature of the water falls from room temperature (15 °C) to freezing point in approximately 15 minutes. However, it is necessary to wait a further $1\frac{1}{2}$ hours before the ice becomes completely frozen. Explain why it is quite usual to wait such a long time for completely solid ice to form in the refrigerator.

(4 marks)

Solution

(i) The tubes in the freezer unit are on the low pressure side of the pump. The liquid entering this low pressure region evaporates and takes up its latent heat. This heat is supplied from the refrigerator compartment, which consequently cools.

(ii) The pump circulates the refrigerant. It reduces the pressure on the freezer side of the system so that the refrigerant evaporates, becoming a vapour. On the high pressure side it compresses the vapour so that it liquefies.

(iii) The vapour liquefies in the part of the tubes attached to the fins. As it liquefies it gives out its latent heat. The metal tube and fins are good conductors of heat and are warmed up by the heat from the condensing refrigerant.

(iv) Heat is continually being withdrawn from the freezing compartment but because the latent heat of fusion of ice is large a lot of heat has to be withdrawn before all the water present has frozen. The heat that must be withdrawn to cool the water from 15 °C to 0 °C is about one-sixth of that needed to freeze the water at 0 °C.

Ice is a poor conductor of heat, so the rate of energy transfer slows down as the ice thickens.

Examination questions

(Numerical answers and hints on solutions will be found at the end of the chapter.)

Question 1

Here are two common types of kettle.

Metal kettle

Plastic jug kettle

(a) Give one advantage of the plastic kettle over the metal one. (1 mark)

(b) For safety, the mains plug has a fuse. This breaks the circuit when the current is too large. Explain why it is in series with the kettle element.
(2 marks)

(c) The metal kettle heats 2 kg of water from 10 °C to 100 °C. How much energy is transferred to the water? [The specific heat capacity of water is 4200 J/kg °C.]

(d) (i) The kettle has an element rated at 3 kW (3000 W) and takes 5 minutes to boil. How much energy is transferred from the electrical supply in this time? (2 marks)

(ii) The kettle is boiled six times a day. Show that the amount of energy transferred in a day is 5.4 MJ (5 400 000 J). (2 marks)

(iii) 1 MJ (1 000 000 J) of electrical energy costs 2p. How much does it cost to use the kettle in this way for a day? (1 mark)

(MEG Nuffield, Jun 95, Intermediate Tier, Q3)

Question 2

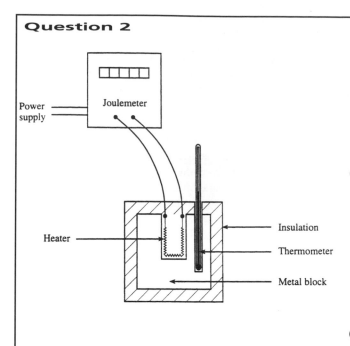

A joulemeter is connected to an electric heater to measure the energy supplied to the heater. The heater is being used to raise the temperature of an insulated metal block.

At the start the reading on the joulemeter was 3.020 kJ and, after 40 s, the reading was 6.620 kJ.

(a) Calculate
   (i) the number of kJ supplied to the heater
      (1 mark)
   (ii) the number of joules supplied to the heater
      (1 mark)
   (iii) the number of joules supplied to the heater in one second. (1 mark)

(b) If the mass of the metal block is 0.5 kg and its specific heat capacity is 450 J/kg °C, calculate the temperature rise of the block after heating. Write down the formula that you use and show your working. (3 marks)
      (WJEC, Jun 95, Intermediate Tier, Q3)

Question 3

The diagram shows part of a heat exchange system used in offices. Waste heat from stale air is extracted and used to heat cool fresh air entering the building. Each hour 625 kg of air is drawn into the building using an electric fan.

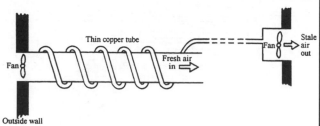

(a) On a particular day the air was drawn into the exchanger at 12 °C and blown into the room at 18 °C.
   (i) Calculate the energy added each hour to the incoming air. The specific heat capacity of air 180 J/kg K.
   (ii) The input power of the electric fan is 12 W. How much does this fan use in 1 hour?
   (iii) Does this system 'save' energy? Explain your answer. (7 marks)

(b) The fan works on a 240 V mains supply. Calculate the current in the fan. (3 marks)
      (ULEAC, Syll B, Jun 95, Higher Tier, Q2)

Question 4

(a) Alcohol and mercury are used as thermometric liquids.
   Give the reason why
   (i) alcohol, rather than mercury, is used to measure a temperature of −60 °C.
   (ii) mercury, rather than alcohol, is used to measure a temperature of 200 °C. (4 marks)

(b) Liquid in glass thermometers are not suitable for measuring very high temperatures.
   State which type of thermometer you would use for each of the following, giving a reason for your answer.
   (i) A steady temperature of 1000 °C
   (ii) The temperature at a point in a flame.
   (iii) Changing temperatures around 1000 °C.

(c) A coil of metal wire is placed in pure melting ice.
   Its resistance is measured and found to be 1.76 Ω.
   The coil is then placed in a beaker of boiling liquid.
   Its resistance is then 2.14 Ω.
   The temperature coefficient of resistance (α) of the metal used for the coil is 0.004/°C.
   Calculate the boiling point of the liquid.
      (NEAB, Jun 95, Higher Tier, Q6)

Answers to examination questions

1. (a) Plastic is a poor conductor of heat so the outside will be cooler and therefore less likely to burn you if you touch it.

   (b) All the current has to flow through the fuse and the current passing through the element and the fuse are the same size.

   (c) $Q = mcT = 2 \text{ kg} \times 4200 \text{ J/kg °C} \times 90 \text{ °C}$
   $= 756\ 000 \text{ J}$

   (d) (i) $\text{Power (W)} = \dfrac{\text{Energy (J)}}{\text{Time (s)}}$

   So, $\text{Energy (J)} = \text{Power (W)} \times \text{Time (s)}$
   $= 3000 \times (5 \times 60 \text{ s})$
   $= 900\ 000 \text{ J}$

   (ii) $\text{Energy used} = 6 \times 900\ 000 \text{ J}$
   $= 5\ 400\ 000 \text{ J (5.4 MJ)}$

   (iii) $\text{Number of MJ used} = \dfrac{5\ 400\ 00}{1\ 000\ 000} = 5.4$

   $\text{Cost} = 5.4 \times 2 = 10.8\text{p}$

2. (a) (i) $6.020 \text{ kJ} - 3.020 \text{ kJ} = 3 \text{ kJ}$

   (ii) $3 \text{ kJ} = 3 \times 1000 \text{ J} = 3000 \text{ J}$

   (iii) $\text{Number of Joules per second} = \dfrac{3000 \text{ J}}{40 \text{ s}}$
   $= 75 \text{ J/s}$

   (b) $Q = mcT$
   $3000 \text{ J} = 0.5 \text{ kg} \times 450 \text{ J/kg °C} \times T$
   Giving $T = 13.3 \text{ °C}$

3. (a) (i) $Q = mcT$
   $= 625 \text{ kg} \times 180 \text{ J/kg K} \times 6 \text{ K}$
   $= 675\ 000 \text{ J}$

   (ii) $12 \text{ W} = 12 \text{ J/s}$
   In 1 hour there are $60 \times 60 = 3600 \text{ s}$
   Energy used by fan in
   1 hr $= 12 \times 3600 = 43\ 200 \text{ J}$

   (iii) No, because you have to supply energy to work the fan and this is not recoverable. Also, the heat exchange will not be 100% efficient and will therefore not be able to convert the heat from the warm stale air to the cold fresh air.

   (b) $\text{Power} = \text{Current} \times \text{voltage}$
   $12 \text{ W} = I \times 240 \text{ V}$
   Current, $I = 0.05 \text{ A}$

4. (a) (i) At -60 °C, mercury would be solid but since the freezing point of alcohol is below -60 °C the alcohol would still be liquid.

   (ii) Alcohol boils below 200 °C whereas mercury would still be liquid.

   (b) (i) A resistance thermometer or thermocouple because they have high melting points and other thermometers would not be able to measure this high temperature.

   (ii) A thermocouple can be made with a small junction which will fit into the flame.

   (iii) A thermocouple. It is quick to respond to changing temperatures.

   (c) $R_t = R_o (1 + \alpha t)$

   $2.14 = 1.76 (1 + 0.004t)$

   Giving $t - 54$ °C

7

Transfer of heat

Topic

WJEC & NEAB	ULEAC Syll A	ULEAC Syll B	MEG	Topic	MEG Salters'	MEG Nuffield	SEG	NICCEA
✓	✓	✓	✓	**Conduction**	✓	✓	✓	✓
✓	✓	✓	✓	**Convection**	✓	✓	✓	✓
✓	✓	✓	✓	**Radiation**	✓	✓	✓	✓
✓	✓	✓	✓	**The thermos flask**	✓	✓	✓	✓
✓	✓	✓	✓	**House insulation**	✓	✓	✓	✓

1 Heat transfer

Heat energy may be transferred from one place to another by conduction, convection and radiation.

2 Conduction

Conduction is the flow of heat energy through a body which is not at uniform temperature, from places of higher temperature to places of lower temperature, without the body as a whole moving.

If a metal rod is placed on a tripod and one end heated with a Bunsen burner, then heat flows down the rod by the process of conduction and the other end begins to warm up. The energy is passed down the rod by the free electrons in the metal and also by the vibrating atoms passing on their energy to adjacent atoms.

Bad conductors are used to insulate roofs, water pipes and storage tanks, and for the handles of saucepans and teapots. Air is a poor conductor of heat, and cellular blankets, string vests, fibre glass and fur coats depend for their insulating properties on the air which is trapped in them. Good conductors (e.g. metals) are used as bases for saucepans and for cooling fins in air cooled engines.

The rate of conduction of heat through a material depends on (a) its nature, (b) its thickness, (c) its area and (d) the temperature difference across its thickness.

3 Convection

Convection is the transfer of heat energy by the circulation of a fluid (a liquid or a gas) due to a temperature difference within it.

The essential difference between convection and conduction is that in convection the less dense hot body rises and takes its heat with it, and in conduction heat flows through the body, which does not move. Hot water systems, coastal breezes from land to sea at night and the hot air rising above radiators and up chimneys are examples of convection.

4 Radiation

Radiation is the transfer of heat energy from one place to another by means of electromagnetic waves. The amount of heat energy radiated per second by a body increases rapidly as its temperature rises.

Heat energy from the Sun reaches us by electromagnetic waves. Radiation is the only means of heat transfer which can take place in a vacuum. Black surfaces are good absorbers and good radiators. Shiny surfaces are poor radiators and poor absorbers (they reflect well).

5 The thermos flask

A thermos (vacuum) flask is designed to reduce heat flow by convection, conduction and radiation. The vacuum

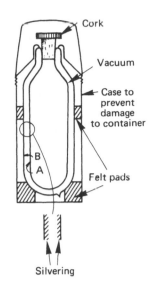

A vacuum flask.

prevents heat flow by conduction and convection. With a hot liquid in the flask the shiny surface on the liquid side of the vacuum (A) is a poor radiator and reduces heat loss by radiation. The shiny surface on the far side (B) reflects back the small amount of radiation that does take place. The main loss of heat is by conduction up the sides and through the cork at the top.

6 House insulation

Insulating materials, such as glass wool or rock wool, in the walls and roofs of houses considerably reduce heat loss by conduction and also by convection. Not only does double glazing produce another layer of poorly conducting glass, but also the air trapped between the glass surfaces is a bad conductor of heat. Plastic strips, coated with shiny aluminium, fitted behind radiators reflect the infra-red radiation back into the room. The loss of heat energy is also reduced if draught proofing material is fitted round doors.

Worked examples

Example 1

Which of the following statements about heat transfer is correct?

A   Conduction takes place both in liquids and in a vacuum

B   Convection takes place in both liquids and solids

C   The amount of heat energy radiated every second by a body increases slowly as its temperature rises

D   Radiation is the *only* way heat transfer can take place in a vacuum

Solution

[Convection and conduction both need a medium and cannot take place in a vacuum. Convection requires that the warmer, less dense part of the medium can rise and take its heat with it, so it can take place only in fluids. See sections 2, 3 and 4. Very little radiation takes place at low temperatures, but the amount of heat radiated from a body does increase very rapidly as the temperature rises.]

Answer **D**

Example 2

The windows of many modern buildings are 'double-glazed' (i.e. have two thicknesses of glass with a small air space in between) to reduce heat losses to the outside. This is mainly because
A   evaporation of moisture from the outside of the windows is reduced
B   convection currents cannot pass through the extra layer of glass
C   radiated heat is not transmitted through the air space
D   air is a very bad conductor of heat
E   glass is a very bad conductor of heat

Solution

[The air trapped between the sheets of glass is a bad conductor of heat and this reduces the heat loss from the house. It is also true that glass is a bad conductor and this fact also reduces the heat loss, but the main reason is the poor conductivity of air. A thick sheet of glass would be expensive and not so effective.]

Answer **D**

Example 3

In cold weather the metal blade of a knife feels colder than the wooden handle because the
A   metal is at a lower temperature than the wood
B   metal is a better conductor of heat than the wood
C   metal has a smaller specific heat capacity than the wood
D   metal has a brighter surface than the wood
E   molecules in the metal are vibrating more vigorously than those in the wood

Solution

[Because the metal is a good conductor of heat, when it comes into contact with skin the heat from the skin can easily pass through the metal blade. Wood is not such a good conductor and the heat cannot easily pass from the hand to the wood.]

Answer **B**

Example 4

Fred put a layer of aluminium cooking foil against the floor in his loft.

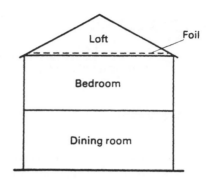

This keeps Fred's house warmer in winter because the foil
A   is a good heat insulator
B   is a good heat reflector
C   prevents convection in the loft
D   is a good heat conductor

Solution

[Shiny foil reflects radiation (see section 4).]

Answer **B**

Example 5

A person sitting on a beach on a calm hot summer's day is aware of a cool breeze blowing from the sea.
    Explain why there is a breeze.   (6 marks)

Solution

The land has a lower specific heat capacity than the sea and in daytime the land is hotter than the sea. The air above the land is warmer and less dense than the air over the sea, and it rises, the cold air from the sea coming in to take its place.

Example 6

The diagram shows some of the main energy losses from a house.

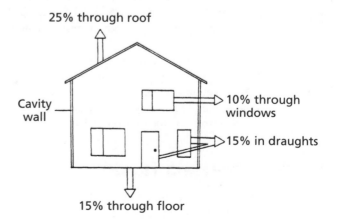

(a) Energy is also lost through the cavity walls.
  (i) What percentage is this of the total energy loss? (1 mark)
  (ii) How can the energy loss through the cavity walls be made less? (1 mark)
(b) The owner decides to improve the insulation of the house by fitting double glazing or insulating the roof. Which of these two methods saves more in heating costs?
  Explain your answer. (2 marks)
(c) Explain how each of the following reduces energy loss.
  (i) fitting aluminium foil behind radiators. (1 mark)
  (ii) blocking off unused fire places and their chimneys. (1 mark)
  (MEG, Jun 88, Intermediate Tier)

Solution

(a) (i) 35% [The total energy losses shown in the diagram are 65%. The rest is lost through the cavity walls.]
  (ii) Fill the cavity with a good insulator [see section 6].
(b) Insulating the roof. 25% of the heat loss is through the roof. Only 10% is lost through the windows.
(c) (i) The shiny aluminium foil is a good reflector of radiant heat. Heat energy which would otherwise escape is reflected back into the room [see section 4].
  (ii) This reduces the heat loss by convection currents flowing up the chimney [see section 3].

Example 7

DOUBLE GLAZING SAVES HEAT!

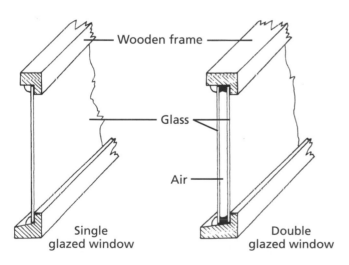

Single glazed window
Double glazed window

The diagrams compare single and double glazed windows of the same area. Heat losses through the single glazed unit are far greater than those through the double glazed unit.
(a) (i) CONDUCTION is one process by which heat

can be transferred from a hot place to a cold place.
  Give **two** others (2 marks)
  (ii) Explain how the double glazing reduces heat loss. (4 marks)
(b) Plastic and wooden frames result in less loss of heat than aluminium frames. Why is this? (2 marks)
(c) A friend suggests that it would be just as good to use single glazing but using glass sheets of double the thickness. Do you agree? Give reasons. (2 marks)

Solution

(a) (i) Convection. Radiation.
  (ii) Air is a very bad conductor of heat. The air between the two sheets of glass greatly reduces the heat loss by conduction. Glass is also a poor conductor of heat and the extra sheet of glass will further reduce the heat loss by conduction.
(b) Plastic and wood are bad conductors of heat. Aluminium is a good conductor of heat [see section 2].
(c) No. Air is a worse conductor of heat than glass. The thick sheet of glass would be heavy and expensive to produce.

Example 8

(a) Heat energy may be transferred by conduction, convection and radiation. Describe three experiments, one for each process, to illustrate the methods of heat transfer. (12 marks)
(b) Some electrical devices, such as power transistors, can become so hot that they do not function properly. In order to prevent this they are fastened in good thermal contact with a 'heat sink', such as a piece of aluminium sheet with aluminium fins as shown in the diagram.

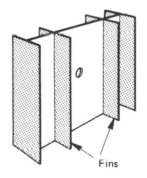

Fins

  (i) What is meant by 'good thermal contact'?
  (ii) Explain how the heat is carried away from the electrical device to the air outside it.
  (iii) Why does the heat sink have fins?
  (iv) Discuss whether the heat sink would operate better if it were placed with its fins horizontal, rather than vertical as shown in the diagram.
  (8 marks)

Solution

(a) A metal rod is placed on a tripod. The rod should be long enough to ensure both ends are well clear of the tripod. A Bunsen burner is placed under one end. Very soon the other end becomes warm; its temperature rises. This can be detected simply by feeling the end of the rod. Heat has been passed down the rod by conduction.

Convection may be demonstrated using the apparatus shown in the diagram. The box has two chimneys as shown. A lighted candle is placed under one of the chimneys and the glass front closed. Very soon smoke from the smouldering rope will be seen passing through the box and out of the chimney above the candle. Convection currents of air are passing through the box.

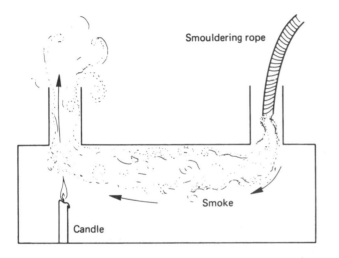

Sit in front of an electric fire which has a shiny reflector behind it. You will feel the radiation reflected on to your body. The heat has not arrived by conduction because air is a bad conductor. Hot air convection currents will rise above the element and circulate around the room but this is not how most of the heat reaches your body. Most of it has arrived as a result of radiation from the heating element falling on your body.

(b) (i) Contact so that heat can flow from one body to the other. Some air between the two bodies would prevent good thermal contact.

(ii) Heat is conducted from the device along the aluminium sheet and through the fins. The air in contact with the aluminium and fins becomes warm and the less dense warm air rises, carrying its heat with it. Denser colder air replaces the warm air and the process continues. At low temperatures very little heat is lost by radiation.

(iii) The fins increase the area of surface in contact with the air, thus increasing the heat loss from the metal surfaces.

(iv) It operates better if placed vertically because the convection currents can flow more freely.

Example 9

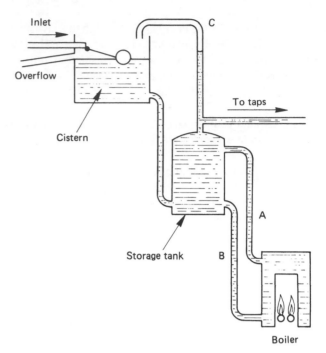

(a) The diagram shows part of a household hot water system.
   (i) Why is pipe A connected between the top of the boiler and the top of the storage tank? (3 marks)
   (ii) Why is pipe B connected between the bottom of the boiler and the bottom of the storage tank? (3 marks)
   (iii) What is the function of pipe C? (2 marks)
   (iv) Suggest, with reasons, what might be added to the hot water system above to make it more efficient. (3 marks)

(b) The temperature of the water inside an aquarium can be controlled by a thermostat which switches an electric heater on and off. Draw a diagram showing how this may be done using a bimetallic strip. (Your diagram must clearly show the construction of the bimetallic strip.)

How may different constant temperatures be achieved using your arrangement? (5 marks)

Solution

(a) (i) and (ii)
   The hot, less dense water in the boiler rises and goes via the pipe A to the top of the storage tank. The denser, colder water from the bottom of the storage tank passes into the boiler. The pipe A is connected to the top of both tanks because hot water is less dense than cold water, and pipe B is connected to the bottom of both tanks because cold water is more dense than hot water.

   (iii) C is an expansion pipe. It is a safety precaution to allow steam to escape should the water boil.

It also allows any dissolved air which is released from the heated water to escape, thus helping to prevent air locks.

(iv) Lagging should be added round all the hot pipes and round the hot water tank. This will reduce the heat lost to the atmosphere and make the system more efficient.

(b) If the screw is screwed down the thermostat will switch off at a higher temperature.

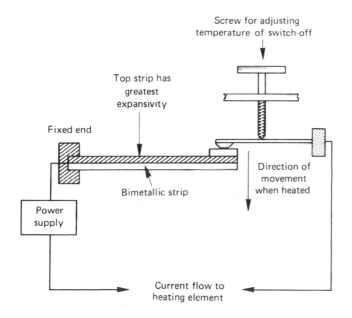

[A similar thermostat has many uses i.c. in domestic boilers and ovens. It is now common practice to heat the domestic hot water supply indirectly using a heat exchanger coil (see diagram below). The dotted pipe inside the tank is the heat exchanger. This ensures

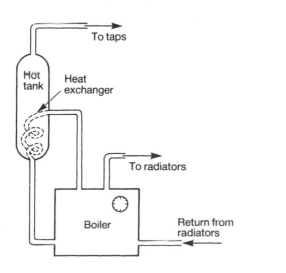

that stagnant water from the radiators cannot enter the domestic supply and that drawing off a large quantity of hot water does not seriously affect the temperature of the radiators.]

Example 10

(a) How would you demonstrate that water is a poor conductor of heat?   (4 marks)

(b) Explain how energy is transferred from the heating element throughout the water in an electric kettle.
                                                                                         (4 marks)

(c)

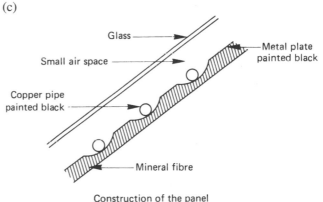

Construction of the panel

The diagram shows the essential features of a solar heating panel. A small electric pump circulates a liquid through the pipes. State briefly why

(i) the pipes and back plate are blackened
(ii) there is a mineral fibre backing to the panel
(iii) the glass sheet increases the energy collected by the panel by a large factor.   (5 marks)

Solution

(a) Put some ice wrapped in a metal gauze (so that it sinks) in a test tube which is nearly full of water. Carefully heat the water at the top of the tube until it boils. The water and ice at the bottom of the tube remain cold because the water (and of course the glass) are poor conductors of heat.

(b) The water in contact with the heating element gets hotter. Hot water is less dense than cold water, so the hot water rises. The denser surrounding cold water flows in to take its place. These circulating convection currents gradually transfer the heat throughout the water in the kettle.

(c) (i) Black surfaces are better absorbers of heat than shiny or light-coloured surfaces. The solar panel works better if as much heat as possible is absorbed from the Sun's rays.

(ii) The mineral fibre is a poor conductor and this reduces the heat escaping from the copper pipes.

(iii) The electromagnetic waves emitted by the copper pipes are of much longer wavelength than the rays arriving from the Sun. They are not transmitted by the glass.

Examination questions

(Numerical answers and hints on solutions will be found at the end of the chapter.)

Question 1

(a)   To reduce heat loss through the wall, some people place shiny metal foil on the wall behind the central heating radiators.

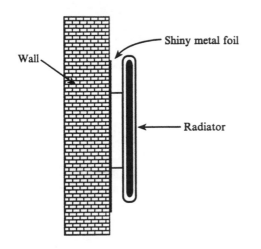

In what way does this reduce heat loss through the wall?

(b)   Mineral wool is frequently used as roof space or loft insulation.

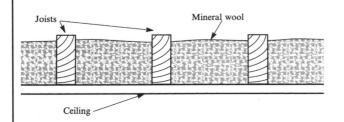

What property of the mineral wool makes it suitable as an insulator?

(NICCEA, Jun 95, Intermediate Tier, Q24)

Question 2

The diagram shows part of a hot water system. An electric immersion heater is to be fitted in the hot water cylinder.

(a)   Jane thinks that the heater should be near the top of the cylinder.
       Chris thinks it should be near the bottom.
       Discuss the advantages and disadvantages of each position.
       (i)   *Heater near the top*
       (ii) *Heater near the bottom*   (4 marks)

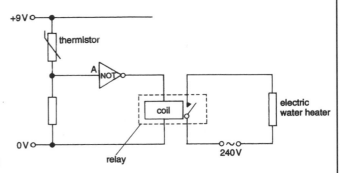

(b)   Describe how the water level in the cold tank is controlled.   (2 marks)

(c)   The circuit below is used to control the electric water heater. It keeps the temperature of the water at 50 °C.

When the temperature of the thermistor is above 50 °C, input *A* is logic 1. Describe what happens in the circuit when the temperature of the thermistor falls.   (3 marks)

(MEG, Jun 95, Intermediate Tier)

Question 3

Susan's Granddad has a greenhouse.

(a)   Complete table 7.1 to show which heating is involved in the energy transfers described.

Choose from: **conduction   convection evaporation   radiation**

Table 7.1

Energy transfer	Heating process
Energy is transferred to the greenhouse directly from the Sun	
Energy is transferred through the metal doors of the greenhouse	
Energy is transferred by air moving through the vents of the greenhouse	

(3 marks)

(b)   In winter Susan's Granddad fixes sheets of clear plastic bubble-wrap to the glass roof and walls of his greenhouse.

bubble of trapped air

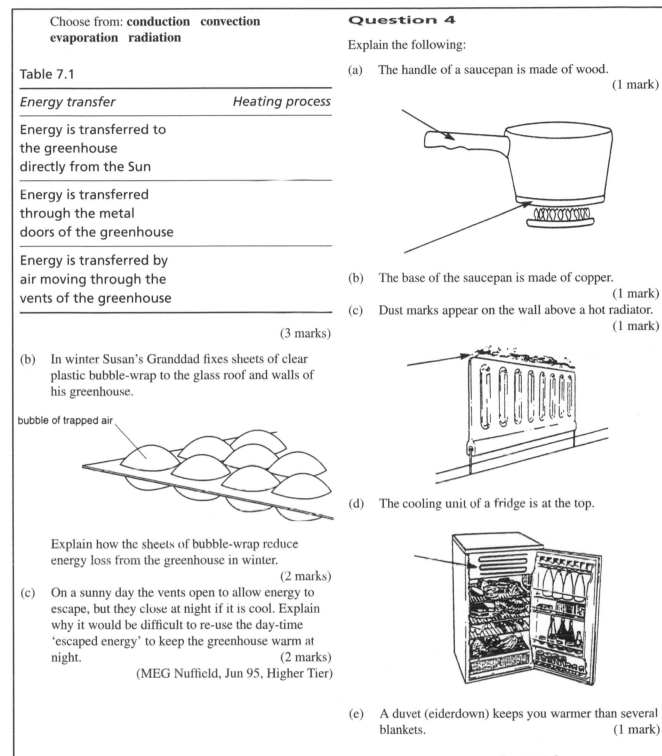

Explain how the sheets of bubble-wrap reduce energy loss from the greenhouse in winter.

(2 marks)

(c)   On a sunny day the vents open to allow energy to escape, but they close at night if it is cool. Explain why it would be difficult to re-use the day-time 'escaped energy' to keep the greenhouse warm at night.

(2 marks)

(MEG Nuffield, Jun 95, Higher Tier)

Question 4

Explain the following:

(a)   The handle of a saucepan is made of wood.

(1 mark)

(b)   The base of the saucepan is made of copper.

(1 mark)

(c)   Dust marks appear on the wall above a hot radiator.

(1 mark)

(d)   The cooling unit of a fridge is at the top.

(e)   A duvet (eiderdown) keeps you warmer than several blankets.

(1 mark)

Answers to examination questions

1. (a) Reflects the infra-red radiated heat away from the wall and back into the room. Reduces the amount of heat loss through the wall near the radiator.

   (b) It contains pockets of trapped air which prevent heat loss by conduction/convection.

2. (a) (i) *Advantages:* The water at the top will get hot quickly and will use less energy if only a small amount of water is needed.
   Disadvantages: Because hot water rises by convection the water lower down would have to be heated by conduction and this would take a long time.

   (ii) *Advantages:* Water rises from the bottom to the top by convection so all the water is eventually heated.
   Disadvantages: Wastes energy if only a small volume of hot water is needed.

   (b) As the hot water is run off using the taps the cold water replaces it in the cylinder. The level in the cold water tank drops, the float falls the valve opens allowing cold water from the mains to enter. As the water in the tank rises, the valve shuts off the water at a certain level.

   (c) When the water cools and reaches 50 °C the input at A is logic level 0. The output from the NOT-gate is logic level 1 so a current flows which turns the relay on which closes the switch turning the heater on. When the temperature falls below 50 °C, it returns logic level 1 and turns the heater off.

3. (a)

Table 7.1

Energy transfer	Heating process
Energy is transferred to the greenhouse directly from the Sun	**radiation**
Energy is transferred through the metal doors of the greenhouse	**conduction**
Energy is transferred by air moving through the vents of the greenhouse	**convection**

   (b) This prevents conduction of heat through the glass to the outside. The air in the bubbles is a poor conductor and because the air is not free to move, convection of the heat from one side of the bubble-wrap to the other will not occur.

   (c) This heat energy is spread out over the whole area of the greenhouse so it would be practically difficult to transfer and store it. The device used would cost more than the cost of the heat saving making it uneconomic.

4. (a) Wood is a good insulator so it will not heat up as much.

   (b) Copper is a good conductor of heat so the heat will be efficiently transferred to the water.

   (c) Air in contact with the radiator heats up, expands, and rises, carrying the particles of dust with it. The dust settles on the wall near the radiator.

   (d) Cool air is denser so it falls and warmer air replaces it. This means the whole fridge will be cooled.

   (e) A duvet is much lighter and therefore contains more trapped air. The air is an insulator and reduces heat losses.

Sources of energy

8

WJEC & NEAB	ULEAC Syll A	ULEAC Syll B	MEG	Topic	MEG Salters'	MEG Nuffield	SEG	NICCEA
✓	✓	✓	✓	**Non-renewable fuels (Fossil and nuclear fuels)**	✓	✓	✓	✓
✓	✓	✓	✓	**Greenhouse effect**	✓	✓	✓	✓
✓	✓	✓	✓	**Renewable resources (Solar, geothermal, tidal, wind, hydroelectric, biomass)**	✓	✓	✓	✓
✓	✓	✓	✓	**Origins of the Earth's energy**	✓	✓	✓	✓
✓	✓	✓	✓	**Storing energy**	✓	✓	✓	✓

1 Non-renewable resources

Most of the energy we use in Britain comes from fossil fuels which include coal, natural gas and oil. Non-renewable resources are resources which are not replaced when we use them and they will eventually run out if used at their present rate.

Fossil fuels

All fossil fuels were originally formed from trees, plants and animals which died and were then buried and compressed under rocks. The illustration shows how these fuels were produced.

Fossil fuels take millions of years to form and are therefore impossible to replace. Fossil fuels are a non-renewable resource and because of this, new, alternative sources of energy need to be found. About 90% of the world's energy comes from fossil fuels, so finding a replacement is not easy.

Problems with using fossil fuels

- Fossil fuels produce carbon dioxide when burned, which adds to the 'greenhouse effect' and causes global warming.
- Fossil fuels contain small amounts of sulphur which is released as sulphur dioxide when burned. This gas dissolves in rain to form acid rain which kills plants and trees.
- They may be used to make plastics and pharmaceuticals so it is a shame to burn them.

The greenhouse effect and the ozone layer

The shorter wavelength infra-red radiation and visible light from the Sun pass through the glass of a greenhouse and are absorbed by the soil and the plants, raising the temperature. The infra-red radiation emitted by the soil and plants is of much longer wavelength than the infra-red radiation emitted by the Sun (the higher the temperature, the shorter is the wavelength of the radiation emitted). The longer wavelength infra-red radiation does not pass through the glass and so energy is not lost from the greenhouse; it is trapped in the greenhouse and the greenhouse temperature rises.

Between 10 km and 15 km above the Earth's surface there is a layer of gases which act like the glass of a greenhouse. The Sun's rays penetrate this layer but the heat radiated back from the Earth does not easily penetrate it. Instead of being lost in space, much of heat radiated back from the Earth is trapped. This layer of gases is increasing and the temperature of the Earth is gradually rising. If the rise in temperature continues this could have disastrous consequences for the Earth. For example, the water in the Earth's oceans would be heated up causing it to expand and cause the sea level to rise, flooding many lower level areas. The melting of the Polar ice would have a similar effect but not to the same extent.

The accumulating layer of gases results from the burning of fossil fuels (e.g. coal, oil and natural gas). Fossil-fuelled power stations, factories and cars are continually producing gases which result in the greenhouse effect.

Above this layer of gases is the ozone layer. This is formed and maintained by the action of ultra-violet light from the Sun on oxygen molecules in the atmosphere. It shields the Earth from the harmful ultra-violet light (too much ultra-violet light is one of the causes of skin cancer). Chemicals which escape into the atmosphere are destroying this protective layer of ozone. One of the most damaging types of gas are the chlorofluorocarbons (CFCs) which are used in aerosols, but CFC-free aerosols are now available. The CFCs are also very effective greenhouse gases.

Uranium (nuclear fuel)

Uranium is the name of the fuel used in many nuclear power stations. Although the uranium ore is dug up from the ground, it is not a fossil fuel and is classed as non-renewable because once it is used up, no more will be made. Plutonium, which is man-made, may also be used as a nuclear fuel.

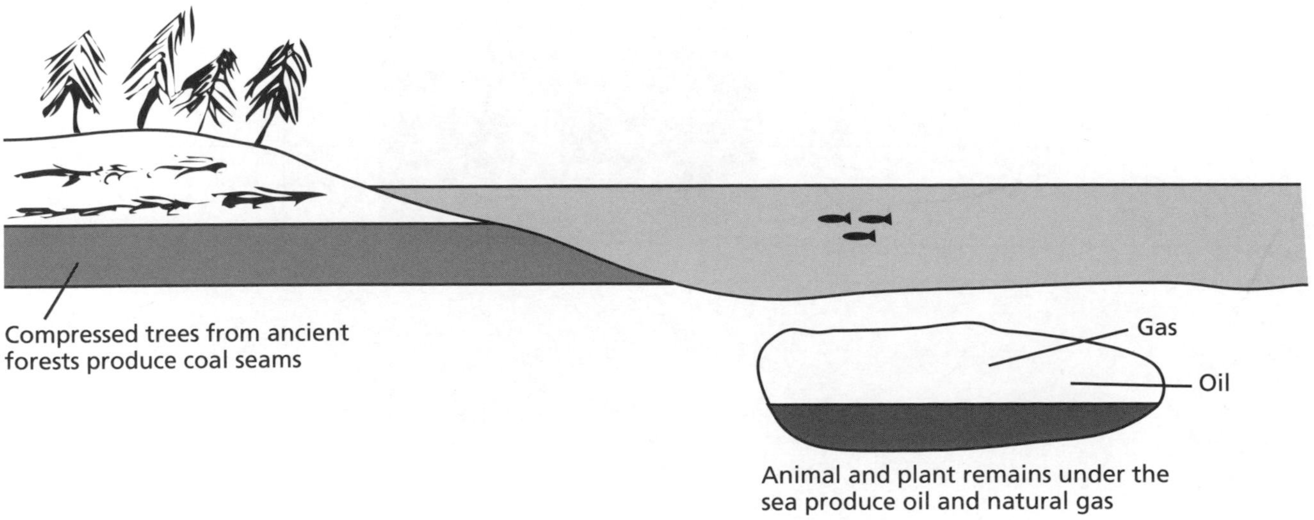

Compressed trees from ancient forests produce coal seams

Gas

Oil

Animal and plant remains under the sea produce oil and natural gas

Problems with nuclear power

- Although nuclear fuel is cheap, the cost of building and operating the reactor is high.
- Nuclear power is 'clean' because it does not produce any polluting or greenhouse gases but the spent fuel and plant is radioactive and needs to be disposed of carefully.
- There have been leaks of radiation in the past and no one wants such a plant on their doorstep.
- The cost of getting rid of radioactive waste and that of decommissioning (closing down) power stations is high.

2   Renewable resources

Renewable resources are those sources of energy which are inexhaustible, which means that they can never be used up because they are constantly being replaced. Wind, waves, solar, biomass, tidal, geothermal and hydroelectric are all examples of renewable energy resources.

Solar power

Believe it or not, in Britain we receive enough solar energy to satisfy all our energy needs. However, harnessing the Sun's rays to produce either hot water for heating or to produce electricity is both difficult and expensive.

Solar power may be used to produce electricity using solar cells to power calculators, watches, satellites etc. or to heat up water using a solar panel.

Solar panels

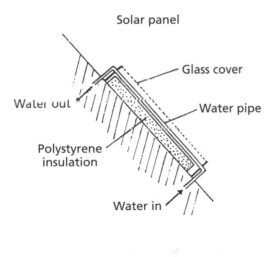

Solar panel

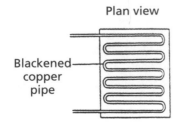

Plan view

The main features of a solar panel are:

- The long bent tube, which is painted matt black to absorb the Sun's rays. The tube is long so that there is a greater surface area exposed to the Sun.
- The glass cover lets the Sun's rays in but won't let them back out again. This trapping of heat is commonly called the 'greenhouse effect'.
- A polystyrene board insulates the back of the panel to prevent heat loss.
- Water travels around the tube by convection, although a pump is often added.
- The hot water is stored in a storage tank which also has an immersion heater in it so that the water may be heated when there is no Sun or when there is a high demand for hot water.

Problems with solar power

- The amount of current produced by solar cells is usually too small to power devices that need large amounts of electricity such as heaters.
- Using solar panels to heat water will produce the hottest water on the hot days when it is least needed.
- Solar panels are usually mounted on the roof of the house which means that the house really needs to point in a certain direction in order to make maximum use of the Sun's rays.

Geothermal energy

Originally the Earth was a fireball and although the outside has now cooled down, the Earth's centre is still at a high temperature. The heat energy to keep the core at this temperature comes from nuclear reactions which take place in the core. This means that the nearer you get to the centre of the Earth the hotter it gets. In some areas, where the Earth's crust is thin, water reaches the hot rocks and changes into steam and this steam finds its way to the surface where it is released as steam jets called geysers. The illustration shows how geysers are produced.

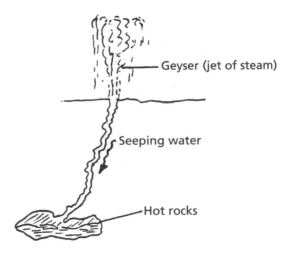

This high pressure steam can be used to drive a turbine which powers a generator to produce electricity. Steam may also be produced for driving turbines by pumping water down boreholes where it comes back up as steam. Energy produced from the Earth's hot interior is called geothermal energy. Geothermal energy is only important in countries such as Iceland and New Zealand where the Earth's crust is thin and only contributes about 0.1% to the total energy consumption of the world.

Tidal power

Tides are due to the gravitational attraction of water to the Moon and to some extent the Sun. In other words, the tides are due to the Moon trying to pull the water in the oceans towards it.

How big a tide is depends mainly on the relative positions of the Earth and the Moon. As the tide comes in the level of water in an estuary rises, sometimes quite considerably. A huge volume of water flows up and down the estuary at high and low tides. If a dam, called a barrage, is built across the mouth of an estuary then it is possible to make the fast flowing water pass through turbines which in turn drive generators during high and low tides. More and more tidal schemes are being used and this is likely to become a significant renewable energy resource. In France there is a tidal power plant across the River Rance, where the level of the water in the estuary rises and falls a distance of 13.5 m (44 feet) during high and low tides and this contains twenty-four 10 MW turbines.

Problems with tidal power

- The barrier across the estuary has an adverse visual impact.
- The estuary is flooded for a longer period and this prevents wading birds from feeding.
- Large areas of land are needed.

Wind power

It is possible to generate electricity using huge windmills.

In remote and exposed parts of the country it is common to see many windmills grouped together in what is called a wind farm. Although wind power will never make a significant contribution to electricity generation in Britain, it may become important in remote areas and offshore islands. One advantage with wind power is that peak demand for electricity usually corresponds with high winds. So windmills produce the most power when it is most needed. Like sunlight, wind is not always with us and so wind power needs to be linked with another, more reliable source of power.

Problems with wind power

- Rows of windmills look unattractive.
- They produce a humming sound which is annoying.

Wave power

Wave power makes use of the bobbing motion of a float on the surface of the sea. The up and down motion of the float is converted into rotation which is then used to drive a generator to produce electricity.

Problems with wave power

- During storms, the floats become damaged and need repair or replacement.
- Waves depend on wind so no wind means no power.

Hydro-electric power (HEP)

In hydro-electric schemes, rivers are dammed to form large lakes. The water is allowed to flow through pipes to the bottom of the dam and is used to power turbines which drive generators and produce electricity.

Because the water is higher on one side of the dam than the other, it possesses gravitational potential energy. When the water rushes down the pipes, the potential energy of the water decreases and is changed into kinetic energy. The water then rushes through a turbine which it spins at high speed which is used to power a generator to produce electricity.

Problems with hydro-electric power

- Although hydro-electric schemes are clean in the sense that they do not give off any pollution causing waste products, large areas of countryside need to be flooded and this can have serious effects on the balance of nature.
- Only really suitable for mountainous regions.

Biomass

In some countries, plants are grown to trap the Sun's energy. The plants are then burned to produce heat directly or they are used to produce other chemicals which may also be burned. Growing trees for wood which is then burned is classed as biomass. Sugar cane is grown in Brazil and then fermented to produce alcohol which is then used as fuel to power cars and tractors. Biomass is classed as a renewable energy source because more plants may be planted to replace those used. One problem is that over-use of the land in this way can lead to the land becoming infertile. Rotting vegetable waste from rubbish dumps or sewage give off a gas called methane which may be burned and used as a fuel.

Problems with biomass

- Growing plants quickly soon renders the soil infertile.
- Large areas of land are needed.

3 The origins of the Earth's energy

Most of the Earth's energy comes from the Sun

Most of the energy we use on the Earth originally came from the Sun. The Sun was responsible for the following list: wind, wave, hydro-electric, solar and biomass. Here's how:

Solar Power: This uses the infra-red heat in the Sun's rays directly.
Wind: The Sun heats air up and changes its density, causing it to move and produce wind.
Hydro-electric: The Sun heats up water and causes it to evaporate. This water falls as rain and fills the dammed reservoir up with water.
Water waves: These are caused by wind which is produced from the heat from the Sun.
Biomass: The Sun is needed to make the plants grow.
Fossil fuels: These were at one time living plants and animals and so the Sun was needed to enable them to live.

Some energy sources not produced by the Sun

Geothermal: the heat produced by the nuclear reactions which take place in the core of the Earth.
Tidal: this source of energy is mainly produced by the Moon, although the Sun does play a minor part.

4 Storing energy

Pumped storage schemes

For any power station to work efficiently it is necessary for the electrical energy to be produced 24 hours per day. The demand for the electricity is not as great during the night as during the day so this electricity is wasted. A pumped storage scheme uses the night time electricity to pump water from a lower reservoir to a higher one. During the day, when the demand for electricity is higher, the water is allowed to flow back down the mountain and drive a turbine to power a generator and produce electricity.

Batteries

These provide transportable electrical energy. Rechargeable batteries are more expensive than non-rechargeable ones, but because they can be recharged they are cheaper in the long run. Their internal resistance (the electrical resistance of the chemicals inside them) is small and they are rapidly discharged and damaged if short-circuited. Their low internal resistance means that less energy is lost as heat in driving currents through the chemicals in the battery. They last less time before running down than ordinary batteries.

Worked examples

Example 1

(a) (i) The diagram shows four stages in the production of electricity at a coal-fired power station. Use words from the list below to complete the labelling of the boxes. One has already been done for you.

boiler   generator   moderator   reactor transformer   turbine

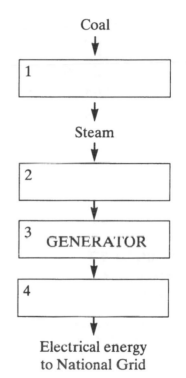

(ii) Write down the energy which takes place in the generator.
.... energy is transferred to .... energy.
(5 marks)

(b) Coal is a fossil fuel.
Describe how fossil fuels were formed. (2 marks)

(c) Coal is a non-renewable energy resource.
Explain what is meant by a non-renewable energy resource. (1 mark)

(d) Some energy resources are renewable
(i) Write down **one** renewable energy resource.
(ii) Describe how this renewable energy resource may be used to produce electricity. (4 marks)
(NEAB, Jun 95, Intermediate Tier)

Solution

(a)   (i)

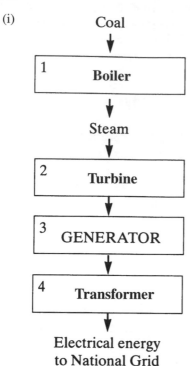

Coal
↓
| 1 | **Boiler** |

↓
Steam
↓
| 2 | **Turbine** |

↓
| 3 | GENERATOR |

↓
| 4 | **Transformer** |

↓
Electrical energy
to National Grid

(ii) Kinetic is transferred to electrical

(b)   Formed millions of years ago from the remains of ancient forests which have decayed and been subjected to high pressures from the layers above.

(c)   Once they have been used they cannot be replaced.

(d)   (i) [Could have any one of the following: wave, tidal, geothermal, wind, solar, hydroelectric and biomass (biofuels such as wood).]

(ii) [If wind were chosen in (d)(i) then we could have:]
The moving air causes the blades of the turbine to rotate and drive a generator which is used to convert the kinetic energy of the wind to electrical energy.

Example 2

(a)   Electricity companies try to encourage people to use more electricity at night and less during the day. They do this by charging a cheaper price for electricity used at night. They call this scheme 'Economy 7'.

The diagram below is taken from an electricity company leaflet. It shows a special hot water tank designed to use electricity on the 'Economy 7' scheme.

(i) Suggest and explain **one** reason why the outlet to the hot taps is at the **top** of the cylinder.
(3 marks)

(ii) How does the insulation 'make sure that the water remains hot all day'?
(1 mark)

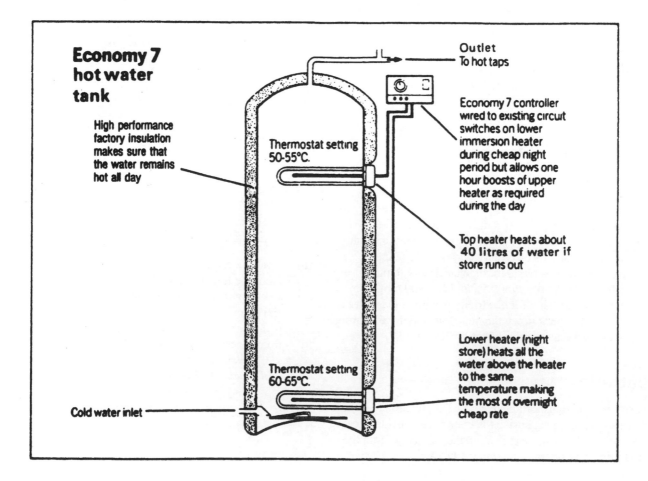

Economy 7 hot water tank

High performance factory insulation makes sure that the water remains hot all day

Thermostat setting 50-55°C.

Thermostat setting 60-65°C.

Cold water inlet

Outlet
To hot taps

Economy 7 controller wired to existing circuit switches on lower immersion heater during cheap night period but allows one hour boosts of upper heater as required during the day

Top heater heats about **40 litres of water** if store runs out

Lower heater (night store) heats all the water above the heater to the same temperature making the most of overnight cheap rate

(b) Power stations are designed to produce large quantities of electrical energy. The diagram below shows a coal-fired power station.

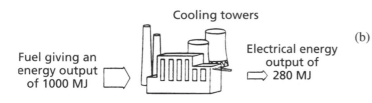

Cooling towers

Fuel giving an energy output of 1000 MJ

Electrical energy output of 280 MJ

   (i) Choose the words from the list in the box below which will complete the sentences which follow. (3 marks)
   Chemical Electrical Gravitational Potential Heat Light Kinetic Nuclear Sound
   In the power station .... energy in the coal is changed into .... energy. This energy turns water into steam. The steam turns turbines and these turn generators. Generators change .... energy into electrical energy.
   (ii) Explain where the energy in coal originally came from. (2 marks)
   (iii) [A] Calculate how efficient the power station is at changing the energy in the coal into electrical energy. (3 marks)
   [B] In what form is most energy lost from this power station? (1 mark)
   [C] What eventually happens to this 'lost' energy? (1 mark)

(c) The chart below shows the major sources of energy used to produce electricity in one country. Just over one-third of this energy comes from coal.

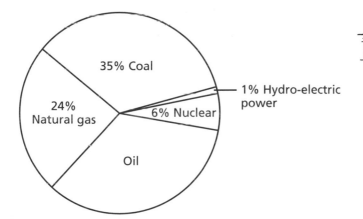

   (i) Calculate the percentage (%) of electrical energy supplied by oil. (1 mark)
   (ii) [A] Name **one** energy source shown in the chart which is 'renewable'. (1 mark)
   [B] Explain what is meant by describing an energy source as being 'renewable'. (2 marks)
   [C] Why is it important that we should be using renewable sources of energy to provide our electricity? (1 mark)
   (SEG, Spec, Intermediate Tier)

Solution

(a) (i) The water will be hottest there because the hot water rises because of convection.
   (ii) The insulation reduces heat losses due to convection and radiation.
(b) (i) Chemical, Heat, Kinetic.
   (ii) The Sun enabled plant life to grow. The energy in the Sun enabled the plant material to be produced by photosynthesis. When the plants died, they were compressed to form coal.

   (iii) [A] Efficiency $= \dfrac{\text{Power out}}{\text{Power in}} \times 100$

   $= \dfrac{(280\ \text{MJ})}{(1000\ \text{MJ})} \times 100$

   $= 28\%$

   [B] Heat energy
   [C] Warms up the surrounding air and so is effectively lost.
(c) (i) 34%
   (ii) [A] Hydro-electric
   [B] A renewable energy source will never run out since it is available all the time
   [C] Fossil fuels will soon run out so alternatives need to be found.

Example 3

An isolated house generates its own electricity by damming a small river to make a lake. The water is used to drive a turbine and generator.

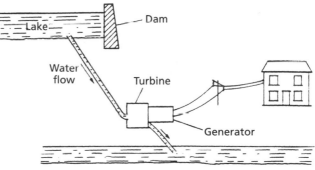

(a) State **two** main energy transformations in this process. (1 mark)
(b) The height of the dam above the turbine is 5 m. Every second 0.5 m³ of water flows through the turbine. The density of water is 1000 kg/m³.
   (i) Use this data to show that the maximum power output is 25 kW. (2 marks)
   (ii) Do you think that an output of 25 kW would be suitable for a house? Justify your answer by estimating the typical power requirements of items that may be used in the house. (3 marks)
   (MEG, Jun 94, Intermediate Tier)

Solution

(a) gravitational potential kinetic electrical
(any **two** of these)

(b) (i) Mass of water travelling through turbines per second

= volume per sec × density of water
= 0.5 m³/s × 1000 kg/m³ = 500 kg/s

GPE lost by water per sec = gravitational field strength × height

= 500 kg/s × 10 N/kg × 5 m
= 25 000 J/s = 25 kW

(ii) Suppose an average house has 10 rooms. Each could be heated by a 1 kW electric fire and lit with a 100 W bulb. This would give a power consumption of 11 kW. In addition there could be an immersion heater to supply hot water, a cooker and various appliances such as irons, kettles, toasters, etc. These are unlikely to be switched on all at the same time and might have a total power of 6 kW. So a total power of 17 kW might be required. In practice the power requirements might be lower than this because some houses use gas heating and gas cookers so 25 kW would be adequate for most houses.

Example 4

The map below shows an industrial region (shaded).

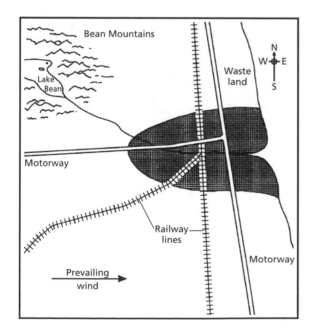

The prevailing wind is from the west. There is a nearby mountainous area, from which a river flows through the region. The major road and rail links are shown.

A power station is to be built to supply electrical energy to the region. The energy will be for a range of domestic and industrial uses.

The choice is between a coal fired power station, wind turbines and a hydroelectric scheme.

Three local groups each support a different option. Choose which option you would support and justify your choice by making reference to the financial, social and environmental implications of your choice compared with those of the alternative systems.

(8 marks)
(NEAB, Jun 94, Science Double Coord, Higher Tier)

Solution

In this answer all three schemes are looked at but to answer the question you would have to consider the best scheme (as far as you are concerned) and compare it with the others.

Hydro-electric scheme

- Renewable source of energy which does not cause pollution.
- High capital costs of constructing the dam and flooding the valley.
- Once constructed only a few people need to be employed.
- Flooding valley kills wildlife and destroys their habitat.
- May not be enough water from the mountain to make the scheme economically viable.

Wind turbines

- Renewable source of energy which does not produce polluting gases although there is some noise and a detrimental visual impact.
- Wind would need to blow all the time to make the scheme viable.
- No employment prospects with this scheme.
- Can give quite cheap electricity although some form of backup would be required.

Coal-fired power station

- Coal when burned produces a lot of heat energy which may be turned into electrical energy. Most coal now comes from abroad but since there are rail links and the site could be placed on the waste land near the sea, transport of the raw material would not be a problem.
- More people are employed in this power station.
- When coal is burned carbon dioxide and sulphur dioxide are produced. The former adds to the greenhouse effect and causes global warming and the latter when dissolved in rain falls as acid rain which kills plants and trees. Smuts from the chimneys could get on local residents' washing.

Example 5

A large wind-powered generator has blades which sweep out a circular area of diameter 50 metres. It faces head-on into a wind of average speed 10 m/s. Calculations show that in 1 second approximately 25 000 kg of air passes through the area swept out by the blades.

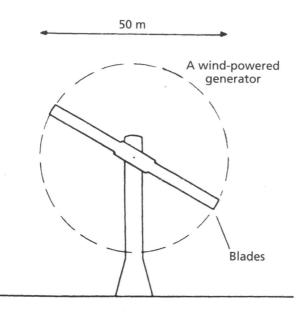

50 m

A wind-powered generator

Blades

(a) Calculate the total kinetic energy of all the air that passes in 1 second through the area swept out by the blades when the wind speed is 10 m/s.
(b) The generator has an efficiency of 10% in converting the kinetic energy of the wind to electrical energy. Calculate the power output of the generator.
(c) If the wind speed doubles, what mass of air now passes in 1 second through the area swept out by the blades?
(d) If the wind speed doubles, how many times bigger will the power output of the generator be?
(e) The power output of a modern coal-fired power

station may be 1000 MW (1×10^6 W). Use the figures in this question to discuss whether or not wind-powered generators are a practical alternative.

Solution

(a) Kinetic energy $= \frac{1}{2}mv^2 = (\frac{1}{2} \times 25\,000 \times 10^2\ \mathrm{J}) = 1\,250\,000\ \mathrm{J} = 1.25\ \mathrm{MJ}$

(b) $\dfrac{10}{100} \times 1.25\ \mathrm{MJ/s} = 0.125\ \mathrm{MW}$

(c) 50 000 g
(d) Kinetic energy $= \frac{1}{2}mv^2$. m and v both double (doubling quadruples v^2). Therefore the kinetic energy is 8 times bigger and the output of the generator is 8 times bigger. [This part is difficult and only grade A* and A grade candidates got it right.]
(e) Power output of the wind-powered generator when wind velocity is 20 m/s is (0.125×8) MW = 1 MW. 1000 wind-powered generators would be needed to produce the same power as one coal-fired power station. The space required would be very large and they would work only when the wind was blowing. The power available would be totally dependent on the wind. They could never replace coal-fired power stations but could be useful for generating small quantities of electrical power on windy days, especially in remote locations.

Examination questions

(Numerical answers and hints on solutions will be found at the end of the chapter.)

Question 1

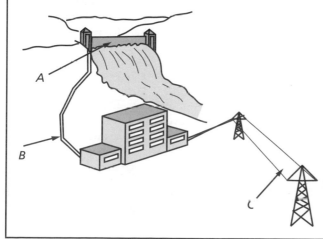

A

B

The diagram above shows a hydro-electric power station. Water is stored in a reservoir beyond the dam. Some of this water flows down the hill through large pipes to the power station.

(i) Name the type of energy possessed by the water in the reservoir at A.

(ii) Name the main type of energy possessed by water flowing down the hill at B.

(iii) Name the machinery contained within the power station building.

(iv) Name the type of energy supplied at C.
(Intermediate Tier)

Question 2

(a) It has been estimated that the Earth receives from

the Sun 10^{17} joules of energy per second and that on average 10^{14} joules per second are being used on the Earth, most of which comes from the burning of fossil fuels. If the Earth is receiving 1000 times as much energy as we use, and it is free, why do we not make use of this energy rather than burn so much fossil fuel?

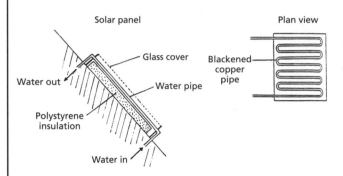

Solar panel   Plan view

The diagram above shows, in simplified form, one method of heating a house using solar energy. It consists of a roof-mounted heat absorbing panel filled with water.

(b)   (i) Explain how the system works, clearly stating the main energy change that takes place.
    (ii) Why is the absorbing panel usually painted black?
    (iii) Suggest ONE reason why the transparent cover increases the efficiency of the system.
    (iv) Explain the purpose of the expanded polystyrene board.
    (v) Give TWO reasons why this system cannot be used as the sole supply of domestic heat in Britain.

(Intermediate Tier)

Question 3

The list below gives some sources of energy.

Coal Oil Natural gas Nuclear
Hydro (water) Wind Wave Solar
Geothermal Tidal

(a)   (i) Name **two** of these sources which are used to generate electricity on a large scale in the United Kingdom.
    (ii) When electricity is generated in power stations much of the input energy is 'wasted'. State **one** reason why electrical energy is so useful that we accept this energy 'loss'.   (3 marks)
(b) Coal is a fossil fuel.
    (i) Give **one** disadvantage of coal.
    (ii) Name **one other** fossil fuel in the list above.
                               (2 marks)
(c) Solar energy is a renewable source of energy.
    (i) Explain what is meant by the term renewable energy.

(ii) Name **one other** renewable energy source in the list above.
(iii) Explain why solar energy is not likely to be used to generate electricity on a large scale in the United Kingdom.   (3 marks)
(d)   (i) What is geothermal energy?
    (ii) Explain briefly how geothermal energy can be extracted.   (3 marks)
           (NEAB, Syll B, Jun 94, Foundation Tier)

Question 4

(a) The pie chart shows five sources of energy used by a country.

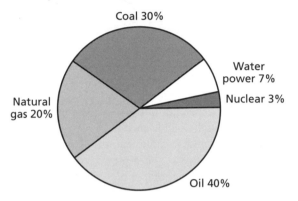

Table 8.1 shows the proportional use and estimated reserves of coal, oil and natural gas.

Table 8.1

	Relative estimated reserves	Relative quantity used/year
Coal	500	1.25
Oil	100	3
Natural gas	90	1.5

(i) Explain why it is always difficult to make accurate predictions of how long reserves will last.   (4 marks)
(ii) Why will the pie chart be likely to be different in about 20 years' time?   (2 marks)
(iii) Explain the economic, environmental and social **benefits** of using nuclear energy as the main source of providing electrical power.
                                 (4 marks)

(b) Pumped storage power stations are used to produce electricity during periods of peak demand. Water is stored in one reservoir and allowed to flow through a pipe to another reservoir at a lower level. The falling water is used to turn turbines which are linked to generators.

In one such station 400 kg of water passes through the turbines every second after falling through 500 m. The gravitational field strength is 10 N/kg.
  Assume that no energy is wasted.

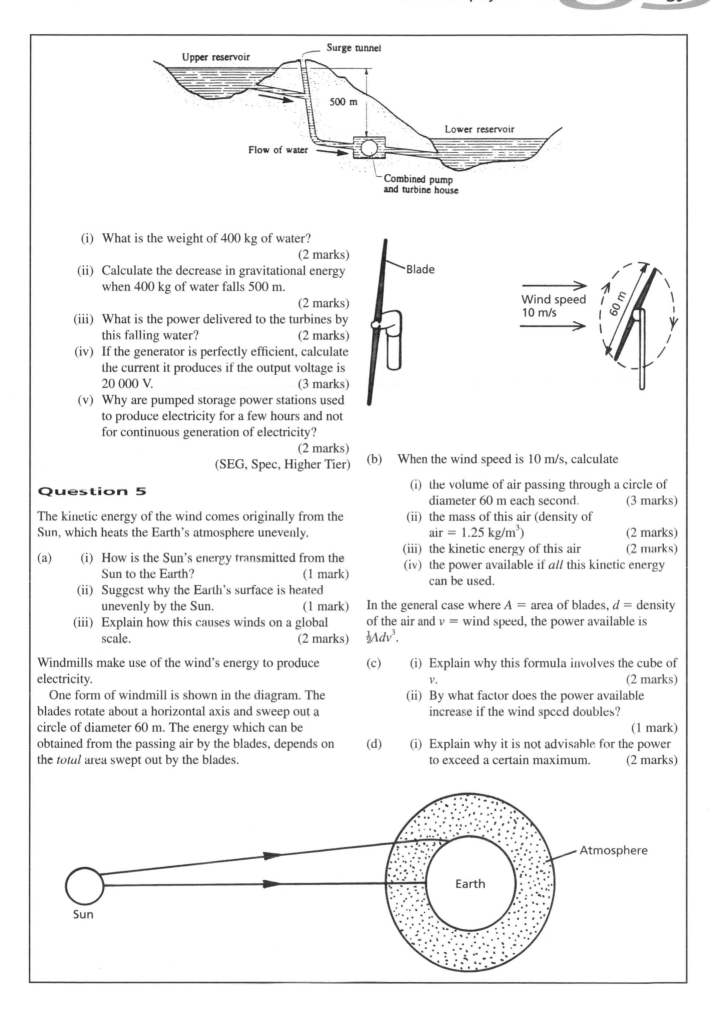

(i) What is the weight of 400 kg of water?
(2 marks)
(ii) Calculate the decrease in gravitational energy when 400 kg of water falls 500 m.
(2 marks)
(iii) What is the power delivered to the turbines by this falling water? (2 marks)
(iv) If the generator is perfectly efficient, calculate the current it produces if the output voltage is 20 000 V. (3 marks)
(v) Why are pumped storage power stations used to produce electricity for a few hours and not for continuous generation of electricity?
(2 marks)
(SEG, Spec, Higher Tier)

Question 5

The kinetic energy of the wind comes originally from the Sun, which heats the Earth's atmosphere unevenly.

(a) (i) How is the Sun's energy transmitted from the Sun to the Earth? (1 mark)
(ii) Suggest why the Earth's surface is heated unevenly by the Sun. (1 mark)
(iii) Explain how this causes winds on a global scale. (2 marks)

Windmills make use of the wind's energy to produce electricity.
   One form of windmill is shown in the diagram. The blades rotate about a horizontal axis and sweep out a circle of diameter 60 m. The energy which can be obtained from the passing air by the blades, depends on the *total* area swept out by the blades.

(b) When the wind speed is 10 m/s, calculate
(i) the volume of air passing through a circle of diameter 60 m each second. (3 marks)
(ii) the mass of this air (density of air = 1.25 kg/m^3) (2 marks)
(iii) the kinetic energy of this air (2 marks)
(iv) the power available if *all* this kinetic energy can be used.

In the general case where A = area of blades, d = density of the air and v = wind speed, the power available is $\frac{1}{2}Adv^3$.

(c) (i) Explain why this formula involves the cube of v. (2 marks)
(ii) By what factor does the power available increase if the wind speed doubles?
(1 mark)

(d) (i) Explain why it is not advisable for the power to exceed a certain maximum. (2 marks)

(ii) Suggest how the power from the windmill could be controlled. (1 mark)

(iii) Give two reasons why all the available wind power is not converted to electrical power. You should state why the loss occurs and how it could be reduced. (4 marks)

(e) Calculations based on another windmill lead to a theoretical maximum power of 500 kW. In practice, such a windmill would have a maximum efficiency of 40%.

(i) Explain the meaning of 'a maximum efficiency of 40%'. (1 mark)

(ii) What is the maximum electrical power which can be produced by this windmill? (2 marks)

(f) State **two** advantages and **two** disadvantages of using wind power to produce electricity. (4 marks)

(MEG, Jun 94, Higher Tier)

The Greenhouse Effect We have the power to help prevent It

Nuclear power
At BNFL, we believe that it has an important role to play in a balanced energy policy for this country. Not only that, it has an important role to play in the environment.
In France and Belgium, for example, they generate more than two-thirds of their electricity from nuclear power. This has helped to reduce their output of carbon dioxide faster than the rest of Europe. In Britain, we could do the same. And we must.
BRITISH NUCLEAR FUELS PLC

The Greenhouse threat would be increased by nuclear expansion taking resources away from real solutions – like energy efficiency

Carbon dioxide is not the only greenhouse gas. And coal fired power stations only produce 10% of the overall greenhouse gases. Replacing all or even part of these with nuclear stations would only address one tenth of the problem.

The massive costs of nuclear power stations would mean that significant expansion is impossible

The major risks associated with nuclear power – the radioactive waste problem and the dangers of catastrophic accidents – would also be increased by nuclear expansion.

Question 6

(a) Explain as fully as you can, the reasons which scientists have suggested for the rise in the amount of 'greenhouse gases' in the atmosphere. (4 marks)

(b) The cuttings represent some viewpoints on the use of nuclear power.

Read all the extracts then evaluate the claim made by BNFL **'In Britain, we could do the same. And we must.'** In your evaluation you should refer to **all** five extracts, explaining the scientific reasons behind each 'claim'. (10 marks)

(NEAB, Jun 94, Science Double Coord, Higher Tier)

Answers to examination questions

1. (i) Gravitational potential energy
   (ii) Kinetic energy
   (iii) Turbine, generator (dynamo)
   (iv) Electrical energy.

2. (a) The solar energy is spread over the whole surface of the Earth and since most of the Earth's surface is covered with water it makes the majority of it impossible to collect.
   Solar cells and panels are expensive.
   The demand for energy is greatest when the Sun's rays are at their weakest.
   Need a back-up source of energy when the Sun's rays are not shining.

   (b) (i) Infra-red radiation from the Sun enters the panel through the glass where it is absorbed by the pipe causing the water inside to heat up. The solar energy from the Sun is therefore changed into heat energy.
   (ii) Black is a good absorber of infra-red radiation.
   (iii) Infra-red radiation is able to pass from the outside to the inside of the panel but not the other way owing to the greenhouse effect. This keeps the heat inside the panel.
   (iv) The expanded polystyrene board is a good insulator and this stops the heat from being lost from the inside to the outside by conduction and convection.
   (v) We can have days (even weeks) without any Sun.
   The Sun's rays are weak compared with hotter countries.

3. (a) (i) Coal and oil (nuclear is still quite small).
   (ii) Electrical energy may be easily moved from place to place using wires. It is also very easy to change into other forms of energy.

   (b) (i) Coal is bulky and therefore difficult to transport.
   Only really suitable for heating.
   (ii) Could have either oil or natural gas.

   (c) (i) Energy source which will never run out since it is constantly being produced.
   (ii) Any one of the following: hydro (water), wind, wave, solar, geothermal or tidal.

(iii) The Sun is not strong enough. It could not be the only source of power because the Sun does not always shine.

(d) (i) Heat from the nuclear reactions which occur in the Earth's core.

(ii) Boreholes are sunk into the hot rocks. Water is pumped down and it comes back up as steam. The steam is used to drive turbines and generate electricity.

4. (a) (i) Reserves are only estimated. New reserves of oil for example are found all the time. Nuclear power is likely to become more significant so this will make oil used only as a portable supply of energy.
More schemes involving renewable sources of energy will be developed.
Reserves become depleted so alternatives need to be found.

(ii) Due to pollution from the burning of fossil fuels, the water and nuclear power will supply a greater percentage of the power. Other sources of power will probably be developed, such as wind, wave, solar and tidal.

(iii) Although the cost of building is high the cost of the electricity produced is low.
No carbon dioxide produced to add to the greenhouse effect and no sulphur dioxide to produce acid rain.
No smoke and ash are produced.
Cheaper electricity benefits the community.
Highly skilled jobs available, usually in areas of high unemployment.

(b) (i) Weight = mass in kg × gravitational field strength in N/kg
= 400 kg × 10 N/kg = 4000 N

(ii) Decrease in GPE = force × distance
= (4000 N) × (500 m)
= 2 000 000 J = 2 MJ

(iii) Power = 2 MJ/s = 2 MW

(iv) Watts = Volts × Amps

Hence, Amps = $\dfrac{\text{Watts}}{\text{Volts}}$

= $\dfrac{2\,000\,000}{20\,000}$ = 100 A

(v) More power is needed to pump the water up to the top than that released when it falls back down. So it is uneconomic.

5. (a) (i) Radiation (remember only radiation is able to travel through a vacuum).

(ii) The Earth is tilted (by 22°) so the Southern Hemisphere gets more Sun than the Northern Hemisphere. Also, points on the Equator are the nearest to the Sun.

(ii) Uneven heating of the air owing to (ii) above means that the air will have different

densities and this makes the air move and cause wind.

(b) (i) In each second, a cylinder of air 10 m long and having diameter 60 m will pass through the windmill.
Volume of a cylinder = $\pi r^2 h$
= $\pi \times (30\,\text{m})^2 \times (10\,\text{m})$
= 28 274 m³
So, volume of air passing per second
= 28 274 m³/s

(ii) Mass of the above volume of air = volume × density
= 28 274 m³ × 1.25 kg/m³
= 35 343 kg/s

(iii) KE = $\frac{1}{2}$ × mass × (velocity)²
= $\frac{1}{2}$ × 35 343 kg × (10 m/s)²
= 1 767 150 J

(iv) Answer to (iii) is the KE per second so since (1 J/s = 1 W) this will be the power.
Power available = 1 767 150 W
= 1.8 MW (n.b. 1 MW = 1 000 000 W)

(c) (i) KE = $\frac{1}{2}mv^2$ but m, the mass passing through the windmill also depends on v, the velocity of the wind. This means that the KE depends on the cube of the wind's velocity.

(ii) Doubling v will mean that the KE will go up according to 2^3. Hence the KE will increase by a factor of eight.

(d) (i) An increase in power would result in an increase in electrical power and this could damage appliances connected in the circuit. Large stresses could be set up on the blades and the tower and this could be dangerous if either of these were to snap.

(ii) The angle of the blades could be altered which would reduce the area on which the wind could act.

(iii) Friction converts some of the kinetic energy of the windmill into heat energy which is lost from the system. The use of ball bearings can help reduce this.
The wind power available is not captured by the blades because the blades do not reduce the wind to zero behind the windmill. Careful aerodynamic design of the blades will help reduce this.

(c) (i) This means that only 40% of the input power is available as useful electrical power so 60% is lost.

(ii) Maximum electrical power = 40% × 500 kW
= 200 kW

(f) *Advantages:* Renewable source of energy
No greenhouse or other polluting gases produced

Disadvantages: Unsightly (disfigures the landscape)
Can produce an annoying hum.

6. Energy efficiency is important but new sources of energy still need to be developed to replaced fossil fuels which will run out over the next 15 to 20 years.

Greater quantities of fossil fuels are being burned owing to more cars on the road and a greater demand for electricity is resulting in greater air pollution. Acid rain and global warming caused by the greenhouse effect. As forests are destroyed the trees and plants are no longer able to cope with the carbon dioxide produced so its concentration has built up in the air. CFCs from aerosols also add to the concentration of greenhouse gases.

All this is in addition to the smoke and ash produced from a fossil fuel burning power station.

Nuclear power does not produce greenhouse gases or acid rain and although the original building costs are high, the electricity costs are quite low.

There have been problems with power stations (mainly in older reactors) regarding escapes of radiation but with careful monitoring these have been reduced. Many people have been killed in coal mines or on oil rigs so there are major risks associated with fossil fuels.

Optics

9

WJEC & NEAB	ULEAC Syll A	ULEAC Syll B	MEG	Topic	MEG Salters'	MEG Nuffield	SEG	NICCEA
✓	✓	✓	✓	**Law of reflection**	✓	✓	✓	✓
✓	✓	✓	✓	**Plane mirrors**	✓	✓	✓	✓
✓	✓	✓	✓	**Refraction**	✓	✓	✓	✓
✓	✓	✓	✓	**Critical angle, prisms, optical fibres**	✓	✓	✓	✓
✓		✓	✓	**Lenses**			✓	✓
✓		✓	✓	**The eye**	✓		✓	✓
✓	✓	✓	✓	**The camera**	✓		✓	✓
✓	✓	✓	✓	**The slide projector**			✓	✓
✓	✓	✓		**The astronomical telescope**				
✓		✓		**Curved mirrors**				✓

1 The law of reflection

The angle of incidence is equal to the angle of reflection. (A normal is a line at right angles to the surface.)

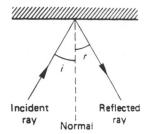

The angle of incidence i *is equal to the angle of reflection* r.

2 Plane mirrors

The image formed by a plane mirror lies on the normal from the object to the mirror and is as far behind the mirror as the object is in front. The image is virtual, laterally inverted and the same size as the object.

A simple periscope may be constructed using two reflecting prisms or two plane mirrors (see example 6).

Plane mirrors are often placed behind a scale over which a pointer passes. By aligning the eye with the pointer and its image the error due to parallax in reading the scale is avoided.

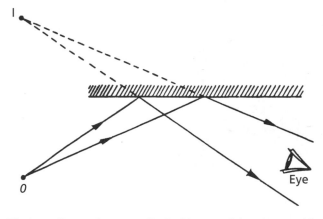

The image lies on the perpendicular bisector of the mirror and is as far behind the mirror as the object is in front.

3 Refraction

When light (or any other wave motion) crosses a boundary between two different media, it is refracted. Refraction results from the change in speed of the wave as it crosses the boundary. The greater the change in the speed, the greater is the refraction of the light.

The next illustration shows a ray of light incident on an air/glass boundary. Some of the light is reflected at the boundary. The light passing into the glass is bent towards the normal. If an object under water or under a glass block is viewed vertically from above then the apparent depth is less than the real depth (see example 12).

When a ray of white light falls on a prism, the different

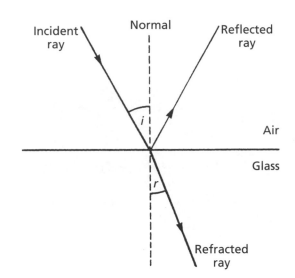

A ray of light passing from air to glass is bent towards the normal.

colours composing the white light are each refracted a different amount and a spectrum is formed. The shorter the wavelength of the light, the greater is the refraction (blue light is refracted more than red light). The greater the amplitude of the waves, the greater is the brightness of the light.

4 Critical angle, prisms, optical fibres

The critical angle for any medium is the angle of incidence of light on the boundary such that the angle of refraction is 90°. If the angle of incidence is greater than the critical angle, total internal reflection occurs. Total internal reflection can only occur when light is passing from an optically denser to an optically less dense medium (i.e. from glass to air).

Isosceles totally internally reflecting prisms are often used instead of plane mirrors because (i) they do not form multiple images (see example 11), and (ii) there is no silvering to wear off. Optical fibres make use of total internal reflection. The outside 'cladding' is less dense than the core, and electromagnetic waves travelling along the fibre are continually being totally internally reflected (in a similar way to the light in a plastic tube: see example 13).

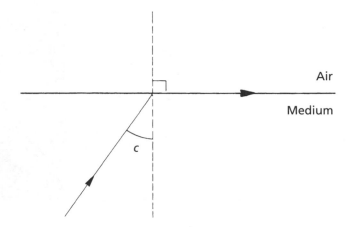

A ray of light incident at the critical angle.

Optical fibres are now being used instead of wires for telephone cables. The fibres can carry more messages and it is less easy to interfere with the transmission, because it travels down the centre of the tube. Security is therefore greater. Signals in optical fibres (usually infra-red laser light) stay strong over larger distances and so 'boosters' can be a long way apart. Doctors use thin flexible fibres with light passing through them to examine the throat and other parts of the body. Optical fibres are also used behind motor car dashboards to distribute light from one bulb to a number of different instrument panel indicators.

5 Lenses

The principal terms used to describe the action of a lens are shown in the diagram opposite. Parallel rays converge to a point in the *focal plane*. The distance from the focal plane to the lens is the *focal length* of the lens. If a lens is used to focus a distant object on a screen, then the distance from the lens to the screen is the focal length.

All objects which are a long way from the lens form inverted real images close to the focal plane, and this fact is made use of in the *camera*.

If the object is close to the principal focus but outside it, the image formed is inverted, real and magnified. A lens is used in this way in a *slide projector*.

If the object is at a distance from the lens which is less than the focal length of the lens, then a virtual, magnified, erect image is formed. When used in this way, the lens acts as a magnifying glass.

These facts are summarised in the table below.

$$\text{Magnification} = \frac{\text{height of image}}{\text{height of object}} = \frac{\text{image distance}}{\text{object distance}}$$

The following formula may be used in lens calculations:

$$\frac{1}{v} + \frac{1}{u} = \frac{1}{f}$$

where u is the object distance to the lens, v is the image distance to the lens and f is the focal length of the lens. When using this formula it is important to note the following: Converging (convex) lenses have positive focal

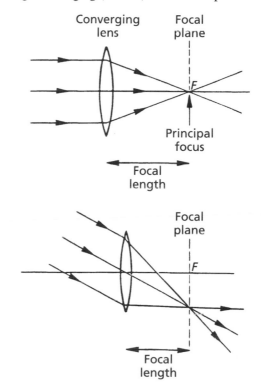

Parallel rays arriving from a distant object and passing through a converging lens.

Object position	Image position	Use
At infinity, or a very large distance	At *F*, or very close to *F*	*Camera*
Outside *F* but close to it	Quite large distance from lens	*Slide projector*
Inside *F*	Same side of lens as object but further from lens	*Magnifying glass*

Images formed by a converging lens.

lengths and diverging (concave) lenses have negative focal lengths. Real image and object distances are positive and virtual image and object distances are negative. [See examples 20 and 21.]

6 The eye, the camera, the slide projector and the astronomical telescope

The eye and the camera both have a converging lens which forms a real, diminished, inverted image of the object on a light sensitive area (the retina in the eye, the film in a camera). A slide projector has a converging projection lens which forms a real, inverted, magnified image on a screen.

The astronomical telescope consists of two convex lenses separated by a distance. The eyepiece lens has a short focal length whereas the other lens, called the objective lens, has a longer focal length. The object (heavenly bodies usually) is so distant that the light rays coming from it arrive at the objective lens almost parallel to each other. They are focused by the objective lens and produce a real inverted image which is then viewed as the object by the eyepiece lens which produces a magnified virtual image of the object.

7 Curved mirrors

Parallel beams of light incident on a concave mirror are focused at the principal focus (see the illustration below). This property of concave mirrors is made use of in micro-wave dishes and radio telescopes: the waves from distant sources are brought to a focus. In headlamp reflectors the source is placed at the principal focus and a parallel beam is produced. Electric fires also make use of curved reflectors.

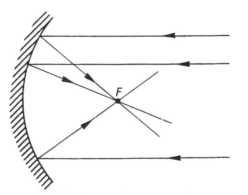

The curved reflector brings parallel rays to a focus at F. On the other hand, if a source is placed at F, a parallel beam results.

Worked examples

Example 1

Which one of the following is not a property of the image of an object placed 12 cm in front of a plane mirror?

A   It is behind the mirror
B   It is 12 cm from the mirror
C   It is laterally inverted

D   It is real
E   A line joining the top of the object to the top of the image is perpendicular to the plane of the mirror

Solution

[No light travels from the object to points behind the mirror. The rays of light reflected at the mirror appear to come from a point behind the mirror. The image is virtual.]

Answer **D**

Example 2

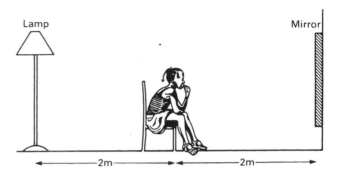

A girl is sitting on a chair 2 m in front of a plane mirror. There is a lamp 2 m behind her. She sees the image of the lamp in the mirror.

What is the distance between the girl and the image of the lamp?

A 2 m   B 4 m   C 6 m   D 8 m

Solution

[The image of the lamp is 4 m behind the mirror. The girl is 2 m in front of the mirror. The distance between the girl and the image of the lamp is 4 m + 2 m = 6 m.]

Answer **C**

Example 3

The diagram shows the image of a watch-face in a plane mirror.

What is the time shown on the watch-face?
A 3:55   B 4:05   C 4:55   D 8:05   E 8:55

Solution

[If you find this difficult, hold a watch in front of a mirror. Remember that the image is laterally inverted. In the

diagram below the dotted lines show the true watch-face time.]

Answer **B**

Example 4

Which of the following shows what happens to a ray of light when it travels from air into glass?

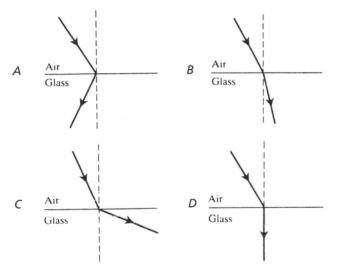

Solution

[When a ray of light passes from air to glass it is bent towards the normal but it does not reach it. See section 3.]

Answer **B**

Example 5

Which one of the diagrams correctly shows the path of the ray through the glass block?

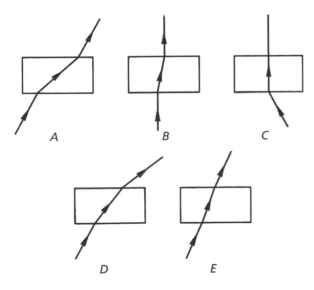

Solution

[When light passes from air to glass it is bent towards the normal. It is bent away from the normal when it passes from glass to air. It emerges parallel to the incident ray.]

Answer **E**

Example 6

Which of the five diagrams below best represents the path of a ray of light through a periscope?

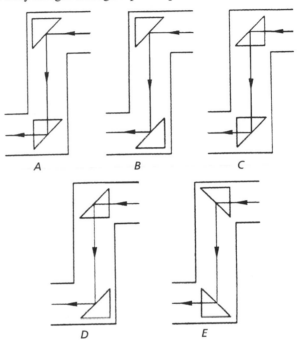

Solution

[Prisms are used to reflect light by making use of total internal reflection (see section 4).]

Answer **C**

Example 7

The diagrams show a ray of light incident on a glass prism.

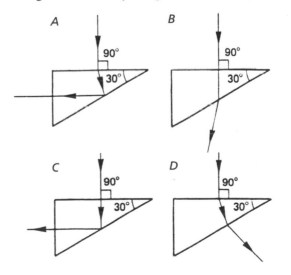

Which diagram shows the subsequent path of the ray through the glass prism correctly?

Solution

[The ray is incident normally on the block (i.e. the angle of incidence is 0°). It therefore passes into the block undeviated. When it reaches the glass–air interface, it is bent away from the normal, as shown in *B*. It is not totally internally reflected, as the angle of incidence at the glass–air interface is less than the critical angle (the critical angle for glass is about 42°).]

Answer **B**

Example 8

The diagrams below show rays of light leaving a point. The point is on an illuminated slide which is in a slide projector. The rays are shown passing through the projection lens and forming an image on a screen. In which diagram will the image on the screen be a clear one? (The diagrams are not drawn to scale.)

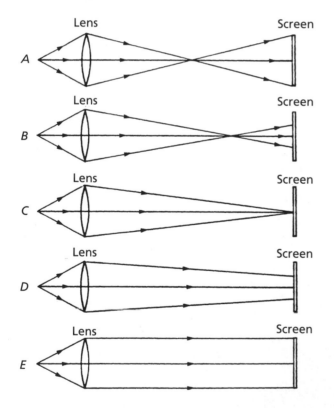

Solution

[For an image to be in focus on a screen the rays of light leaving a point on the object must all pass through a point on the screen.]

Answer **C**

Example 9

(i) A girl stands at a distance of 2 m in front of a plane mirror, and a boy stands at a distance of 3 m in front of the same mirror. How far from the boy is the girl's image in the mirror? (3 marks)
(ii) The girl stands still and the mirror is moved away from her at 3 m/s. At one instant the girl and her image are 6 m apart. How far apart will they be 2 s later? (5 marks)

Solution

(i) The girl's image is 2 m behind the mirror. The boy is 3 m in front of the mirror. Therefore the girl's image is 5 m from the boy.
(ii) When they are 6 m apart the girl is 3 m in front of the mirror and her image is 3 m behind the mirror. 2 s later she is 9 m in front of the mirror and her image is 9 m behind the mirror. They are 18 m apart.

Example 10

(a) A pin is placed in front of a plane mirror. Describe an experiment you would do to locate the position of the pin's image. (9 marks)
(b)

(Not drawn to scale)

The diagram shows a girl standing in front of a plane mirror. *H* is the top of her head, *E* her eyes and *F* her feet. The girl is 140 cm high and her eyes are 10 cm below the top of her head. Draw a ray diagram (which need not be to scale) showing
  (i) a ray of light which travels from the top of her head to her eyes and
  (ii) a ray of light which travels from her feet to her eyes. On your diagram the girl may be shown as a straight line with *H*, *E* and *F* shown as dots on the line.
(c) What is the minimum length of the mirror that would be required in order to enable her to see a full length image of herself in the mirror? (11 marks)

Solution

(a) The plane mirror is placed on a sheet of paper with its surface vertical and the object pin *O* placed in front of the mirror. With the eye at *A* two pins A_1 and A_2 are

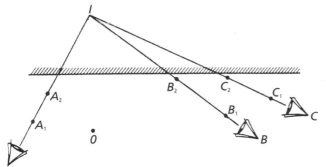

placed so that they are in line with the image *I*. The procedure is repeated with the eye at *B* and *C*. The mirror is removed and the lines A_1A_2, B_1B_2 and C_1C_2 are drawn. They intersect at the position of the image *I*.

(b)

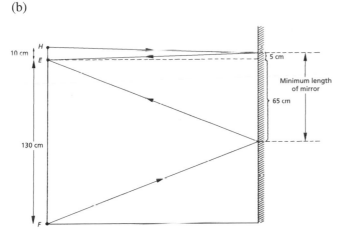

(c) As shown in the diagram the minimum length of the mirror is 70 cm.

Example 11

An object is placed in front of a thick sheet of glass.

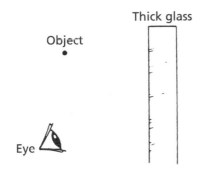

(a) An image, I_1, is formed by reflection from the front surface of the glass.
  (i) Mark and label the exact position of I_1.
  (ii) Draw a ray diagram to show how this image is seen by the eye.

(b) A second image, I_2, will also be seen by the same observer.

(i) What causes the formation of this second image?
(ii) Mark on the diagram the position of this second image, I_2. (4 marks)

Solution

(a) [On the scale of the diagram *O* is 2.2 cm in front of near surface of glass, so I_1 is 2.2 cm behind the near surface of glass. Draw lines from either side of the pupil directed towards I_1. When they hit the near surface they go to the object. Don't forget to dot the virtual rays and to put arrows on light rays from the object going into the eye.]

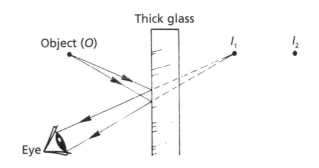

(b) (i) Reflection at the rear surface of the glass of light refracted at the first interface.
    (ii) [The object is 3 cm in front of this rear surface and the image I_2 is nearly 3 cm behind it. It would be exactly 3 cm if there were no refraction at the front surface of the glass.]

Example 12

(a) When you look vertically down into a pond, the pond appears to be shallower than it really is. Draw a ray diagram to illustrate this phenomenon. Mark clearly on your diagram the position of the bottom of the pond and where the bottom appears to be. (5 marks)

(b) The diagrams show a ray of light incident on a glass prism. In each case complete the diagrams, showing the subsequent path of the ray. (5 marks)

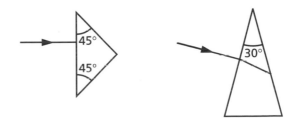

Solution

(a)

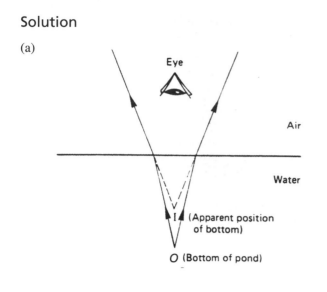

(b)

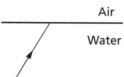

[For (b), in the first diagram the ray is incident on the glass–air interface at an angle greater than the critical angle and is totally internally reflected. In the second diagram the angle of incidence on the glass–air interface is much less than the critical angle (the critical angle is about 42° for glass) and the ray passes into the air, being bent away from the normal. Some light will be reflected.]

Example 13

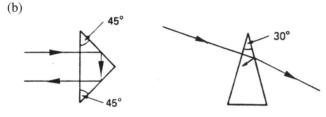

(a) The diagram shows a ray of light in water incident on a water–air boundary. Draw sketches to show what would happen to the ray of light when the angle of incidence is
   (i) about 15°
   (ii) about 60°. (5 marks)

(b) The diagram shows a light beam incident on a curved transparent plastic tube. Explain why the light will stay in the tube and come out at the other end. (4 marks)

Solution

(a) [When the angle of incidence is 15° the light passes into the air, being refracted away from the normal. Some light will be reflected. At an angle of incidence of 60°, total internal reflection occurs and no light passes into the air. Total internal reflection occurs whenever the angle of incidence in the glass exceeds 42°.]

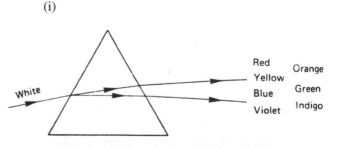

(b) The diagram shows the path of the ray. Each time it is incident on the plastic–air surface the angle of incidence is greater than the critical angle and it is totally internally reflected.

Example 14

(i) Draw a diagram showing how a ray of white light entering a 60° glass prism is refracted and dispersed. (4 marks)
(ii) Beyond the red end of the spectrum there is some invisible radiation. How would you detect this radiation? (5 marks)
(iii) State, putting them in order of increasing wavelength, five regions of the electromagnetic spectrum. (4 marks)

Solution

(i)

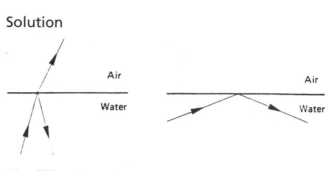

(ii) Infra-red radiation may be detected by using a phototransistor [a semiconductor device] connected in series with a battery and a milliammeter. The reading on the milliammeter is a measure of the infra-red radiation falling on the phototransistor. The phototransistor is moved through the spectrum and into the region beyond the red end. In this position the reading on the milliammeter increases.

(iii) Gamma rays, X-rays, ultra-violet light, infra-red light and radio waves (see section 3).

Example 15

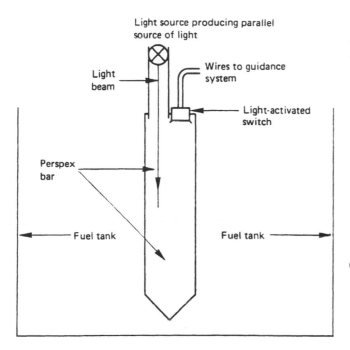

Light source producing parallel source of light

The diagram shows a design for an 'optical fuel gauge' suitable for use in a rocket fuel tank. When the tank runs low, the light-activated switch is triggered.

Air is less 'optically dense' than perspex. Perspex is less 'optically dense' than rocket fuel.

(i) Complete the diagram showing the paths of the light rays when the tank is (a) FULL; (b) EMPTY; Label the rays you draw (a) and (b).
(5 marks)

(ii) Why is the term 'CRITICAL ANGLE' of importance in this application? (3 marks)

(iii) What is the term given to the process by which light changes direction in order to reach the light-activated switch? (1 mark)

Solution

(i) *See the illustration at the top of the next column.*

(ii) When the light passes from a denser to a less dense medium and the angle of incidence in the denser medium is greater than the critical angle, the ray is totally internally reflected as shown above, and the light-activated switch is triggered. [When there is fuel round the perspex the light is passing into a denser medium and is bent towards the normal as shown in (a) above.]

(iii) Total internal reflection.

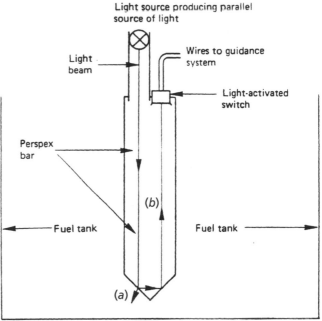

Light source producing parallel source of light

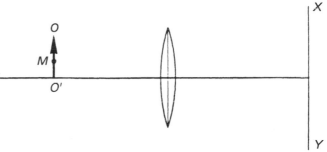

Example 16

(a) The diagram shows an object OO' placed in front of a converging lens. The image is formed on the line XY.

(i) Draw three rays leaving O (the arrow head at the top of the object) and passing through the lens. Show clearly the position of the image. (6 marks)

(ii) Show the path of two rays leaving M (the mid-point of OO'), which pass through the lens and travel to the image. (3 marks)

(b) Suppose the lens were dropped and broken into two approximately equal pieces. What effect, if any, would this have on the brightness and size of the image formed in (a)(i)? (3 marks)

Solution

(a) (i) and (ii)

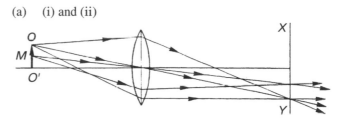

[The ray through the centre of the lens passes through undeviated. Where this meets the line *XY* is the position of the image. Once this position is fixed any other ray leaving *O* goes through this same point on the image. All rays leaving *M* go through the mid-point of the image.]

(b) The rays of light passing through the remaining half will travel the same path as when the whole of the lens was present. The image will therefore be the same size. Since only half the light will reach the image, the image will be less bright.

Example 17

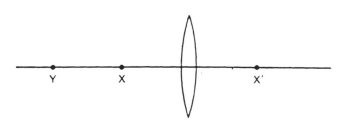

The diagram shows a point object *X* on the principal axis of a converging lens. The image is formed at *X'*.

    (i) Draw three rays from *Y*, another point on the principal axis of the lens, and show a possible path for each of these rays after they pass through the lens. Mark the position of the image *Y'*. (4 marks)

    (ii) Draw another diagram showing the paths of rays coming from a distant object and passing through the lens. (3 marks)

Solution

    (i)

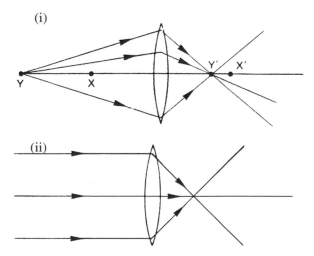

    (ii)

[As the object distance increases, the image distance decreases. When the object is a long way from the lens, the rays arrive at the lens nearly parallel. They pass through the principal focus, which is closer to the lens than *Y'*.]

Example 18

A camera has a lens of focal length 50 mm.

(a) When one takes a photograph of a distant object, where should the film be placed? (2 marks)

(b) What adjustments can be made to adapt the camera to take photographs in bright sunlight? (4 marks)

Solution

(a) Rays from a distant object are brought to a focus in the focal plane of the lens and the film must therefore be placed at the focal plane of the lens, i.e. 50 mm from the lens.

(b) The amount of light entering the camera can be reduced. This may be done by (i) decreasing the aperture or (ii) decreasing the exposure time. Another way is to use a film of lower sensitivity.

Example 19

(a) State *one* similarity and *one* difference in the images formed by a slide projector and a camera. (2 marks)

(b) The diagram shows a camera being used to photograph a distant object. Complete the diagram by continuing the rays to show how the image is formed. (2 marks)

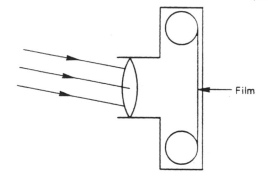

(c) The same camera is now used to photograph an object which is about 3 m from the lens. What adjustment must be made to the lens in order that a sharp image is formed on the film? (2 marks)

Solution

(a) Both images are inverted. [Remember that the slide is put in the projector upside down.] The slide projector image is magnified. The camera image is diminished.

(b)

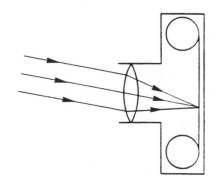

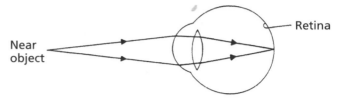

(c) The lens must be moved further from the film. [If the lens were left in the same position, the image would be formed behind the film, so the rays reaching the film would not be focused and a blurred image would be formed on the film.]

(a) Describe how the eye adjusts to focus light from a **more distant** object. (2 marks)

Example 20

An object is positioned 10 cm from the centre of a convex (converging lens) of focal length 20 cm. Calculate the position of the image.

Solution

Using the lens equation with $u = +10$ cm (real object) and $f = +20$ cm (converging lens):

$$\frac{1}{v} + \frac{1}{u} = \frac{1}{f}$$

$$\frac{1}{v} + \frac{1}{10} = \frac{1}{20}$$

Giving image distance to the lens, $v = -20$ cm. [Note that the negative sign indicates that the image is virtual and on the same side as the object. The lens is being used as a magnifying glass since the object is placed inside the focal length of the lens.]

Some people, who can see objects clearly, have an eye defect which stops them focusing rays from distant objects correctly.

(b) Name this defect. (1 mark)

(c) On the diagram below, show how an eye with this defect would focus the rays from a distant object. (1 mark)

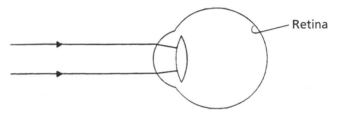

(d) On the diagram below, show how a suitable lens can be used to correct this defect. (4 marks)

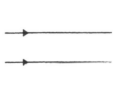

(MEG, Jun 95, Intermediate Tier)

Example 21

An object is positioned 12 cm from the centre of a concave lens of focal length 18 cm. Find the position and nature of the image produced by this lens.

Solution

Using the lens equation with $f = -18$ cm [negative because the lens is concave (diverging)] and $u = 12$ cm gives

$$\frac{1}{v} + \frac{1}{u} = \frac{1}{f}$$

$$\frac{1}{v} + \frac{1}{12} = -\frac{1}{18}$$

Solving this gives image distance, $v = -7.2$ cm

Hence, a virtual image [we know this because of the negative image distance] is formed 7.2 cm from the lens on the same side as the object. Image is upright and diminished. [To determine this you would need to either remember or be able to construct a ray diagram.]

Example 22

The next illustration shows how a human eye focuses light from a **near** object.

Solution

(a) Light from a distant object does not need to be turned through as great an angle by the lens, so the lens does not need to be as fat. The ciliary muscles make the lens thinner thereby increasing the focal length of the lens.

(b) Short sight

(c)

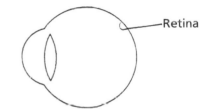

(d)

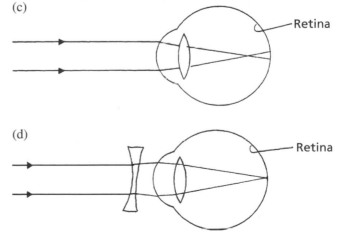

Examination questions

(Numerical answers and hints on solutions will be found at the end of the chapter.)

Question 1

The diagram shows a section through the human eye.

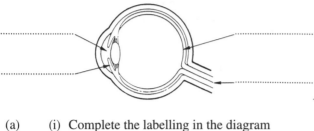

(a) (i) Complete the labelling in the diagram above.
     (ii) Complete table 1.

Table 1

Part of the eye	Description
	Sensitive to light
	Carries signal to brain
	Alters the size of the pupil

(7 marks)

(b) The diagram shows one type of eye defect.

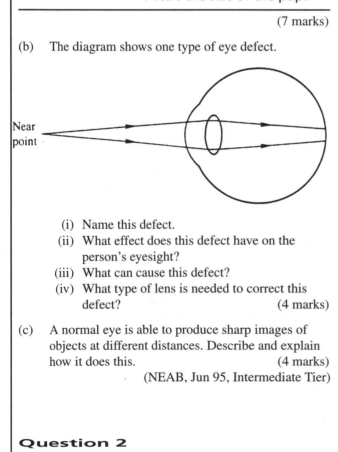

     (i) Name this defect.
     (ii) What effect does this defect have on the person's eyesight?
     (iii) What can cause this defect?
     (iv) What type of lens is needed to correct this defect? (4 marks)

(c) A normal eye is able to produce sharp images of objects at different distances. Describe and explain how it does this. (4 marks)
(NEAB, Jun 95, Intermediate Tier)

Question 2

(a) Complete the path of a ray of light as it passes through and comes out of the glass block shown below. (3 marks)

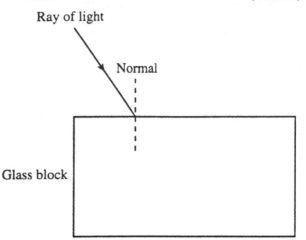

(b) Explain why the light changes direction when it enters the block. (1 mark)
(WJEC, Jun 95, Intermediate Tier, Q15)

Question 3

(a) The diagram below shows wavefronts of light in air arriving at a glass block.

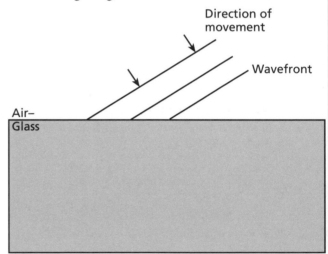

     (i) State what happens to the speed of the wavefront as it enters the glass block. (1 mark)
     (ii) Complete the diagram to show the direction of the wavefronts once they have entered the glass. (1 mark)
     (iii) Explain what causes the wavefront to change direction as it enters the glass. (2 marks)

(b) The diagram below shows a simple camera which has been focused on a distant object.

     (i) Continue the rays entering the lens to show how an image is produced on the film. (2 marks)
     (ii) What changes need to be made to the camera to take a picture of a close object? Explain your answer. (3 marks)

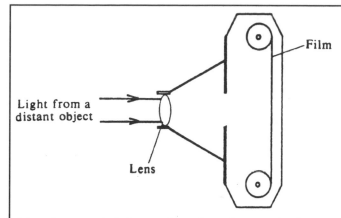

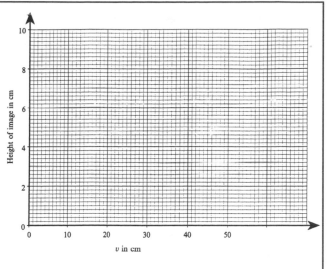

(c) A camera is being used to take a photograph of a passing car which is moving at a steady speed of 20 m/s. To take the photograph, the shutter on the camera is opened for 0.05 s. The car is then 10 m from the lens in the camera and the film is 5.0 cm behind the lens.

    (i) Calculate the distance the car travels during the time the shutter is open. (2 marks)

    (ii) Explain how this causes a blurred image to be produced on the film. (2 marks)

    (ULEAC, Syll A, Jun 95, Intermediate Tier, Q4)

Question 4

(a) The working of a slide projector was investigated by a student using the arrangement below.

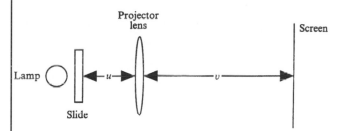

The distance u, between the slide and the lens was varied. For each value of u the screen was moved until a sharp image of the slide was produced on the screen. Each time the distance v, between the lens and the screen was measured and also the height of the image was measured. The values obtained are shown in table 2.

Table 2

v in cm	15	25	30	33	40
Height of image in cm	1.0	3.0	4.0	4.6	6.0

    (i) What type of lens is used in the projector? (1 mark)

    (ii) On the grid below plot a graph to show how the height of image (y-axis) depends on the distance v (x-axis), between the lens and the screen. (6 marks)

(iii) Using the graph complete table 3.

Table 3

v in cm	Height of image in cm
	5
	10
10	

(3 marks)

The focal length of the lens used in this slide projector was 10 cm.

    (iv) Use the graph to find the height of the image when it is 20 cm from the lens. (1 mark)

    (v) What is the height of the slide used in this investigation? (2 marks)

(b) The full scale diagram below shows an object in front of a converging lens. A sharp image is formed on the screen. (6 marks)

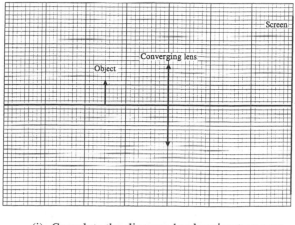

    (i) Complete the diagram by drawing two rays from the top of the object to show how the image is formed.

    (ii) On the diagram mark and label the principal focus. (2 marks)

(iii) Mark and measure the focal length of the lens. (3 marks)

(iv) What two changes would you make to show the formation of a more magnified image of the same object using the same lens?(2 marks)

(c) A magnifying glass also uses a converging lens, but it works in a different way.

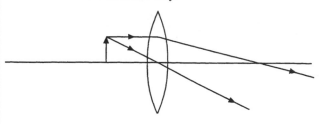

Describe the differences in terms of:

(i) the position of the object (1 mark)

(ii) the position of the image. (1 mark)

(iii) Could this type of image be produced on a screen?

Explain your answer. (2 marks)

(NICCEA, Jun 95, Higher Tier P1, Q3)

Answers to examination questions

1. (a) (i) Working around the diagram from the top left in a clockwise direction we have: Pupil, Retina, Optic nerve and Iris.

   (b) (i) Long sight.

   (ii) The eye cannot see near objects clearly. They can only see distant objects clearly.

   (iii) Any one of: the eye is too short, the muscles have lost their elasticity or the eye lens is too thin.

   (iv) A converging (convex) lens.

   (c) (i) The ciliary muscles alter the shape of the lens and therefore the focal length. The eye lens is fat when near objects are viewed and thin when distant objects are viewed.

   (ii)

Table 1

Part of the eye	Description
Retina	Sensitive to light
Optic nerve	Carries signal to brain
Iris	Alters the size of the pupil

2. (a)

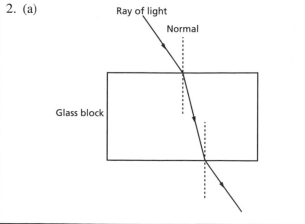

(b) Light travels slower in glass than in air and this causes the light to bend towards the normal as it enters the block. On leaving the other side of the block the light bends away from the normal and is therefore parallel to the incident ray.

3. (a) (i) The speed decreases.

   (ii)

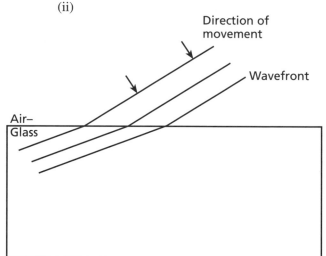

   (iii) Light travels slower in glass than in air and this means that one end of the wavefront is slowed before the other which causes the light to travel a smaller distance in the glass compared with the air. The effect is to bend the light.

(b) (i) the rays on the diagram should come together and meet on the film.

   (ii) The lens is moved so that it is further away from the film. The divergence of the rays for a near object are greater so that more room is needed between the lens and the film for the rays to converge.

(c) (i) Distance = Speed × Time = 20 m/s × 0.05 s = 1 m

   (ii) For a 1 m movement in the object, there will

be a 0.5 cm movement in the image which means the light will be spread over this distance, thus causing blurring.

4. (a) (i) A converging (convex) lens.

(ii)

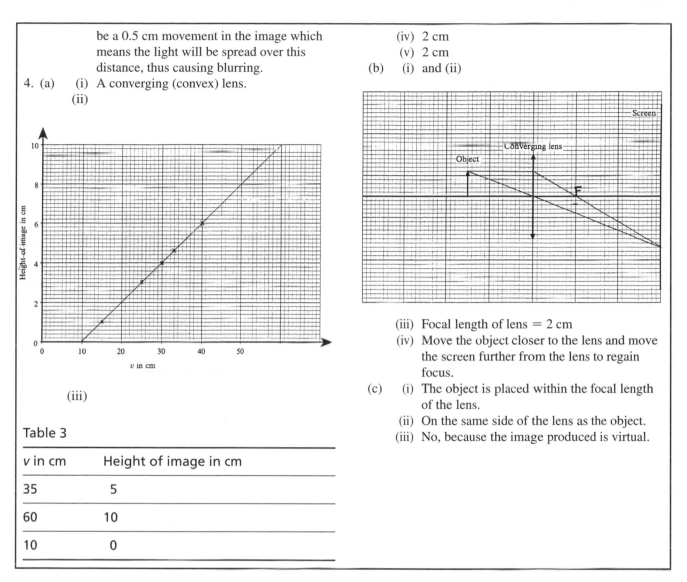

(iii)

Table 3

v in cm	Height of image in cm
35	5
60	10
10	0

(iv) 2 cm
(v) 2 cm

(b) (i) and (ii)

(iii) Focal length of lens = 2 cm
(iv) Move the object closer to the lens and move the screen further from the lens to regain focus.

(c) (i) The object is placed within the focal length of the lens.
(ii) On the same side of the lens as the object.
(iii) No, because the image produced is virtual.

10

Wave motion

WJEC & NEAB	ULEAC Syll A	ULEAC Syll B	MEG	Topic	MEG Salters'	MEG Nuffield	SEG	NICCEA
✓	✓	✓	✓	**Energy, speed, frequency and wavelength**	✓	✓	✓	✓
✓	✓	✓	✓	**Longitudinal and transverse waves**	✓	✓	✓	✓
✓	✓	✓	✓	**Reflection and refraction**	✓	✓	✓	✓
✓	✓	✓	✓	**Sound and its properties**	✓	✓	✓	✓
✓	✓	✓	✓	**Noise**	✓	✓	✓	✓
✓	✓	✓	✓	**Resonance**	✓	✓	✓	✓
✓	✓	✓	✓	**The electromagnetic spectrum**	✓	✓	✓	✓
✓	✓	✓	✓	**Diffraction and interference**	✓	✓	✓	✓
✓	✓	✓	✓	**Polarisation**	✓	✓	✓	✓

1 Wave motion

Energy, speed, frequency and wavelength

In all wave motion, a disturbance (energy) travels through a medium without the medium moving bodily with it. The wave transmits energy from the source to the receiver.

The *amplitude* of the waves is their biggest displacement from the undisturbed level (see example 6), and is a measure of the energy of the waves. In light a greater amplitude results in greater brightness, and in sound a greater amplitude results in a louder note. The *frequency* is the number of oscillations or cycles of the wave motion occurring every second (unit: hertz (Hz)). The *wavelength* is the distance between successive crests or successive troughs (see example 6).

The speed, v, of any wave is related to the frequency, f, and the wavelength, λ, by the equation $v = f\lambda$. The speed of a wave depends on the medium in which it is travelling. All electromagnetic waves (e.g. radio, light and X-rays) travel at the same speed in a vacuum – namely, 3×10^8 m/s.

Longitudinal and transverse waves

In a transverse wave, oscillations are perpendicular to the direction of propagation of the wave (water waves and electromagnetic waves are transverse). In a longitudinal wave the medium oscillates along the direction in which the wave travels (sound waves are longitudinal).

Reflection and refraction

Waves can be reflected and refracted. Refraction results from a change in speed as the wave crosses a boundary between two different media; the wavelength of the wave also changes, but the frequency does not change (see examples 7 and 8 and question 5).

2 Sound and its properties

Every source of sound has some part which is vibrating. Sound needs a medium in which to travel and cannot travel through a vacuum. Sound waves consist of a series of alternate compressions and rarefactions travelling away from the source. The pitch of a note depends on its frequency; if the frequency increases, the pitch of the note goes up. The harder you strike a drum, the greater is the amplitude of vibration and the louder is the note heard.

The human ear is sensitive to sounds with frequencies ranging from about 20 Hz to 20 000 Hz.

Resonance

Any object, or part of an object, has its own frequency of vibration, called its natural frequency which it would like to vibrate with.

These objects or their parts may be forced to vibrate by other vibrations. For instance:

*   A bridge can be forced to vibrate by vibrations due to the wind.
*   A glass vibrates due to the note received from an opera singer's voice.
*   A loose fitting on a car is forced to vibrate due to vibrations from the engine.

If the frequency of the forcing vibrations is the same as the natural frequency of the object, then the amplitude of vibrations of the object can eventually become large enough to damage the object. This is called resonance and the following can occur due to resonance:

*   A bridge can fall down when the frequency of vibrations due to the wind is equal to the natural frequency of the bridge.
*   A glass can break if the vibration of the note from the singer is equal to the natural frequency of the bridge.
*   Parts of a car work loose and eventually fall off if not secure.

Noise

Noise is really unwanted sound and there are laws which protect people from noisy neighbours, aircraft noise, noise from machinery and road noise.

Noise has many effects, the main ones being:

*   It increases stress and can make people ill.
*   It can cause damage to hearing and can cause deafness.
*   It is distracting and makes it difficult to concentrate, and this can lead to accidents.

Hard surfaces reflect sound and can make the noise inside buildings worse, so soft furnishings (carpets, curtains, etc.) are used to absorb the sound.

3 The electromagnetic spectrum

Electromagnetic waves are characterised by oscillating electric and magnetic fields. In a vacuum they all travel at the speed of light. They are transverse waves. The spectrum of these waves includes γ-rays (gamma rays), X-rays, ultra-violet rays, visible light, infra-red rays, microwaves, and radio waves. The list is in order of increasing wavelength (decreasing frequency), γ-rays having the shortest wavelength (largest frequency). The main features of electromagnetic waves are summarised in table 10.1 on the next page.

Table 10.1

Name and approximate wavelength in continuous spectrum	Source	Detection	Properties and uses
γ-rays $\lambda = 10^{-14}$ m	Cobalt 60 Radioactive isotope	G-M tube Photographic film	Very penetrating Used in treatment of cancer and for sterilisation Can be harmful
X-rays $\lambda = 10^{-9}$ m	X-ray tube	Photographic film Fluorescent screen	X-ray photography X-rays pass through skin and are stopped by bone Study of crystal structure
Ultra-violet $\lambda = 10^{-8}$ m	U-V lamp Sun	Photographic film Photo cells Fluorescent chemicals	Absorbed by glass Helps tanning of skin but causes sunburn (too much damages body cells)
Visible light $\lambda = 0.4 \times 10^{-6}$ to 0.7×10^{-6} m	Sun Lamps	Eye Photographic film	Focused by the eye Essential to photosynthesis and plant growth
Infra-red $\lambda = 10^{-4}$ m	Sun Fires and hot objects	Skin Semiconductor devices (e.g. phototransistor)	Makes skin feel warm Used in photography, when dark Detected by instruments that 'see' in the dark
Microwaves $\lambda = 10^{-2}$ m	Microwave ovens Microwave transmitters	Microwave receiving dish	Absorbed by water molecules Microwave communication
Radio $\lambda = 10^{2}$ m	TV and radio transmitters	Aerials and radio or TV sets	Radio TV Radar

Communications systems

Links between satellites and ground stations usually use microwaves, and telephone systems use infra-red laser light passing through optical fibres.

Microwave ovens

In these, the microwaves penetrate the food and are absorbed by the water molecules in the food, thus heating the food from within. They are therefore faster and more economical than conventional ovens. The heat energy spreads through the food by radiation (the microwaves) and by conduction. Microwaves pass through air, glass and plastic without causing any heating and are reflected by metals. The steel walls of a microwave oven reflect the microwaves back into the food and also prevent the microwaves escaping from the oven.

Security systems

The breaking of an invisible infra-red beam can be used to trigger an alarm system.

4 Diffraction and interference

If a plane wave passes through a small gap, it spreads out round the edges of the gap. The effect is known as diffraction (see example 15).

Young's slits was a classic experiment which demonstrated the constructive and destructive interference of light from a double slit. It was an important experiment in the establishment of the wave theory of light.

A similar experiment with sound waves uses two loudspeakers connected to the same signal generator and placed about 50 cm apart. The variation in intensity resulting from the interference of the waves from the two sources may be heard by walking along a line parallel to

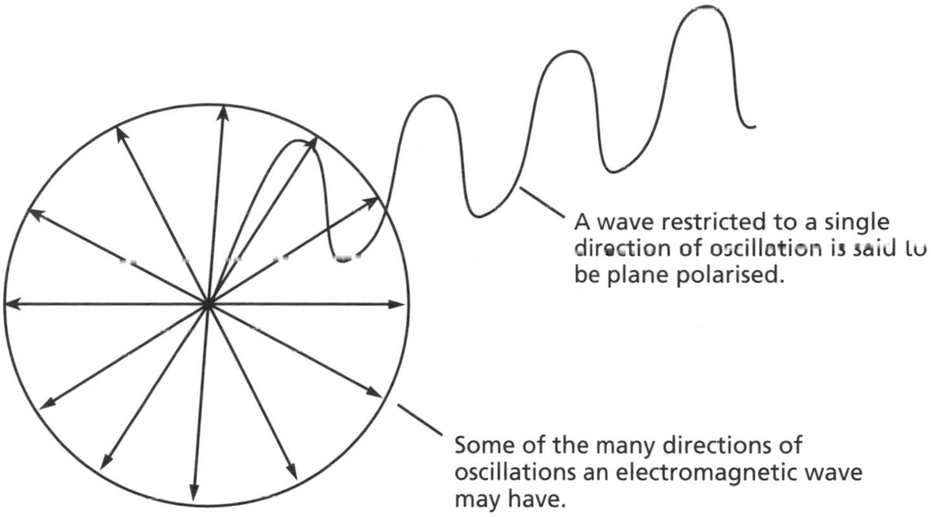

A wave restricted to a single direction of oscillation is said to be plane polarised.

Some of the many directions of oscillations an electromagnetic wave may have.

the line joining the two speakers. There are some places where a loud note is heard. These are where compressions from the two sources arrive together, and rarefactions arrive together. Where a compression arrives with a rarefaction, no sound is heard (see example 15 and question 6).

5 Polarisation

Only transverse waves may be polarised. Transverse waves such as light often consist of frequency oscillations in many directions as shown below.

Polarisation involves restricting the directions of oscillations to just one, when the wave is said to be plane polarised. A Polaroid is a material which when light passes through it produces plane polarised light. In restricting the light to vibrations in a single plane, light in other planes is not allowed through and this lowers the intensity of the light. This is why Polaroid materials are used in sunglasses.

Worked examples

Example 1

Which one of the following statements, concerning the vibrations of a string on a violin which is being played, is true?

A   A longitudinal wave is set up on the string.
B   A transverse wave is set up on the string.
C   The frequency of the note produced is higher the longer the vibrating length of the string.
D   The velocity of the sound waves in the air depends on the tension of the string.

Solution

[The wave travels along the string and the particles of the string vibrate from side to side, i.e. perpendicular to the

direction of travel of the wave. Therefore the wave must be a transverse wave. See section 1, 'Longitudinal and transverse waves'.]

Answer **B**

Example 2

Radio	Infra-red	Visible Light	Ultra-violet	X-rays	Gamma rays

The diagram illustrates the electromagnetic spectrum.

Reading from left to right, the quantity which is increasing is

A   amplitude
B   frequency
C   velocity
D   wavelength

Solution

[All electromagnetic waves have the same speed in a vacuum. The amplitude determines the intensity of the wave. Radio waves have long wavelengths and gamma-rays short wavelengths. Therefore the frequency is increasing from left to right in the diagram. See also sections 1 and 3.]

Answer **B**

Example 3

Which one of the following pairs of waves contains one example of a longitudinal wave and one example of a transverse wave?
A   radio and X-rays
B   infra-red and ultra-violet
C   sound and radio
D   waves on a ripple tank and light

Solution

[All electromagnetic waves are transverse waves. Water waves are transverse waves. So all the waves in the question are transverse except sound.]

Answer **C**

Example 4

The diagram (not drawn to scale) shows the crests of waves spreading out from a point source.

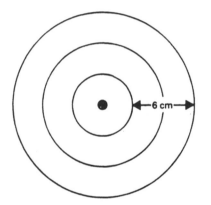

The wavelength of the waves is
A 2 cm B 3 cm C 4 cm D 6 cm E 12 cm

Solution

[The wavelength is the distance between *successive* crests. Two crests are 6 cm apart. The wavelength is 3 cm.]

Answer **B**

Example 5

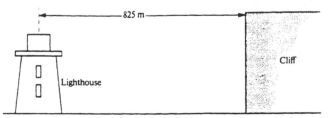

When the foghorn on the lighthouse in the diagram sounded a short blast, the echo was heard at the lighthouse 5 seconds later. The speed of sound was
A 165 m/s B 330 m/s C 825 m/s D 4125 m/s

Solution

[The sound travels to the cliff and back. The distance travelled in 5 s is (825 m) × 2 = 1650 m.

$$\text{Speed} = \frac{\text{distance travelled}}{\text{time taken}} = \frac{1650 \text{ m}}{5 \text{ s}} = 330 \text{ m/s}]$$

Answer **B**

Example 6

(a)   Draw a labelled diagram to illustrate the meaning of the words *wavelength* and *amplitude*.   (3 marks)
(b)   What is meant by frequency?   (2 marks)
(c)   Water waves are made by a dipper moving up and down 3 times every second. If the velocity of the waves is 12 cm/s, what is the wavelength of the waves?   (3 marks)

Solution

(a)
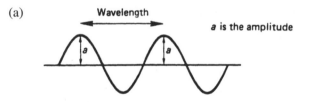

(b)   Frequency is the number of complete oscillations (or cycles) in one second.
(c) $v = f\lambda$ [see section 1]
        12 cm/s = 3 Hz × λ

$$\lambda = \frac{12}{3} = 4 \text{ cm}$$

Example 7

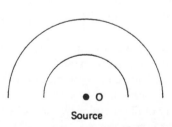

(a)   The diagram shows waves spreading out from a point source O and travelling towards a plane reflecting surface. Complete the diagram, showing what happens to the waves as they arrive at, and leave, the reflecting surface. Show the position of the image of the source on your diagram.   (5 marks)

(b) Plane waves in a ripple tank have a frequency of 6 Hz. If the wave crests are 1.5 cm apart, what is the speed of the waves across the tank? (3 marks)

Solution

(a)

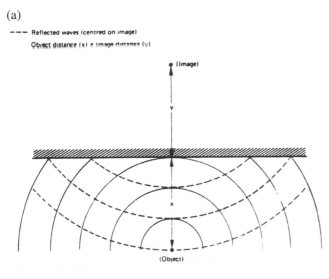

- - - Reflected waves (centred on image)

Object distance (x) = Image distance (y)

(b) $v = f\lambda$ [see section 1]
$v = 6\,\text{Hz} \times 0.015\,\text{m} = 0.09\,\text{m/s}$

Example 8

(a) Draw two diagrams showing plane waves crossing a straight boundary and passing into a medium in which their speed is greater, (i) when the wavefronts of the incident wave are parallel to the boundary and (ii) when the wavefronts make an angle with the boundary. (5 marks)

(b) What change occurs in (i) the frequency, (ii) the speed and (iii) the wavelength, as a result of refraction when light passes into an optically less dense medium. (3 marks)

(b) The frequency remains constant. The speed and wavelength increase.

Example 9

This question is about cooking food in a microwave cooker and in a conventional electrical one.

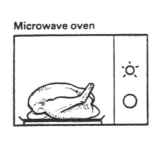

Microwave oven

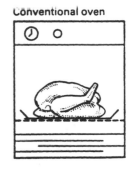

Conventional oven

(a) The microwave cooker uses electromagnetic waves of wavelength 0.12 m.
   (i) Underline the part of the electromagnetic spectrum which includes this wavelength:
      gamma rays ultra-violet visible infra-red radio. (1 mark)
   (ii) Complete the following sentence about the conventional cooker:
      The radiation from the heating elements comes mainly from the .... part of the electromagnetic spectrum. (1 mark)
   (iii) If the speed of the microwaves is 300 000 000 m/s (3×10^8 m/s), find their frequency. (3 marks)

(b) The following table compares the performance of the ovens when cooking a chicken. The conventional oven is first warmed up before putting the chicken in.
   (i) Complete table 10.2 on the next page to find the total energy used by each cooker to cook a chicken.

Solution

(a)

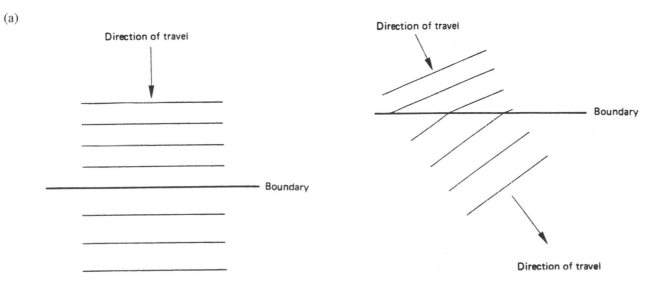

Table 10.2

	Microwave	Conventional
Time to warm up/hours	0	0.5
Input power whilst warming up/kW	0	1.2
Energy to warm up/kWh	0	
Input power whilst cooking/kW	1.5	0.6 (average)
Time to cook chicken/hours	0.4	2.0
Energy used whilst cooking/kWh		
Total energy used/kWh		

(4 marks)

(ii) The energy absorbed by each chicken is 0.2 kWh.
Find the efficiencies of the cookers:
microwave ....
conventional .... (3 marks)

(c) (i) Why could a shiny inside surface reduce energy loss from the **microwave** cooker?

(ii) How do you think energy loss from the **conventional** cooker is minimised?

(3 marks)
(ULEAC, Intermediate Tier)

Solution

(a) (i) radio
(ii) infra-red
(iii) $v = f\lambda$ [section 1] $\Rightarrow 3 \times 10^8$ m/s $= f \times 0.12$ m

$$\Rightarrow f = \frac{3 \times 10^8}{0.12} \text{ Hz} = 2.5 \times 10^9 \text{ Hz}$$

Frequency $= 2.5 \times 10^9$ Hz.

(b) (i)

Table 10.2

	Microwave	Conventional
Time to warm up/hours	0	0.5
Input power whilst warming up/kW	0	1.2
Energy to warm up/kWh	0	
Input power whilst cooking/kW	1.5	0.6 (average)
Time to cook chicken/hours	0.4	2.0
Energy used whilst cooking/kWh	0.6	1.2
Total energy used/kWh	0.6	1.8

[Energy to warm up conventional oven = power × time = 1.2 kW × 0.5 h = 0.6 kWh. Whilst cooking the microwave oven uses 1.5 kW × 0.4 h = 0.6 kWh. The conventional oven uses 0.6 kW × 2.0 h = 1.2 kWh. The total energy is found by adding the energy to warm up to the energy used whilst cooking.]

(ii) Efficiency $= \dfrac{\text{power output}}{\text{power input}}$

For microwave oven,

$$\text{efficiency} = \frac{0.2 \text{ kWh}}{0.6 \text{ kWh}} = \frac{1}{3} \text{ (or } 33\tfrac{1}{3}\%\text{)}$$

For conventional oven,

$$\text{efficiency} = \frac{0.2 \text{ kWh}}{1.8 \text{ kWh}} = \frac{1}{9} \text{ (or } 11\tfrac{1}{9}\%\text{)}$$

(c) (i) Shiny surfaces reflect the microwaves travelling towards the sides of the cooker back towards the food.

(ii) Lagging the sides. Preventing hot air escaping into the atmosphere by ensuring that the oven door is well sealed (gas expansion is allowed for by letting cooler air at the bottom of the oven flow out through a ventilator). A thermostat limits the use of energy to that which is needed to maintain the oven's cooking temperature.

Example 10

A girl stands 100 metres from a large vertical wall. She claps her hands at a steady rate and adjusts the rate until the echo from the wall returns at the same time as the next clap. When this is achieved, a friend times her clapping and finds that she makes 50 claps in 30 s. Use these figures to calculate the speed of sound in air. (3 marks)

Solution

Time between claps $= \dfrac{30}{50}$ s

In this time the sound travels to the wall and back, i.e. 200 m.

$$\text{Speed of sound} = \frac{200 \text{ m}}{30/50 \text{ s}}$$

$$= 333 \text{ m/s}$$

Example 11

A sonar device at the bottom of a ship emits a very short pulse of sound. The pulse travels vertically downwards. The echo is received back after 0.5 s.
(a) How is the echo produced?
(b) How far does the sound pulse travel in 0.5 s? [The speed of sound in sea water is 1500 m/s.]
(c) How deep is the sea under the ship?

Solution

(a) By reflection at the sea bed.
(b) Distance travelled = 1500 m/s × 0.5 s = 750 m
(c) 375 m
     [The sound travels to the bottom of sea and back again.]

Example 12

Table 10.3 gives the speed of water waves in metres per second for various wavelengths and for four different depths of water.

Table 10.3

		Wavelength/m					
		0.001	0.01	0.1	1	10	100
Depth/m	0.1	0.67	0.25	0.40	0.93	0.99	0.99
	1	0.67	0.25	0.40	1.25	2.95	3.13
	10	0.67	0.25	0.40	1.25	3.95	9.33
	100	0.67	0.25	0.40	1.25	3.95	12.5

(a) What is the speed of a wave of wavelength 0.1 m at a depth of 10 m?   (1 mark)
(b) For a wavelength of 0.1 m or less, what is the relationship between speed and depth?   (2 marks)
(c) Describe, in general terms, how the speed varies with the depth, for a wavelength of 100 m.   (2 marks)
(d) Describe, in general terms, how the speed varies with the wavelength, for a depth of 100 m.   (2 marks)
(e) What is the frequency of a wave of wavelength 10 m at a depth of 10 m?   (3 marks)

Solution

(a) 0.40 m/s.
(b) The speed is independent of the depth. [For example,

at wavelength 0.1 m at every depth the speed is 0.4 m/s.]
(c) As the depth increases, the speed increases.
(d) The speed increases as the wavelength increases, increasing very rapidly at large wavelengths.
(e) $v = f\lambda$ [see section 1].
     [At a wavelength of 10 m and a depth of 10 m the speed is 3.95 m/s.]
     3.95 m/s = $f \times$ 10 m

$$f = \frac{3.95 \text{ m/s}}{10 \text{ m}} = 0.395 \text{ Hz}$$

Example 13

(a) The diagram shows an electric trembler bell suspended inside a bell jar from which the air can be removed. When the lowest pressure has been reached, only a very faint ringing can be heard.

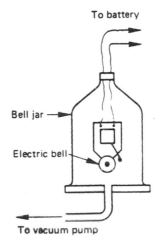

To battery

Bell jar

Electric bell

To vacuum pump

     (i) What does the experiment illustrate about the transmission of sound?   (1 mark)
     (ii) Why is a faint ringing heard?   (2 marks)
     (iii) Why is the Moon sometimes referred to as 'the silent planet'?   (2 marks)

Solution

(a) (i) Sound cannot travel in a vacuum.
     (ii) Vibrations travel up the wires suspending the bell. These vibrations can be heard as a faint ringing.
(b) The Moon has no atmosphere and there is therefore no gas through which the sound can travel. The only way sound can travel is by vibration on the surface.

Example 14

(a) A microphone is connected, via an amplifier, to a cathode ray oscilloscope. Three different sounds are made, one after the other, near the microphone. The traces produced on the screen are shown in the diagrams below. During the experiment the controls of the oscilloscope were not altered.

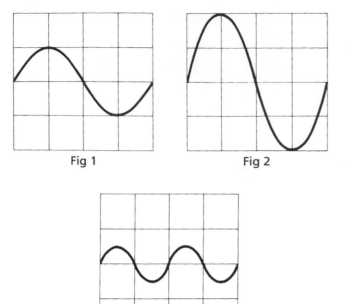

Fig 1   Fig 2

Fig 3

(i) Which trace results from the sound having the highest frequency? (2 marks)

(ii) Which trace results from the loudest sound? (2 marks)

(b) If in Fig. 1 the time base on the oscilloscope is set at 1 ms/cm and the lines on the grid on the screen are 1 cm apart, what is the frequency of the note which emits the sound giving the trace shown in Fig. 1? (4 marks)

Solution

(a) (i) Fig. 3. [In Fig. 1 and Fig. 2 a whole cycle takes four grid squares. In Fig. 3 a whole cycle only takes two grid squares.]

(ii) Fig. 2. [The amplitude is greatest in Fig. 2.]

(b) [There is one complete cycle on the screen.

$$4\,\text{ms} = \frac{4}{1000}\,\text{s} = \frac{1}{250}\,\text{s}]$$

The time for 1 cycle is 4 ms.

$\Rightarrow$ It takes $\dfrac{1}{250}$ s for 1 cycle

$\Rightarrow$ There are 250 cycles every second

Frequency = 250 Hz

Example 15

(a) Draw diagrams to illustrate what happens when plane waves are incident on a slit,

(i) when the width of the slit is large compared with the wavelength of the waves. (3 marks)

(ii) when the width of the slit is small compared with the wavelength of the waves. (3 marks)

(b) A student set up a demonstration using two loudspeakers connected to the same oscillator, which was producing a note of fixed frequency. The loudspeakers were placed at *A* and *B* and they emitted waves which were in phase. An observer walked along the line *PQRS*. A loud note was heard at *Q* and a faint note at *R*.

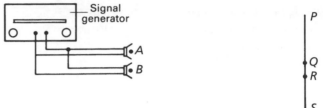

(i) On the same axes sketch two graphs showing how the displacement of the vibrating air molecules varies with the time for the disturbance at *Q*, one for the waves from *A* and one for the waves from *B*. On the same axes sketch a third graph showing the displacement at *Q* for both sets of waves arriving together. (4 marks)

(ii) Repeat (i) for waves arriving at *R*. What is the relationship between the distances *AR* and *BR*? (5 marks)

Solution

(a)

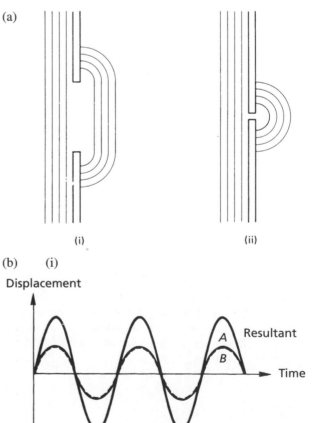

(i)   (ii)

(b) (i)

Displacement

Resultant
A
B
Time

Waves arriving from *A* and *B* in phase

(ii)

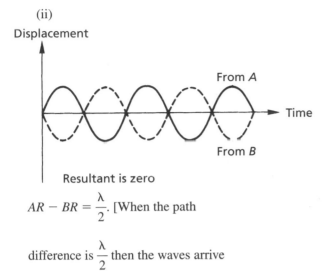

$AR - BR = \dfrac{\lambda}{2}$. [When the path

difference is $\dfrac{\lambda}{2}$ then the waves arrive

at R 180° out of phase, i.e. in anti-phase.]

Example 16

Light is a type of electromagnetic radiation.
(a) Write down one type of electromagnetic radiation with a **wavelength**
   (i) longer than that of light.
   (ii) shorter than that of light.
(b) Complete table 10.4 to show the name and use of some types of electromagnetic radiation.

Table 10.4

Type of radiation	Use
	Sending information to and from satellites
	Making toast
	Producing shadow pictures of bones
Gamma rays	

(2 marks)

(c) Microwaves are used for cooking vegetables.
   (i) Explain, in terms of their effect on the plant cells, how microwaves can be used for cooking.
   (ii) Explain why the door of a microwave oven must have a good seal.   (4 marks)
(d) The diagram below shows some Polaroid sunglasses.

Polaroid film

Polaroid sunglasses are often sold with a small piece of Polaroid film attached to them. The purpose of the Polaroid film is to enable people to test whether the sunglasses are made of genuine Polaroid.
   The test is carried out by putting the sunglasses on and holding the Polaroid film in front of them.
   The film is turned through 360°, keeping it parallel to the lenses.
   Describe what you will see during the test if the sunglasses are
   (i) Polaroid
   (ii) not Polaroid   (5 marks)
(e) Explain the observations seen during the test when the sunglasses are Polaroid.   (3 marks)
(NEAB, Jun 95, Higher Tier, Q7)

Solution

(a) (i) One of radio waves, infra-red or microwaves.
     (ii) One of ultra-violet, X-rays or gamma rays.
(b) In order working down the table: microwaves, infra-red, X-rays.
   Uses of gamma rays could be: sterilising equipment, treatment of cancer, detecting welding flaws in pipes and detecting leaks in underground pipes.
(c) (i) Microwaves are absorbed by the water in the cells causing them to vibrate with greater amplitude and therefore heat up. Conduction heats up the rest of the cells which are not at the surface of the food.
     (ii) Microwaves could escape and damage human cells by causing the water in the cells to heat up. The heat will then be released into the body and this will cause damage.
(d) (i) The intensity of the light changes. There will be two positions 180° apart where there will be no light let through and mid-way between these will be positions where there will be maximum brightness. Between the maximum brightness and minimum brightness there will be an angle of 90°.
     (ii) The light intensity will be reduced (because the small piece will not allow as much light through) but now the intensity will not vary with the angle of rotation.
(e) Polaroids only allow light through them if the light's vibrations are in the same plane as the orientation of the Polaroid. When the Polaroids have directions parallel to each other, then light is transmitted but when one of them is rotated through 90° no light is able to pass through.

Examination questions

(Numerical answers and hints on solutions will be found at the end of the chapter.)

Question 1

A ship is using ultrasound waves to check the depth of the sea.

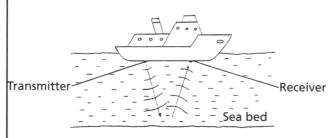

A short pulse of ultrasonic waves is emitted by a transmitter under the ship.

(a) What are ultrasonic waves? (2 marks)
(b) Why does the pulse return to the ship? (2 marks)
(c) The pulses emitted by the ship are shown on a chart as traces *X*. The return pulses are also displayed on the chart as traces *Y* which follow each emitted pulse *X*.

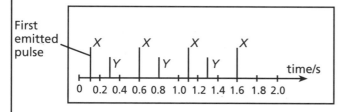

What is the time interval between the emitted pulses? (1 mark)
(d) What is the time between a pulse leaving the ship and returning to it? (1 mark)
(e) Use the equation to calculate the distance a pulse travels between leaving the ship and returning to it. The speed of ultrasonic waves in sea water is 1500 m/s.
distance = speed × time (2 marks)
(f) How deep is the water under the ship? (1 mark)
(MEG, Jun 94, Intermediate Tier)

Question 2

(a) The diagram shows the structure of the human ear.

Complete the sentences below:

Sounds entering the ear cause a thin sheet of skin called the .... to vibrate. These vibrations pass through three small bones called ...., which increase the strength of the vibration. The vibrations are converted to electrical impulses in the .... These electrical impulses pass through the .... to the brain. The function of the semicircular canals is .... (5 marks)

(b) In a flute, a column of air is made to vibrate at its natural frequency so that a loud note is produced.
    (i) What name is given to this effect?
    (ii) A column of air in a flute produces a note of wavelength 64 cm. The speed of sound in air is 330 m/s. Calculate the frequency of the note produced. (5 marks)
(ULEAC, Syll B, Jun 95, Intermediate Tier P2, Q5)

Question 3

(a) (i) The diagram below shows part of a telephone.

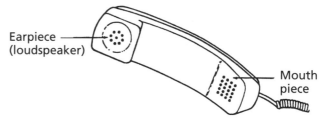

Choose the words from the list below which complete the sentence which follows.
marks)

chemical   electrical   gravitational   heat
light   nuclear   sound

The microphone in the mouthpiece changes .... energy into .... energy.
    (ii) The earpiece contains a loudspeaker. The diagram below shows a sectional view of a loudspeaker.
    [A] When an electric current flows through the wire of the coil the cardboard cone vibrates. Explain why. (3 marks)
    [B] Explain how the sound produced by the loudspeaker reaches the ear of a listener. (3 marks)

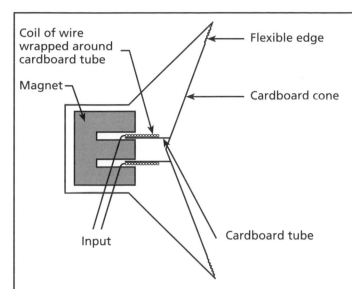

Coil of wire wrapped around cardboard tube

Magnet

Input

Flexible edge

Cardboard cone

Cardboard tube

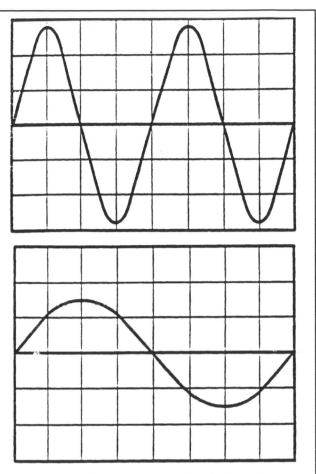

(b) Sound waves patterns can be shown as traces on the screen of an oscilloscope. The diagrams below show traces for two sound waves, A and B.

   (i) Which sound wave has the highest pitch? Give a reason for your answer. (2 marks)

   (ii) Which sound wave is the loudest? Give a reason for your answer. (3 marks)

(c) (i) What equation should be used to find the **speed** of a wave from its **frequency** and **wavelength**? (1 mark)

   (ii) A sound wave has a wavelength of 0.25 metres (m). Use your equation to calculate its frequency. The speed of sound is 330 metres per second (m/s). (4 marks)

(d) Sounds above 20 000 Hz are known as ultrasound. Some singers who can sing notes of very high frequencies are supposed to be able to break wine glasses when they make these high notes. People suffer great pain if 'kidney stones' grow in their kidneys. These stones are formed from calcium compounds. Doctors in some hospitals are trying to break kidney stones using beams of ultrasound focused at the stones. Explain why the ultrasound beam could cause the break up of the kidney stones. (3 marks)

(e) Smoke alarms are designed to produce a sound signal when smoke is detected. The graph below shows the different hearing abilities of a person aged 13 and a person aged 65.

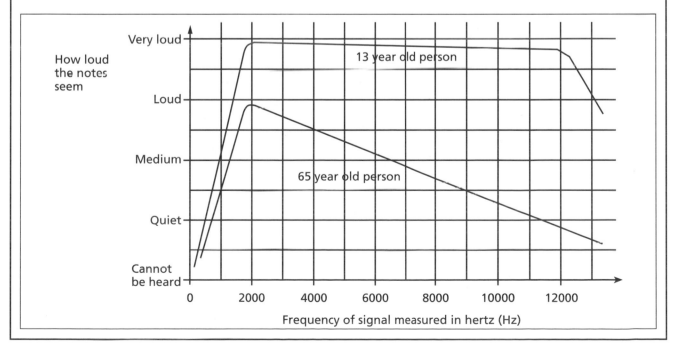

How loud the notes seem

Very loud

13 year old person

Loud

Medium

65 year old person

Quiet

Cannot be heard

0   2000   4000   6000   8000   10000   12000

Frequency of signal measured in hertz (Hz)

Table 10.5										
Number of engine revolutions per second in Hz	1.0	2.0	3.0	4.0	5.0	6.0	8.0	10.0	12.0	14.0
Amplitude of vibration in mm	1.0	3.5	6.0	3.5	1.0	0.5	0.5	0.5	0.5	0.5

(i) Use the information in the graph to suggest the best frequency for the signal produced by the smoke alarm. Give the reason for your choice. (2 marks)

(ii) Most smoke alarms set off a sound signal rather than a visual one, such as a flashing light.
[A] Give four different advantages of sound signals rather than visual ones. (4 marks)
[B] Explain why an alarm with both sound and visual signals is sometimes necessary. (2 marks)

(SEG, Spec, Intermediate Tier)

Question 4

(a) Resonance usually occurs when a child is pushed on a swing. Explain how this happens and what the result is. You may add to the drawing to help you with your answer. (3 marks)

(b) A modern bridge is designed to have its natural frequency of vibration outside the range of any vibrations which could affect the bridge.

(i) Explain why a modern bridge is designed in this way. (3 marks)

(ii) Give **two** examples of a possible cause of vibrations which could affect a bridge. (2 marks)

(c) (i) What is the equation which links frequency, speed and wavelength? (1 mark)

(ii) An ultrasonic wave has a frequency of

30 000 hertz (Hz) and a speed of 300 metres per second (m/s).
What is the wavelength of this wave? (1 mark)

(SEG, Jun 95, Science Double Award, Higher Tier)

Question 5

(a) The wing mirror of a lorry vibrates when the engine is running.

An investigation of how the amplitude of this vibration varied with the number of engine revolutions per second (frequency) was carried out. The measurements taken are shown in table 5.

(i) On the grid below plot a graph of amplitude of vibration (*y*-axis) and the number of engine revolutions per second (*x*-axis).

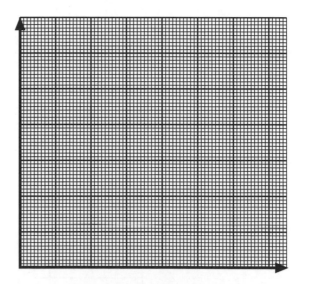

(ii) What is the characteristic (natural) frequency of vibration of the wing mirror?
Why did you choose this value? (2 marks)

(iii) When the lorry is moving, the number of engine revolutions per second is never less than 12 Hz. Explain why it is important that the characteristic frequency of the wing mirror is much less than 12 Hz. (4 marks)

(iv) Give **one** other example of resonance. (2 marks)

(b) (i) State two features that electromagnetic waves have in common. (2 marks)

(ii) Complete table 10.6 by stating a different effect each of the electromagnetic waves listed, has on the human body. The first one has been done for you.

Table 10.6

Electromagnetic wave	Effect on the human body
Visible light	stimulates nerves in the retina
Gamma rays	
Infra-red	
Ultra-violet	

(3 marks)

(iii) Write the waves shown in part (ii) in order of **increasing** wavelength. (2 marks)

(iv) Calculate the frequency of infra-red waves of wavelength 1.0×10^{-6} m. The speed of infra-red wave is 3×10^8 m/s. Show clearly how you obtain your answer. (3 marks)

(c) Distances can be measured using ultrasound waves. The principle of the method used in each case is illustrated below.

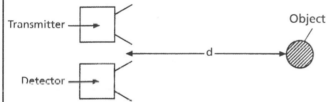

(i) What quantity has to be measured? What quantity has to be known? Write down the formula that is used to calculate the distance *d*. (3 marks)

(ii) This method can be used with ultrasound and electromagnetic waves to measure the distances shown in the table 10.7. For each distance state the type of wave used and give a reason for its use.

Table 10.7

Distance	Wave used	Reason
Airport to plane		
Length of a room		
Earth to the Moon		

(3 marks)
(NICCEA, Jun 95, Higher Tier)

Question 6

(a) Both diagrams below show water waves approaching the sea shore.

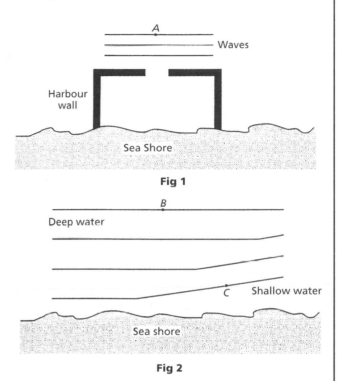

Fig 1

Fig 2

(i) On the figures, mark carefully the direction of the waves at the points *A*, *B* and *C*. (3 marks)

(ii) Complete Fig. 1 by drawing the waves inside the harbour. What name is given to this effect?
(3 marks)

(iii) What name is given to the change in direction of the waves shown in Fig. 2?
(1 mark)

(iv) State a possible cause of this change in direction. (2 marks)

(b) (i) What is meant by a polarised wave? (2 marks)

(ii) Complete the diagram below by adding the piece of apparatus needed so that the light reaching the eye is polarised. (1 mark)

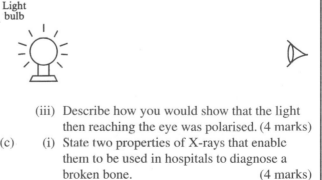

(iii) Describe how you would show that the light then reaching the eye was polarised. (4 marks)

(c) (i) State two properties of X-rays that enable them to be used in hospitals to diagnose a broken bone. (4 marks)

In the 1950s many shoe shops used X-rays to help customers see how well their feet fitted the shoes they were thinking of buying. The diagram shows some of the features of one of these machines.

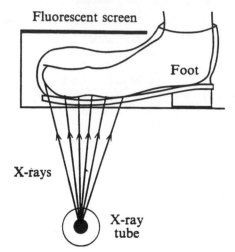

Fluorescent screen

Foot

X-rays

X-ray tube

The fluorescent screen glows when X-rays strike it. The more X-rays striking the screen the brighter it glows.

(ii) Explain why the bones of the feet showed on the screen. (3 marks)

(iii) From your knowledge of the affects of X-rays state why this use of X-rays was not a good idea. (3 marks)

(d) (i) Diagrams (a) and (b) below show a ray of light travelling from glass into air. When the angle of incidence of the ray, in the glass, is greater than 42° total internal reflection takes place.

Complete the two diagrams to show what happens to the rays of light shown.

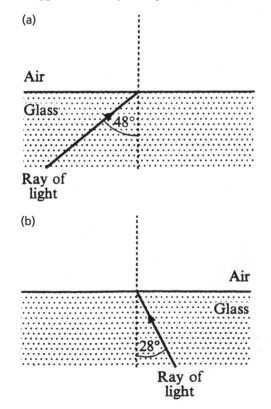

(a)

Air

Glass

48°

Ray of light

(b)

Air

Glass

28°

Ray of light

(ii) Cars and bicycles are fitted, at the rear, with plastic reflectors. These are shaped as shown below. They use total internal reflection so that drivers, following behind, see the car or bicycle in front.

Complete the diagram below to show the paths taken by the rays of light shown. (2 marks)

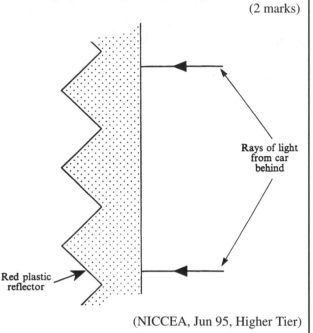

Rays of light from car behind

Red plastic reflector

(NICCEA, Jun 95, Higher Tier)

Question 7

The diagram shows a loudspeaker that provides sound for a TV.

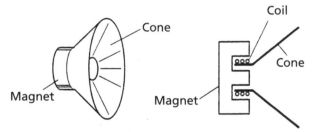

Cone

Coil

Cone

Magnet

Magnet

(a) Explain how sounds are produced when an alternating current flows through the wire coil. (3 marks)

(b) The diagram shows a sound wave produced by the loudspeaker.

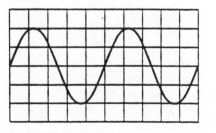

It takes 1/100 (0.01) of a second for the trace to be produced. What is the frequency of the sound from the loudspeaker? (3 marks)

(c) When the loudspeaker produces loud sounds of a particular frequency the casing of the TV vibrates.
 (i) What is this effect called? (1 mark)
 (ii) Explain why it happens. (2 marks)
 (iii) Suggest how these vibrations could be reduced without reducing the volume. (1 mark)

(d) (i) The electromagnetic waves which carry the signal to your home are polarised.
 What do you understand by this? (2 marks)
 The diagram shows how the picture quality depends on the orientation of the receiving aerial.

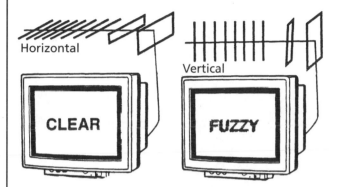

 (ii) Suggest what this tells you about the orientation of the **transmitting** aerial. (1 mark)
 (MEG Nuffield, Jun 95, Higher Tier, Q4)

Question 8

(a) The following diagram shows an experiment. The lamp gives out red light of a single wavelength. Some of the light passes through two narrow slits S_1 and S_2. A series of alternate bright and dark bands is produced on the screen.
 The diagram is not to scale.

 (i) What is the name of the effect which occurs at each of the slits? (1 mark)

 (ii) What is the colour of the **bright** bands? (1 mark)

 (iii) [A] What is the name of the effect which causes the formation of a bright band? (1 mark)
 [B] Explain how a bright band is formed. Draw one or more diagrams if this will help you explain. (2 marks)

 (iv) [A] What is the name of the effect which causes the formation of a **dark** band? (1 mark)
 [B] Explain how a dark band is formed. Draw one or more diagrams if this will help you to explain. (2 marks)

 (v) The distance from one bright band to the next is called the band separation. In this experiment it was 1 millimetre (mm).
 In the following equation **all the lengths are measured in metres (m)**. Use the equation to calculate the wavelength of red light.

 band separation =

$$\frac{\text{wavelength} \times \text{distance from slits to screen}}{\text{distance between slits}}$$

 Show clearly how you get to your answer. (4 marks)

(b) A special light source produces a beam of vertically polarised light waves. The beam then passes through a polarised filter which has a vertical plane of polarisation.

 (i) What are polarised waves? (1 mark)
 (ii) What will happen if you turn the filter through an angle of 90°? (1 mark)
 (iii) What will happen if your turn the filter through an additional angle of 180°? (1 mark)
 (iv) Give **two** uses for a polarised filter. (2 marks)
 (SEG, Jun 95, Science Double Award, Higher Tier)

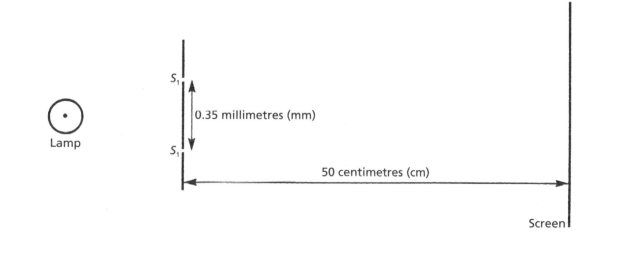

Answers to examination questions

1. (a) High frequency sound waves.
   (b) The ultrasound is reflected off the hard surface of the sea bed.
   (c) 0.5 s
   (d) 0.2 s
   (e) Distance = speed × time = 1500 m/s × 0.2 s = 300 m
   (f) [Answer in (e) is the distance to the sea bed and back.]
   Depth of water under the ship = 0.5 × 300 m = 150 m

2. (a) The words to be placed in the spaces, in order, are: Ear Drum, Ossicles (Hammer, anvil and stirrup), cochlea, auditory nerve.
   The function of the semicircular canals is to detect the speed of movement of the head.
   (b) (i) Resonance.
       (ii) Frequency = $\dfrac{\text{Velocity}}{\text{wavelength}}$ =
       $\dfrac{330 \text{ m/s}}{0.64 \text{ m}}$ = 516 Hz

3. (a) (i) Sound, electrical.
       (ii) [A] When the current goes through the coil it becomes an electromagnet and the interaction (repulsion and attraction) of the magnetic field with the magnetic field of the permanent magnet causes the coil to move and make the cardboard cone vibrate.
       (iii) [B] The vibration of the speaker cone causes the air particles to vibrate and pass energy from one particular to the next as they collide. High pressure regions where the particles are close together are called compressions and low pressure regions where the particles are further apart are called rarefactions.
   (b) (i) Sound wave B because there are more wavelengths in the same space.
       (ii) Sound wave B because it has the largest amplitude. [The distance from the central horizontal line to either a peak or a trough.]
   (c) (i) Speed = frequency × wavelength.
       (ii) Frequency = $\dfrac{\text{Speed}}{\text{wavelength}}$
       = $\dfrac{330 \text{ m/s}}{0.25 \text{ m}}$ = 1320 Hz
   (d) The kidney stones will have their own natural frequency which is the frequency that they would like to vibrate with. When ultrasound with same

frequency as the natural frequency of the stones is directed at the stones, resonance occurs and they vibrate and build up their amplitude. Eventually the vibrations become so large that the stone breaks up.
   (e) (i) 2000 Hz because this frequency is the loudest for both ages.
       (ii) [A] Sounds will wake you up when asleep. Sounds can be heard all around the house no matter which room you are in. During a fire, sounds can be heard whereas a light may not be able to be seen.
       Sounds may be heard by someone who is blind.
       [B] Some people may be deaf and not hear the sound.
       They may be wearing headphones and not hear the sound so a visual signal can still be seen.

4. (a) The swing has its own natural frequency of vibration and when someone provides pushes with the same frequency resonance occurs and the amplitude of vibration of the swing starts to build up and the swing goes higher and higher.
   (b) (i) Vibrations, such as those produced by wind, could be the same as the natural frequency of the bridge and resonance could occur and the amplitude could build up causing the bridge to collapse. By making sure that the bridge has a natural frequency away from other forcing frequencies the above can never occur.
       (ii) Wind.
       Vibrations from vehicles passing across it.
   (c) (i) Speed = frequency × wavelength.
       (ii) Wavelength = $\dfrac{\text{speed}}{\text{frequency}}$ =
       $\dfrac{300 \text{ m/s}}{30\,000 \text{ Hz}}$ = 0.01 m

5. (a) (i) *See figure opposite.*
       (ii) 3 Hz since this is the frequency which produces the largest amplitude of vibration.
       (iii) This stops resonance occurring which could have caused the wing mirror to vibrate with such a large amplitude of vibration that it could break.
       (iv) Breaking up kidney stones in the body using ultrasound.
   (b) (i) They all travel at the same speed in a vacuum. They travel in straight lines. [You could also have: they may be refracted, reflected, polarised and so on.]

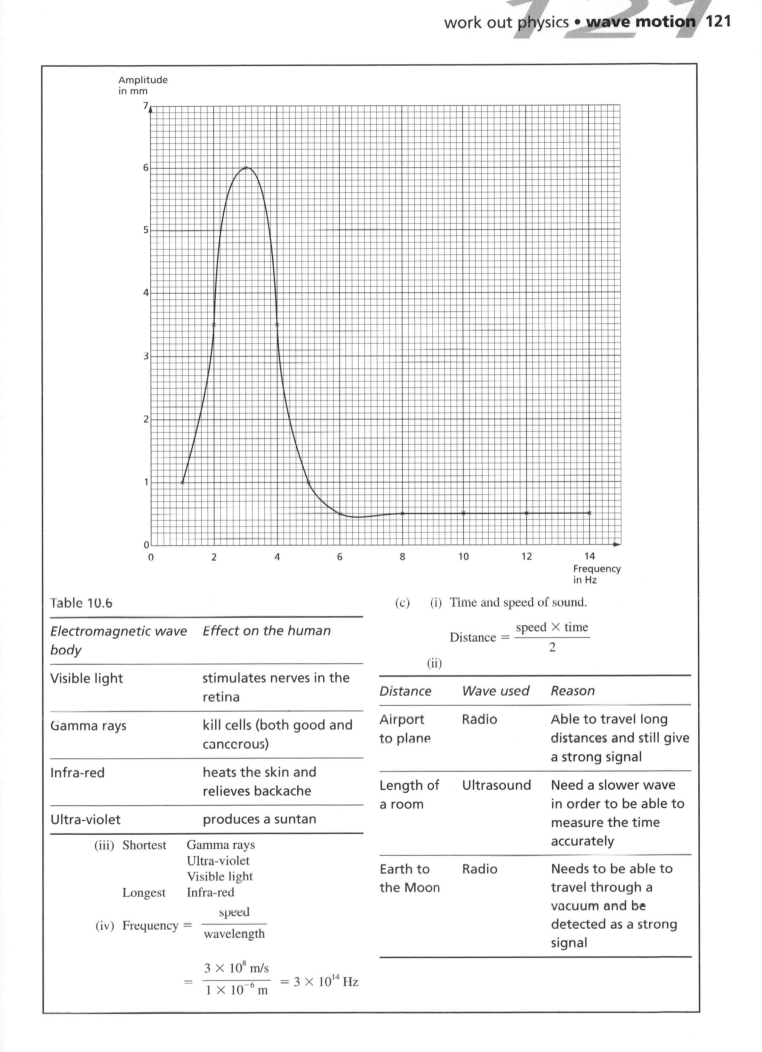

Table 10.6

Electromagnetic wave body	Effect on the human
Visible light	stimulates nerves in the retina
Gamma rays	kill cells (both good and canccrous)
Infra-red	heats the skin and relieves backache
Ultra-violet	produces a suntan

(iii) Shortest   Gamma rays
              Ultra-violet
              Visible light
     Longest   Infra-red

(iv) $\text{Frequency} = \dfrac{\text{speed}}{\text{wavelength}}$

$$= \frac{3 \times 10^{8}\,\text{m/s}}{1 \times 10^{-6}\,\text{m}} = 3 \times 10^{14}\,\text{Hz}$$

(c) (i) Time and speed of sound.

$$\text{Distance} = \frac{\text{speed} \times \text{time}}{2}$$

(ii)

Distance	Wave used	Reason
Airport to plane	Radio	Able to travel long distances and still give a strong signal
Length of a room	Ultrasound	Need a slower wave in order to be able to measure the time accurately
Earth to the Moon	Radio	Needs to be able to travel through a vacuum and be detected as a strong signal

6. (a) (i)

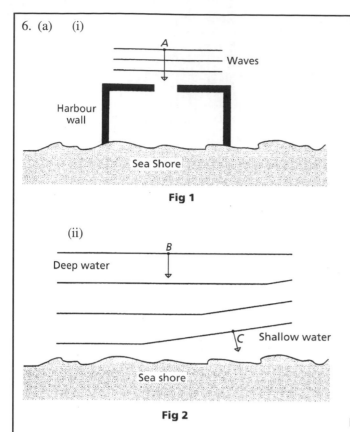

Fig 1

(ii)

Fig 2

Diffraction.
  (iii) Refraction.
  (iv) The change in depth.
  (b)   (i) A wave which has had its vibrations restricted to a single plane.
       (ii) A sheet of Polaroid needs to be added to the diagram between the light and the eye.
       (iii) Use a second sheet of Polaroid and rotate it. The intensity changes with two dark and two bright positions when rotated through 360°.
  (c)   (i) Can travel easily through tissue but are stopped by bone.
           They travel in straight lines and produce an image on photographic film.
       (ii) X-rays are absorbed by the bones and therefore will show up as dark areas on the screen. Soft tissues allow X-rays through and these will be the bright areas on the screen.

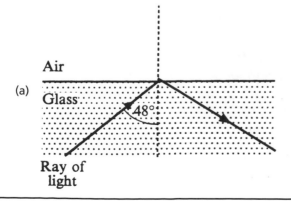

(iii) X-rays are harmful since they can cause cancer. Doses should therefore be restricted to essential uses.
  (d)   (i) In diagram (a) the light is reflected at 48° to the normal back into the glass.
           In diagram (b), the ray is refracted and moves away from the normal when it comes out of the glass.
       (ii)

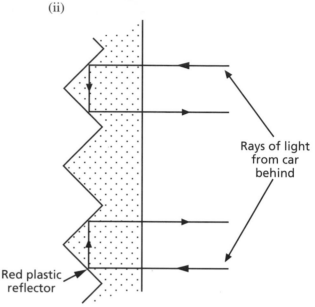

7. (a) Alternating current produces a changing magnetic field in the coil and this field interacts with the magnetic field produced by the permanent magnet causing the coil to move backwards and forwards each time the current in the coil changes direction. The vibrations of the coil cause the speaker cone to vibrate and this gives rise to a sound wave with the same frequency as the alternating current.

  (b) [From the graph we can see that there are two complete wavelengths in 0.01 s so the time for one wavelength (called the period of the wave) is $\dfrac{0.01\ \text{s}}{2} = 0.005\ \text{s}$].

  $\text{Frequency} = \dfrac{1}{\text{Period}} = \dfrac{1}{0.005\ \text{s}} = 200\ \text{Hz}$

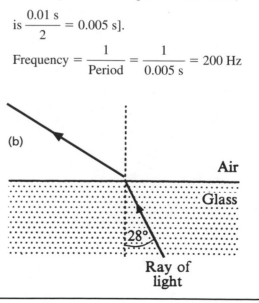

(c) (i) Resonance.

(ii) Vibrations from the speaker cone have the same frequency as the natural frequency of the casing, causing it to vibrate in unison.

(iii) Make the casing so that it has a natural frequency away from those frequencies likely to be produced by the speaker.

(d) (i) The waves have their directions of oscillation restricted to a single plane.

(ii) The waves are plane polarised in a horizontal direction so for a clear picture, the aerial needs to be orientated in this direction.

8. (a) (i) Diffraction.

(ii) Red (because the source is red).

(iii) [A] Constructive interference.
[B] The waves from the two slits arrive in phase so their peaks and troughs coincide. The mixing of the two waves gives a larger wave and this produces the bright band.

(iv) [A] Destructive interference.
[B] The waves arrive out of phase with each other so when one wave has a peak, the other wave has a trough. The effect is to cancel the waves out. A dark band is produced.

(v) Rearranging the equation gives

$$\text{wavelength} = \frac{\text{band separation} \times \text{distance between slits}}{\text{distance from slits to screen}}$$

$$= \frac{(1 \times 10^{-3}\,\text{m}) \times (0.35 \times 10^{-3}\,\text{m})}{0.5\,\text{m}}$$

$$= 7 \times 10^{-7}\,\text{m}$$

(b) (i) Polarised waves are waves where the vibrations have been restricted to a single plane.

(ii) No waves will be transmitted through the filter.

(iii) Again no waves will be allowed through the filter.

(iv) Sunglasses and for looking at stress in plastics.

11

Electrostatics

WJEC & NEAB	ULEAC Syll A	ULEAC Syll B	MEG	Topic	MEG Salters'	MEG Nuffield	SEG	NICCEA
✓	✓	✓	✓	**Charge and charge flow**	✓	✓	✓	✓
✓	✓	✓	✓	**Conductors and insulators**	✓	✓	✓	✓
✓	✓		✓	**The cathode ray oscilloscope**	✓	✓	✓	

1 Charge and charge flow

There are two sorts of charge, positive and negative. A polythene rod rubbed with wool becomes negatively charged, and a cellulose acetate rod rubbed with wool becomes positively charged. Like charges repel and unlike charges attract.

A negatively charged polythene rod attracts an uncharged piece of paper, because the near side of the paper becomes positively charged, owing to the presence of the rod. This positive charge is attracted by the negative charge on the rod.

The magnitude of the force between charges decreases rapidly as the distance between them is increased.

One way of removing pollution in the form of smoke (dust particles) from the atmosphere is by electrostatic precipitation. The smoke is passed through a strong electric field, which produces ions (an atom that has gained or lost one or more electrons, and is therefore a charged atom) which adhere to the dust particles, giving them a charge. The charged particles are attracted towards an earthed plate and collect on it. They are periodically removed by striking the plate with a mechanical hammer. The dust particles fall into collectors for disposal.

Petroleum fuel pipes should be earthed to avoid the build up of static charge and the consequent danger of a spark igniting the fuel.

2 Conductors and insulators

Materials which allow electricity to pass through them are called conductors whilst those which don't are called insulators. Both conductors and insulators may be charged with static electricity, but a conductor must be insulated from the ground. For this reason, a piece of metal held in the hand cannot be charged. Metals, such as copper, solder, aluminium and silver, are the best conductors whilst PVC, rubber and porcelain are the best insulators. Some materials, called semiconductors, only conduct electricity under certain circumstances. Semiconductor materials include silicon and germanium.

3 The cathode ray oscilloscope

A *cathode ray oscilloscope* (CRO) has a heated metal cathode that emits electrons from its surface (thermionic emission). A beam of electrons is produced by positioning a positively charged anode with a small hole in it, close to the filament. The beam is passed through *X*-plates which produce an electric field that deflects the beam horizontally and *Y*-plates which produce an electric field that deflects it vertically. The input terminals connect the

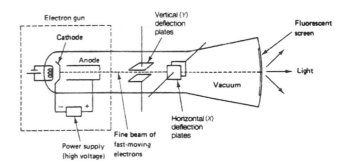

Diagram showing the essential features of a cathode ray tube. The heated cathode emits electrons which are attracted by the positive anode. Some pass through a hole in the anode, producing a beam of electrons.

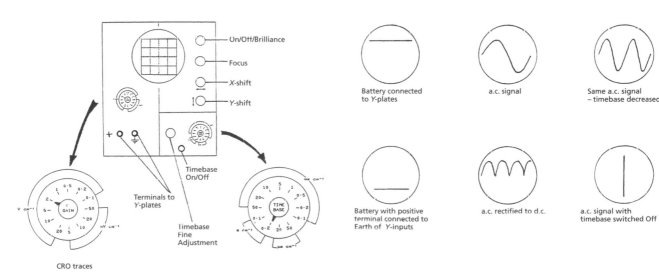

Diagram of the front panel of a cathode ray oscilloscope and showing some CRO traces.

applied potential difference across the *Y*-plates. When the beam strikes the fluorescent screen at the end of the tube a spot of light is produced. When an alternating current is connected across the *Y*-plates, a vertical line appears on the screen. The waveform is displayed if the timebase (which is connected across the *X*-plates and which moves the spot across the screen horizontally) is then switched on.

In a TV tube an electron beam 'scans' the screen – that is, moves across the screen in a series of lines from left to right. The greater the number of electrons that strike any one spot on the screen, the brighter the picture at that point.

Worked examples

Example 1

An oscilloscope screen appears as shown below.

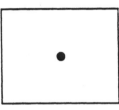

In order to change the appearance to that shown below, which one of the following alterations must be made?

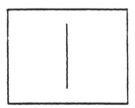

A Apply d.c. to the *Y*-plates.
B Switch on the timebase.
C Apply a.c. to the *Y*-plates.
D Apply a.c. to the *X*-plates.

Solution

[The spot is moving up and down vertically very fast. For this to happen an alternating current must be connected across the *Y*-plates. For further help with this question see section 2.]

Answer **C**

Example 2

The diagram shows the trace on an oscilloscope screen resulting from a coil rotating at a steady speed in the field of an electromagnet.

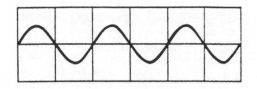

The strength of the magnetic field is increased. The speed of rotation of the coil and the settings on the oscilloscope are unchanged. Which of the diagrams below shows the new trace observed on the screen?

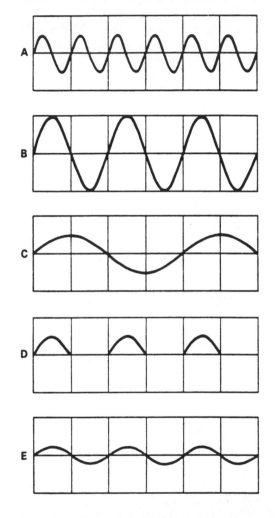

Solution

[The coil is rotated at the same speed and therefore the frequency remains the same. So one complete cycle must still take up two complete grid squares. The strength of the magnetic field is increased and hence the rate of change of flux is increased. This results in an increased e.m.f. (electromotive force) and hence an increase in the peak value on the oscilloscope.]

Answer **B**

Example 3

Richard holds a vibrating tuning fork in front of a microphone connected to the input terminals of an oscilloscope. He observes the trace shown.

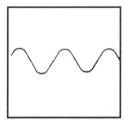

Which of these traces would represent the trace produced by a fork of higher frequency providing a louder note? Assume that the adjustments of the oscilloscope remain unchanged.

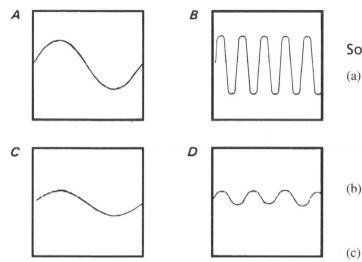

(ULEAC, Intermediate Tier)

Solution

[A louder note will produce a greater peak value. A note of higher frequency will produce more complete cycles in the same time, i.e. more complete cycles in the time the spot takes to travel across the screen.]

Answer **B**

Example 4

(a)

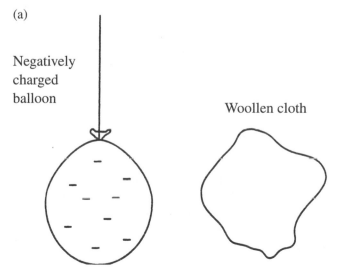

Negatively charged balloon

Woollen cloth

A balloon, after rubbing with a woollen cloth, is found to be negatively charged and is attracted to the cloth.
   Explain
   (i) why the balloon becomes negatively charged when it is rubbed.   (2 marks)
   (ii) why the balloon is attracted to the cloth.
                                                        (1 mark)
(b)   Name **one** practical use of static electricity.   (1 mark)
(c)   (i) State **one** situation where static electricity is dangerous.   (1 mark)
      (ii) Explain what precautions can be taken to reduce the danger.   (1 mark)
                    (WJEC, Jun 95, Intermediate Tier, Q4)

Solution

(a)   (i) Friction causes electrons to be transferred from the woollen cloth to the balloon leaving the cloth deficient in electrons and therefore having a positive charge whilst the balloon is left with a negative charge.
      (ii) The cloth and the balloon have opposite charges and opposite charges attract.
(b)   Paint or insecticide spraying, laser printing or photocopying, electrostatic precipitators to remove smuts in smoke in power station chimneys (any one of these).
(c)   (i) When discharging oil from a supertanker.
      (ii) Make sure that the oiltanker and the storage tanks are earthed so that no spark can jump between them and cause an explosion.

Example 5

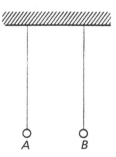

(a)   Two small balls coated with metallic paint are suspended by insulating threads as shown in the diagram. Describe what you would observe if both balls were given a positive charge. How does the effect depend on the distance *AB*?   (3 marks)
(b)   Draw a diagram showing all the forces acting on one of the balls.   (3 marks)
(c)   In order to reduce the noise level of his machinery a manufacturer of nylon thread put rubber matting on the floor underneath his machinery. Small bits of fluff were found to stick to the nylon thread. Explain the reason for this. Someone suggested that the problem could be overcome by installing a humidifier to keep the air moist. Explain whether you think this was a reasonable suggestion.   (5 marks)

Solution

(a) The balls would repel each other and both threads would be at an angle to the vertical. The smaller the distance between the threads the greater is the angle the threads make with the vertical.

(b)

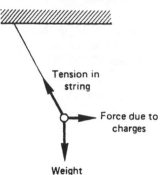

(c) Friction during the manufacture of the thread will cause the thread to become charged. The mat prevents charge accumulating on the machine flowing to earth. If the charge on the thread is negative bits of fluff close to it will have a positive charge induced on the side nearest to the thread. The unlike charges attract and the fluff sticks to the thread.

Dry air is a good insulator but damp air is quite a good conductor. Moist air in contact with the thread would enable the charge to leak away, so the suggestion is a reasonable one.

Example 6

The diagram shows a waveform displayed on the screen of a cathode ray oscilloscope when an a.c. voltage is connected across the Y-plates. The Y gain is set at 2 V/cm and the timebase at 5 ms/cm. The graticule has a 1 cm grid.

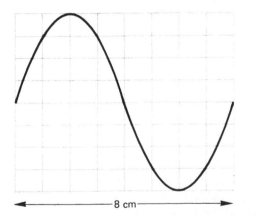

← 8 cm →

(a) What is the peak voltage of the waveform?
(b) What is the frequency of the a.c. voltage?
(c) Sketch on the graticule below the trace that would be seen on the screen if
  (i) the timebase setting were changed to 10 ms/cm; label your trace **A**.
  (ii) the timebase setting were left at 5 ms/cm but the frequency of the applied voltage were halved; label your trace **B**.

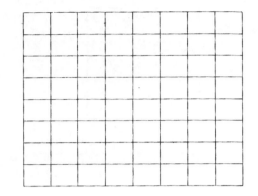

Solution

(a) 2 V/cm × 4 cm = 8 V
(b) One cycle traces out 8 cm on the graticule and, hence, it takes

$$5 \text{ ms/cm} \times 8 \text{ cm} = 40 \text{ ms} = \frac{40}{1000} \text{ s} = \frac{1}{25} \text{ s}$$

If each cycle takes $\frac{1}{25}$ s, then it happens 25 times every second.

Frequency = 25 Hz

(c) [If the timebase takes 10 ms instead of 5 ms to go 1 cm, it is travelling at half the speed. Twice as many cycles will appear on the screen. If the timebase is left at 5 ms/cm and the frequency is halved, then in the time it takes the spot to cross the screen there will be half a cycle on the screen.]

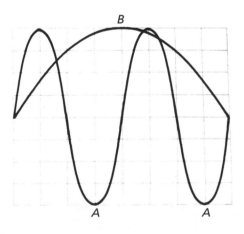

Example 7

Sketch the pictures seen on the screen of a cathode ray oscilloscope when the oscilloscope is adjusted so that the spot is in the middle of the screen and
(a) a battery is connected across the Y-plates. (2 marks)
(b) the output terminals from a transformer connected to the mains are connected across the Y-plates.
(2 marks)

(c)　the transformer is left connected to the *Y*-plates and the timebase is switched on.　(2 marks)

(d)　As in (c) but the speed of the timebase is doubled.　(2 marks)

Solution

[In (b) the spot will be moving up and down many times a second, so the trace will appear as a straight line. In (d) the timebase is moving twice as fast, so that the oscillation will only have time to go through half the number of oscillations.]

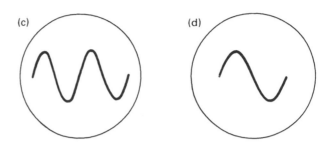

Examination questions

(Numerical answers and hints on solutions will be found at the end of the chapter.)

Question 1

Michael's teacher explains to him that rubbing a plastic rod with a cloth leaves the plastic rod with a **positive** charge.

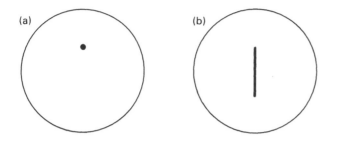

Plastic rod　　　　　　　　　　　　　　Duster

(a)　Explain, in terms of the movement of electrons, how the plastic rod becomes positively charged.

(b)　Michael carries out an experiment with two light, charged balls which hang from a pin by silk threads. When the balls come close together, Michael observes that they repel each other.

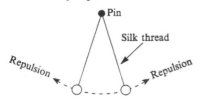

What can Michael conclude about the type of charge on each ball?　(3 marks)

NICCEA, Jun 95, Intermediate Tier)

Question 2

Electrostatic paint spraying is widely used in the motor industry. The fine needle tip of a spray gun gives a negative charge to all the small droplets of paint sprayed out from the gun. This spray of paint falls on the car body which must be given a positive charge during spraying.

This method provides an even coat of paint and also allows difficult areas to be painted easily.

(a)　Explain why the car body has to be given a positive charge during spraying.　(1 mark)

(b)　Suggest one reason why this method is a better way of painting a car than painting with a brush.

(1 mark)

(WJEC, Jun 94, Intermediate Tier)

Question 3

A lightning conductor helps protect tall buildings from being struck by lightning. The lightning conductor is a very thick strip of copper which connects some sharp metal points above the top of the building to a metal plate buried in the ground.

Thunderclouds are electrically charged. A large **negatively** charged thundercloud will cause free electrons to move in the lightning conductor to produce a large charge, opposite to that of the thundercloud, at the sharp metal points. This large charge, at the metal points, produces positively and negatively charged air atoms which are called ions. These ions, depending on their charge, will either be attracted to the thundercloud, or the lightning conductor. Those attracted to the lightning conductor will cancel out some of the charge in the cloud making a lightning strike less likely.

(a) State

  (i) the sign of the charge produced at the metal points by the thundercloud described in the passage.

  (ii) how this charge is produced. (1 mark)

(b) Explain why the positive ions are attracted to the thundercloud. (1 mark)

(c) Give **one** reason why the lightning conductor is connected to the earth plate by a **very thick strip** of a good conductor such as copper. (1 mark)

  (WJEC, Jun 92, Higher Tier)

Question 4

You wash and dry your hair.
You then comb it with a plastic comb.
As you move the comb away from your head some hairs are attracted to the comb.

(a) What has happened to the comb to make it attract the hairs? (1 mark)

(b) If the comb is now held above some small pieces of dry tissue paper what is likely to happen? (1 mark)

(c) If you rub your hands all over the comb it will no longer attract your hair.
  Explain why. (1 mark)

  (NEAB, Jun 95, Intermediate Tier, Q3)

Question 5

(a) The diagram below shows the construction of a simple cathode ray tube.

In the electron gun, electrons from the heater filament are accelerated by the potential difference between C and A.

  (i) What form of energy does the electron gain as it moves from C to A? (1 mark)

  (ii) State the principal energy transfer taking place when an electron strikes the fluorescent screen. (2 marks)

  (iii) State TWO ways in which the brightness of the image on the fluorescent screen could be altered. (2 marks)

(b) (i) The potential difference between C and A is 5000 V. Calculate the energy gained by an electron of charge -1.6×10^{-19} coulomb when it reaches A. (2 marks)

  (ii) What is then the speed of the electron, given that the mass of an electron is 9.1×10^{-31} kg? (3 marks)

(c) When a potential difference of 200 V is placed across the Y-plates, the vertical deflection of the dot in the screen is 10 mm.

  (i) What would be the deflection if the potential difference across the Y-plates were changed to 350 V? (2 marks)

  (ii) State what the X-deflection plates are normally used for in an oscilloscope. (1 mark)

(d) The diagram below shows the path followed by the electron beam when a steady potential difference is applied across the Y-plates.

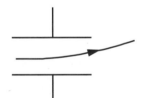

  (i) What is the shape of the path followed by the electron beam called? (1 mark)

  (ii) Describe and explain how the path changes

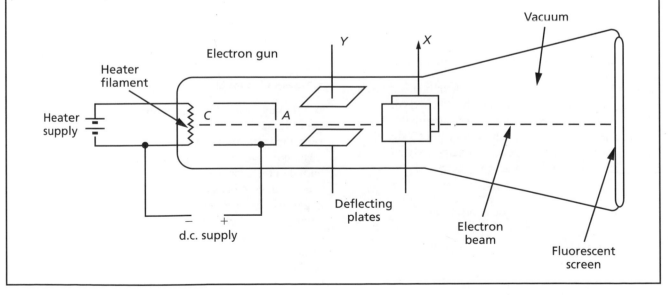

when the potential difference between C and A is reduced from 5000 V to 4000 V.   (2 marks)

(e) When the same steady potential difference is applied across the Y- and X-plates at the same time, the deflection seen on the screen is as shown below.

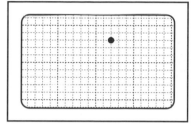

Explain why there is a difference in the horizontal and vertical displacements of the beam, even though the plates are identical and have the same potential difference across them. (Assume that the dot was originally in the centre of the screen.)   (2 marks)

(ULEAC, Syll A, Jun 95, Higher Tier, Q2)

Answers to examination questions

1. (a) Friction transfers some of the electrons from the rod to the duster. The duster gains electrons and becomes negatively charged whereas the duster is deficient in electrons and is therefore positively charged.

   (b) He can only say that the signs of the charges on the balls are the same so they could be both positively or both negatively charged since like charges attract.

2. (a) So that the negatively charged paint particles are attracted to it. [Opposite charges attract.]

   (b) Since the fine paint droplets are repelled because they all have the same charge, they produce a more even coating on the surface.

3. (a) (i) Positive

      (ii) The negatively charged thundercloud repels the electrons in the wire down to the ground leaving the top of the lightning conductor with a positive charge.

   (b) Unlike charges attract each other.

   (c) When the lightning strikes the conductor a very large current needs to flow to ground and if the conductor were thin, the wire would burn out and no longer protect the building. The wire needs to be thick and therefore have a low resistance.

4. (a) Friction transfers electrons so the hairs and the comb become charged with static electricity. They are charged oppositely so like charges attract.

   (b) Small uncharged objects, like tissue paper are attracted.

   (c) The charge will have been earthed and this removes the charge.

5. (a) (i) Kinetic energy.
      (ii) Kinetic energy to light energy.
      (iii) By increasing the number of electrons (i.e. by increasing the current).
           By increasing the velocity of the electrons (i.e. by increasing the p.d. between C and A).

   (b) (i) Energy gained = Electronic charge
           $\times$ Voltage = eV
           $= (1.6 \times 10^{-19}$ C$)$
           $(5000$ V$) = 8 \times 10^{-16}$ J

      (ii) Energy gained = Kinetic energy
           $8 \times 10^{-16} = \frac{1}{2} \times$ mass $\times$ velocity^2
           $8 \times 10^{-16} = \frac{1}{2} \times 9.1 \times 10^{-31} \times v^2$
           $v = 4.2 \times 10^7$ m/s

   (c) (i) 200 V deflects 10 mm so 50 V deflects 2.5 mm.
           Hence 350 V will deflect
           7×2.5 mm = 17.5 mm.

      (ii) To produce a timebase which will move the electron beam horizontally at a steady pre-determined speed across the screen from left to right.

   (d) (i) Parabolic

      (ii) The electrons will move slower so they will spend more time between the Y-plates so the deflection in the screen will be greater.

   (e) The X-plates are nearer to the screen so although there will be the same displacement between the plates there will be less displacement on the screen.

12

Electrical circuits

WJEC & NEAB	ULEAC Syll A	ULEAC Syll B	MEG	Topic	MEG Salters'	MEG Nuffield	SEG	NICCEA
✓	✓	✓	✓	**Basic units**	✓	✓	✓	✓
✓	✓	✓	✓	**Laws for circuits**	✓	✓	✓	✓
✓				**Potential divider**				
✓		✓	✓	**Ring main and lighting circuit**	✓		✓	✓
✓		✓		**Conductors, semiconductors and insulators**				✓
✓	✓	✓	✓	**Electrical power and energy**	✓	✓	✓	✓
✓	✓	✓	✓	**Earthing and wiring**	✓		✓	✓

1 Basic units

When an ammeter reads 1 ampere (A), then 1 coulomb (C) of charge is flowing every second.
1 A = 1 C/s. If a current I passes for a time t, then the charge, Q, which flows is given by $Q = It$.

The potential difference (in volts) between two points is the work done in joules in moving 1 coulomb of charge between them:

$$\text{Potential difference} = \frac{\text{work done}}{\text{charge moved}}$$

1 V = 1 J/C

The e.m.f. (electromotive force) is the total energy supplied by a source to each coulomb of charge that passes through it, including any energy that may be lost as heat in the source itself as a result of the source's internal resistance. It is measured in J/C i.e. volt.

Both e.m.f. and potential difference are sometimes referred to as *voltage*.

The potential difference in volts across the terminals of a cell or generator is the energy which is delivered to the external circuit by each coulomb of charge.

Resistance (ohms)

$$= \frac{\text{potential difference across the object (volts)}}{\text{current through the object (amperes)}}$$

or $R = \dfrac{V}{I}$

You must learn to write the above equation as $V = IR$

and $I = \dfrac{V}{R}$.

An object has a resistance of 1 ohm if a potential difference across it of 1 volt results in a current of 1 ampere passing through it. Most metal conductors have a fixed resistance if their temperature is kept constant. Doubling the length of a conductor doubles its resistance; doubling the cross-sectional area halves its resistance.

2 Laws for circuits

If the total e.m.f. of a circuit is E volts, then

$E = I \times$ (total resistance of circuit)

Ohm's Law

Ohm's Law states that the current in a conductor is proportional to the potential difference across it provided the temperature is kept constant.

If a conductor obeys Ohm's Law, a graph of potential difference against current is a straight line through the origin. The gradient of the graph is the resistance of the conductor (see the first diagram). An example of a non-ohmic conductor is a diode (the symbol is shown in the next diagram). A graph of current against voltage for a diode is shown in the third diagram.

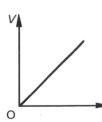

A graph of current plotted against voltage for a component that obeys Ohm's Law.

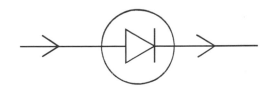

Current can only pass when it is in the direction of the arrows.

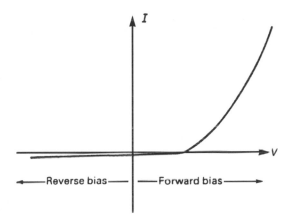

A graph of current plotted against voltage for a diode. A diode has a low forward resistance and a very high reverse resistance.

Laws for series circuits

  (i) The same current passes through each part of the circuit.
 (ii) The applied potential difference is equal to the sum of the potential difference across the separate resistors: $V = V_1 + V_2 + V_3$.
(iii) The total resistance is equal to the sum of the separate resistances: $R = R_1 + R_2 + R_3$.

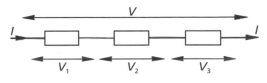

Resistors in series. $V = V_1 + V_2 + V_3$.

Laws for parallel circuits

  (i) The potential difference across each resistor is the same.

(ii) The total current is equal to the sum of the currents in the separate resistors:
$I = I_1 + I_2 + I_3$.

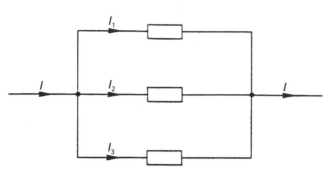

Resistors in parallel. $I = I_1 + I_2 + I_3$.

(iii) The combined, or total, resistance of a number of resistors in parallel is less than the value of any of the separate resistors and is given by

$$\frac{1}{R} = \frac{1}{R_1} + \frac{1}{R_2} + \frac{1}{R_3}$$

For two resistors in parallel

$$\frac{1}{R} = \frac{1}{R_1} + \frac{1}{R_2} \text{ or } R = \frac{R_1 \times R_2}{R_1 + R_2} \ ,$$

i.e. $R = \dfrac{\text{product}}{\text{sum}}$

3 Potential divider

See the illustration opposite. If a p.d. of V_{in} is applied across two resistors R_1 and R_2 in series,

then current in R_1 and $R_2 = \dfrac{V_{in}}{R_1 + R_2}$

p.d. across $R_2 = V_{out} = \left(\dfrac{V_{in}}{R_1 + R_2} \right) \times R_2$

By adjusting the values of R_1 and R_2, V_{out} can be made to vary between 0 and V_{in}.

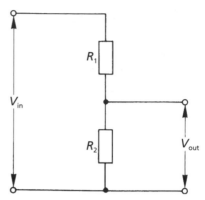

A potential divider. If the resistance of R_2 is increased, then V_{out} increases. If the resistance of R_1 is twice the resistance of R_2, then the p.d. across R_1 is twice the p.d. across R_2.

4 Ring main and lighting circuits

The mains supply is delivered to a house by means of two wires called the live and the neutral (see the illustration below). Every appliance is connected in parallel with the supply. Lighting circuits usually have a 5 A fuse. In the ring

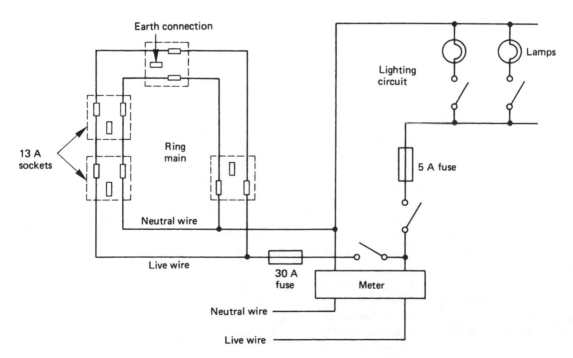

In both lighting and ring main power circuits, the appliances are connected in parallel across the supply. A third wire (not shown) joins all the earth sockets of the ring main.

main, the live, the neutral and the earth wires run in a complete ring round the house.

5 Conductors, semiconductors and insulators

An insulator (e.g. rubber or plastic) is a substance which will not conduct electricity. Conductors (e.g. copper) are good conductors of electricity. In between are the semiconductors (e.g. germanium), which are neither good conductors nor good insulators.

6 Electrical power and energy

Power (watts) = potential difference (volts) × current (amperes) or Power $P = (V \times I)$ joule/second

Substituting, using the definition of resistance given above, namely

Potential difference (V) = current (I) × resistance (R)

we have Power = $(I^2 \times R)$ joule/second

Thus, the energy dissipated in a resistor is $I^2 \times R$ joule/second. Hence, the total energy dissipated in t seconds is given by

Energy dissipated = $I^2 \times R \times t$ joule

A meter supplied by the electricity company records the energy consumption in kilowatt-hours. One kilowatt-hour (kWh) is the total amount of energy supplied to a 1 kW appliance when it is connected to the supply for 1 h. As a 1 kW appliance is supplied with energy at a rate of 1000 J/s, a total of 3.6×10^6 J of energy will be supplied in 1 h (3600 s); thus

1 kWh = 3.6×10^6 joule

The cost of running an electrical appliance may be calculated from the equation

Cost = (number of kilowatts) × (hours) × (cost of a kilowatt-hour).

7 Safety: earthing and wiring

Appliances with metal casing have an earthed wire connected to the metal casing. This is a safety device. For example, if the element in an electric fire breaks and the live end touches the metal case, the connection to earth has a very low resistance, and the current surges to a large value. The fuse in the circuit is designed to melt and break the circuit when the current reaches a predetermined value. The large current that results when the live wire touches the case melts the fuse, the circuit is broken and is thus safe. In many modern circuits the fuse is replaced by a circuit-breaker. 'Double insulated' appliances have their metal parts surrounded by thick plastic, making it impossible for

the user to touch a metal part. When wiring double insulated appliances, no earth lead is necessary. Damp surroundings can be dangerous as moisture greatly reduces the resistance of the skin. In a three-pin plug, the earth wire is yellow and green, the live wire is brown and the neutral wire is blue (see question 2).

A residual current device (RCD), also called an earth leakage circuit-breaker (ELCB), is a main switch automatically operated by a small current passing (leaking) to earth. An important characteristic of an RCD is that only a very small earth current is needed to operate it. It is easily reset and, unlike a fuse, no replacement is needed. RCDs should always be used as a safety precaution when electrical equipment used out of doors, especially in damp conditions. It protects the user from possible electrocution if the cable is cut or if a breakdown occurs in the insulation.

Worked examples

Example 1

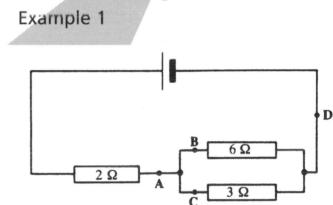

An electrical circuit is connected as shown in the diagram. At which point, **A**, **B**, **C** or **D** is the current smallest?

Solution

[The current at **A** and **D** is the same. On arriving at the junction the current through **A** divides and most of it passes through the lower resistor. The smallest current passes through the higher resistor, i.e. through **B**.]

Answer **B**

Example 2

In the following circuit the lamps P and Q are identical. A_1 and A_2 are ammeters.

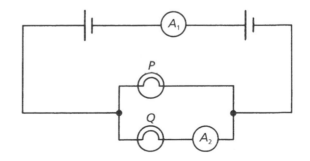

The ammeter A_2 reads 0.30 A. What does the ammeter A_1 read?

A 0 A B 0.15 A C 0.30 A D 0.60 A E 0.90 A

Solution

[The lamps are identical and the same current passes through each. The 0.60 A divides and 0.30 A passes through each lamp.]

Answer **D**

Example 3

The graphs show the potential difference across a component plotted against the current in the component. Which of the graphs would be obtained for a coil of copper wire?

Solution

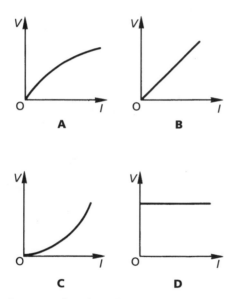

[A coil of copper wire obeys Ohm's Law and the current is proportional to the potential difference. Hence, the graph is a straight line through the origin.]

Answer **B**

Example 4

In the circuit that follows, the ammeter reads 2 A and the voltmeter reads 5 V.

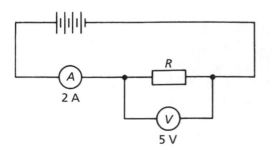

The value of the resistance R is

A 0.4 Ohm B 2.5 Ohm C 3.0 Ohm D 10.0 Ohm

Solution

$$[R = \frac{V}{I} \text{ (see section 1)} = \frac{5\,\text{V}}{2\,\text{A}} = 2.5\,\text{Ohm}]$$

Answer **B**

Example 5

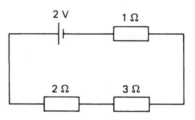

The potential difference, in V, across the 3 Ohm resistor, is

A $\frac{1}{9}$ B $\frac{1}{2}$ C 1 D $\frac{6}{5}$ E 2

Solution

[The total resistance is 6 Ohm. $V = IR \Rightarrow 2\,\text{V} = I \times 6\,\text{Ohm}$, where I is the current in the circuit. Hence, $I = \frac{1}{3}\,\text{A}$. For the 3 Ohm resistor $V_{3\,\text{Ohm}} = I \times 3\,\text{Ohm} = (\frac{1}{3} \times 3)V = 1\,\text{V}$.]

Answer **C**

Example 6

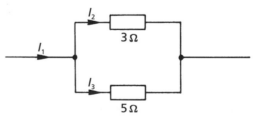

The diagram shows two resistors connected in parallel. Which of the following statements is correct?

A $I_1 = I_2 - I_3$
B $I_1 = I_3 - I_2$
C $5I_2 = 3I_3$
D $3I_2 = 5I_3$

Solution

[In a parallel circuit the potential difference across each resistor is the same; therefore $I_2 \times 3\,\text{Ohm} = I_3 \times 5\,\text{Ohm}$. The current divides in the same ratio as that between the resistances, the larger current passing through the smaller resistance.]

Answer **D**

Example 7

Three resistors, each of resistance 4 Ohm, are to be used to make a 6 Ohm combination. Which arrangement will achieve this?

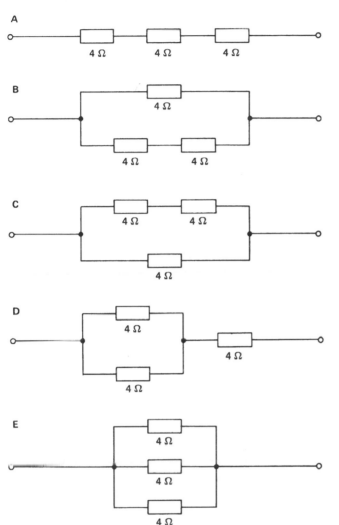

Solution

[When answering this type of question, don't waste time calculating the effective resistance of every combination. Glance at the various answers and see if you can spot the correct one, then check it by calculation. In this case a glance at **D** will tell you that this is the correct answer. The two 4 Ohms in parallel have an effective resistance of 2 Ohm. This 2 Ohm is connected in series with 4 Ohm and hence the total resistance is 6 Ohm.]

Answer **D**

Example 8

An electricity board charges 6p per kilowatt/hour (kWh) of electrical energy supplied. what is the total cost of operating 4 light bulbs, each rated at 100 W, for 5 hours?

A   3p
B   6p
C   12p
D   120p

Solution

[4 bulbs = 400 W = 0.4 kW. Energy used 0.4 kW × 5 h = 2.0 kWh. Cost = 2.0 kWh × 6p = 12p (see section 6).]

Answer **C**

Example 9

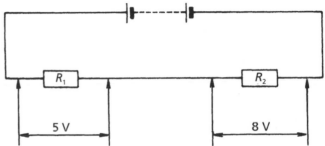

The circuit shows a battery connected to two resistors, R_1 and R_2, in series. There are potential differences of 5 V across R, and 8 V across R_2.
What is the potential difference across the battery?

A   −3 V
B   3 V
C   13 V
D   40 V

(ULEAC, Intermediate Tier)

Solution

[When resistors are in series the sum of the separate potential differences is equal to the applied potential difference (see section 2, 'Laws for series circuits').]

Answer **C**

Examples 10 and 11

Gary has made a circuit with three 6 V lamps and two diodes.

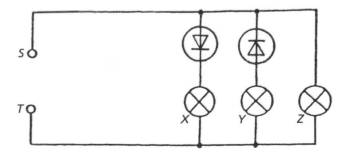

10. He connects a 6 V a.c. supply to S and T.
    Which lamp or lamps will be lit?
    A All three
    B Z only
    C X and Z only
    D Y and Z only
11. He replaces the a.c. supply with a 6 V battery. He
    connects the positive to S and the negative to T.
    Which lamp or lamps would now be lit?
    A All three
    B Z only
    C X and Z only
    D Y and Z only

    (ULEAC, Intermediate Tier)

Solution 10

[Diodes only pass current in one direction. Current will
pass through X and Y when it is in the direction of the arrow
on the diode symbol.]

Answer **A**

Solution 11

[S is positive so the current will pass through X but not
through Y.]

Answer **C**

Example 12

(i) The diagram shows a potential divider circuit.
    The e.m.f. of the cell is 12 V and it has
    negligible internal resistance.

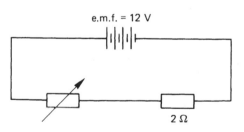

e.m.f. = 12 V

2 Ω

What happens to the potential difference across
the 2 Ohm resistor as the resistance of the
variable resistor is increased?
Explain your answer.
(ii) When the variable resistor is set at 4 Ohm, what
is
    (a) the current in the 2 Ohm resistor?
    (b) the potential difference across the 2 Ohm
        resistor?

    (ULEAC, Intermediate Tier)

Solution

(i) The potential difference across the 2 Ohm
    resistor decreases. The current passing through it

will decrease because the circuit resistance has
increased, and hence the p.d. across it, which is
given by $V = IR$, will decrease.
(ii) (a) Total resistance of circuit
$$= (2 \text{ Ohm} + 4 \text{ Ohm})$$
$$= 6 \text{ Ohm}.$$

$$\text{Current flowing} = \frac{V}{R} \text{ [see section 1]}$$

$$= \frac{12 \text{ V}}{6 \text{ Ohm}}$$

$$= 2 \text{ A}$$

[In a series circuit the current is the same
everywhere. The current in every part of the
circuit is 2 A.]
(b) Potential difference across 2 Ohm is given
by
$V = IR$ [see section 1]
$V = 2 \text{ A} \times 2 \text{ Ohm} = 4 \text{ V}$
[Notice that the p.d. across the 4 Ohm
resistor is 8 V and that the ratio of the p.d.s
is equal to the ratio of the resistors.]

Example 13

A 12 V battery is used, together with a potential divider, to
provide a variable voltage across a 12 V car bulb. Sketch
the circuit you would use to measure the current in the
lamp for various voltages across it. Sketch the graph you
would expect to obtain.

(8 marks)

Solution

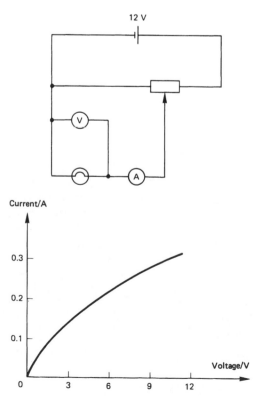

[As the slider is moved from the right-hand end of the variable resistor to the left-hand end, the voltage across the bulb varies from 12 V to 0 V (this arrangement is known as a potential divider). Remember that voltmeters are connected across appliances and ammeters in series.]

Example 14

This question is about paying for electricity.

(a) At home the Webb family use electricity for heating and lighting only. They have three heaters rated at 2 kW, 1.6 kW and 1 kW. They have five lights: three are rated at 100 W, one at 60 W and one at 40 W.

   (i) What is the total power of the three heaters?
                                      (1 mark)

   (ii) What is the total power (in watts) of the five lights? (1 mark)
       What is the total power of the five lights in **kilowatts**? (1 mark)

(b) They use all the lighting and heating for 6 hours each day.

   (i) How much electrical energy (in kWh) does the family use in one day? (2 marks)
   (ii) The bill is for 90 days. How much electrical energy should the bill charge for? (1 mark)

(c) This is the family's electricity bill.

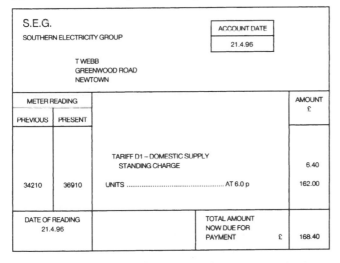

METER READING			AMOUNT £
PREVIOUS	PRESENT		
		TARIFF D1 – DOMESTIC SUPPLY STANDING CHARGE	6.40
34210	36910	UNITS .... AT 6.0 p	162.00

S.E.G.
SOUTHERN ELECTRICITY GROUP

ACCOUNT DATE
21.4.96

T WEBB
GREENWOOD ROAD
NEWTOWN

DATE OF READING 21.4.96 — TOTAL AMOUNT NOW DUE FOR PAYMENT £ 168.40

   (i) Use the meter readings to calculate the number of units charged. (1 mark)
   (ii) Is the bill correct? Give your reasoning. (1 mark)
   (iii) Would you expect the Webb's next electricity bill to be larger, smaller or much the same? Give your reasoning. (2 marks)

(d) Give one piece of advice to the Webb family to help them reduce the cost of keeping their house warm. (1 mark)

Solution

(a)   (i) 4.6 kW [2 kW + 1.6 kW + 1 kW]

    (ii) 400 W [(100 W × 3) + 60 W + 40 W]
       1 kW = 1000 W ⇒ 400 W = 0.4 kW

(b)   (i) Electrical energy used in a day = 5 kW × 6 h
       = 30 kWh.

    (ii) 30 kWh × 90 = 2700 kWh

(c)   (i) 2700 units [36 910 − 34 210]

    (ii) Cost = 2700 × 6p = 16 200p = £162
       The bill is correct.

    (iii) Smaller. This bill is for the months before April, i.e. in winter. The next one will cover the summer months.

(d) Reduce the heat loss (one way would be to insulate the roof).

Example 15

   (i) A current of 2 A is passing through a circuit. How much charge flows past any point in 5 s? (2 marks)

   (ii) 120 J of work is done when 10 C passes through a bulb. What is the p.d. across the bulb? (2 marks)

   (iii) A 60 W 240 V lamp is connected to a 240 V mains.

     (a) What current does it take? (2 marks)
     (b) What is its resistance? (2 marks)
     (c) How much does it cost to keep it on for 50 hours if the cost of a kWh is 6p? (3 marks)

Solution

   (i) $Q = It$ [see section 1. Remember that 1 A = 1 C/s]
      $Q = (2 \text{ C/s})(5 \text{ s}) = 10 \text{ C}$

   (ii) $\text{p.d.} = \dfrac{\text{work done}}{\text{charge moved}}$

      [see section 1; remember that p.d. is energy per coulomb]

      $= \dfrac{120 \text{ J}}{10 \text{ C}} = 12 \text{ V}$

   (iii) (a) Power $= (V \times I)$ [see section 6]
         60 W $= 240 V \times I$

         $I = \dfrac{60 \text{ W}}{240 \text{ V}} = 0.25 \text{ A}$

     (b) $R = \dfrac{V}{I}$ [see section 1]

         $R = \dfrac{240 \text{ V}}{0.25 \text{ A}} = 960 \text{ Ohm}$

     (c) Cost = (number of kilowatts) × (hours) × (cost of kilowatt-hour) [see section 6]

         $= \left(\dfrac{60}{1000} \text{ kW}\right)(50 \text{ h})(6\text{p/kWh}) = 18 \text{ p}$

Example 16

    (i) Why is a three-pin mains plug fitted with a fuse?
                         (2 marks)
    (ii) What fuse would you put in a plug if the plug is
        connected to an appliance labelled 1 kW 250 V?
                         (3 marks)

Solution

    (i) The fuse is a safety device. If the current exceeds
        a certain value, the fuse will melt, breaking the
        circuit and disconnecting the appliance from the
        supply.
    (ii) Power $= (V \times I)$ [see section 6]

$$1000\ \text{W} = 250\ \text{V} \times I$$

$$I = \frac{1000\ \text{W}}{250\ \text{V}} = 4\ \text{A}$$

        A 5 A fuse would be fitted. [Fuses fitted to
        household appliances are usually 3 A, 5 A or
        13 A.]

Example 17

A 3 V battery of negligible internal resistance is connected
in series with a 3 Ohm resistor and an ammeter as shown in
the diagram.

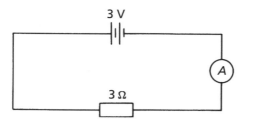

  (a)   What does the ammeter read?   (2 marks)
  (b)   A 6 Ohm resistor is connected across the 3 Ohm
        resistor. What is the new reading on the ammeter?
                         (4 marks)

Solution

  (a) $I = \dfrac{V}{R} = \dfrac{3\ \text{V}}{3\ \text{Ohm}} = 1\ \text{A}$

  (b) For resistances in parallel,

$$R = \frac{\text{product}}{\text{sum}}$$

    [see section 2: 'Laws for parallel circuits']

$$= \frac{3 \times 6}{3 + 6}\ \text{Ohm} = \frac{18}{9}\ \text{Ohm} = 2\ \text{Ohm}$$

$$I = \frac{V}{R} = \frac{3\ \text{V}}{2\ \text{Ohm}} = 1.5\ \text{A}$$

[Notice that when a resistance is added in parallel, the
resistance of the circuit decreases and the current
increases.]

Example 18

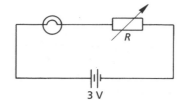

  (a)   The diagram shows an incomplete circuit for an
        experiment to investigate how the resistance of a
        torch bulb varies with the current flowing through it.
        (i) Add to the circuit diagram an ammeter for
             measuring the current through the bulb and a
             voltmeter for measuring the p.d. across the
             bulb.
        (ii) State clearly how you would obtain the
             readings needed to carry out the investigation.
        (iii) How would you calculate the resistance of the
             bulb?
        (iv) If the bulb is 2.5 V and takes 0.25 A at its
             working temperature, calculate the resistance of
             the bulb at the working temperature.
        (v) The resistance of the bulb when the filament is
             cold is 5 Ohm. Sketch the graph you would
             expect to obtain if you plotted resistance
             against current for the bulb.
  (b)   The diagram shows an electrical circuit containing a
        3 V battery, an ammeter of negligible internal
        resistance and three resistors with resistances shown.

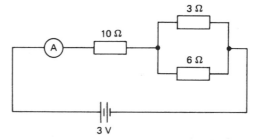

What is (i) the resistance of the parallel combination,
(ii) the reading on the ammeter and (iii) the potential
difference across the 3 Ohm resistor?

Solution

  (a)   (i)

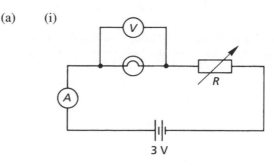

(ii) The resistance R is set at its highest value and the readings on the ammeter and the voltmeter are recorded. The value of the resistance R is decreased and the reading on each meter again recorded. continuing in this way a series of readings are obtained. The readings are continued until the bulb is burning brightly.

(iii) The resistance for each pair of readings is calculated by using the equation $R = V/I$, that is by dividing the voltmeter reading by the ammeter reading.

(iv) $R = \dfrac{V}{I} = \dfrac{2.5\ V}{0.25\ A} = 10\ Ohm$

(v)

Resistance/Ω

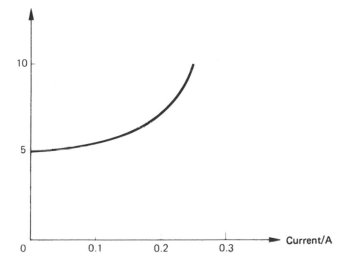

[The same circuit is used to investigate the $\dfrac{voltage}{current}$ relationship for other components such as a resistor or a diode.]

(b) (i) $R = \dfrac{product}{sum}$

[see section 2: 'Laws for parallel circuits']

$= \dfrac{3 \times 6}{3 + 6}\ Ohm = \dfrac{18}{9}\ Ohm = 2\ Ohm$

(ii) $I = \dfrac{V}{R} = \dfrac{voltage\ of\ cell}{total}$

resistance $= \dfrac{3\ V}{12\ Ohm} = 0.25\ A$

(iii) p.d. across 3 Ohm and
6 Ohm $= IR = 0.25\ A \times 2\ Ohm = 0.5\ V$

Examination questions

(Numerical answers and hints on solutions will be found at the end of the chapter.)

Question 1

The diagram below shows part of a lighting circuit in a house. The circuit is protected by a 5 A fuse. There is a current of 0.4 A in each lamp when it is turned on.

(a) (i) How are the lamps wired in the circuit?
(1 mark)

(ii) What is the current at **P** when ALL the lamps are on? (2 marks)

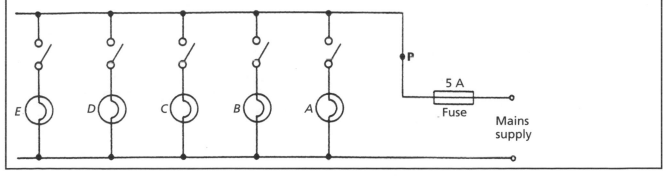

(iii) Is the 5 A fuse suitable for this circuit? Explain your answer. (2 marks)

(iv) What is the purpose of the fuse? (1 mark)

(b) The diagram below shows a way of switching a lamp on and off using a pair of two-way switches.

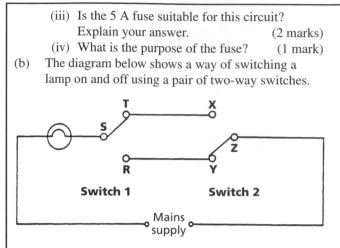

(i) If **switch 1** connects to **S** and **T**, how could the lamp be turned on using **switch 2**? (1 mark)

(ii) How could the lamp then be switched off using **switch 1**? (1 mark)

(iii) Where would this switching arrangement for the lamp be useful in the house? (1 mark)
(ULEAC, Syll A, Jun 95, Intermediate Tier)

Question 2

The diagram shows the inside of a three-pin plug.

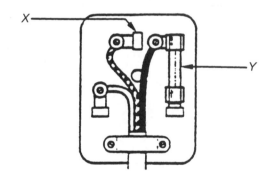

(a) (i) Choose words from the box to label the parts X and Y.

black, blue, brown, cable grip, conductor, copper, earth pin, fuse, green and yellow, insulator, live pin, neutral pin, plastic, red, steel, wire

(ii) Complete the following sentences using words from the box.

In a mains plug the colour of the wire attached to the neutral terminal is .... The fuse is always connected to the .... terminal. The wires are covered in .... which is a good .... (6 marks)

(b) In a discussion at school Sue said her hairdryer had only two wires to connect to the plug. Sam said this was unsafe and all electrical equipment needed three wires to connect to the plug.

(i) Was Sam correct?

(ii) Explain your answer.

(iii) Which wire was missing from Sue's hairdryer?
(4 marks)

(c) The label on the hairdryer is shown below.

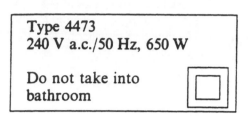

Type 4473
240 V a.c./50 Hz, 650 W

Do not take into bathroom

(i) What is the power rating of the hairdryer?

(ii) What rating of fuse should Sue fit into the plug? (2 marks)
(ULEAC, Syll B, Jun 95, Intermediate Tier)

Question 3

Here is a circuit installed in the boot of Joe's car.

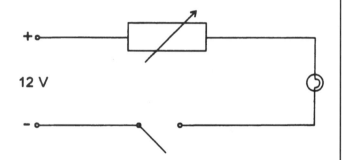

(a) (i) The light bulb comes on when the boot is opened. Complete table 12.1 with the words OPEN and CLOSED.

Table 12.1

State of boot	State of switch
Open	
Closed	

(2 marks)

(ii) Joe decides to test the bulb by including an ammeter in the circuit. Redraw the circuit in the space below, with the ammeter measuring the current in the bulb. (2 marks)

(b) The variable resistor has a knob on it. When Joe rotates the knob clockwise, he notices that the ammeter reading goes up.

(i) What happens to the brightness of the bulb? (1 mark)

(ii) What happens to the resistance of the variable resistor? (1 mark)
(iii) What happens to the voltage across the bulb? (1 mark)
(MEG Salters', Jun 95, Intermediate Tier, Q1)

Question 4

The illustration above shows the mains wiring to an electric heater.

(a) The heater shown in Fig. 12.30 has an earth wire attached to the case.
If the live wire accidentally touches the case of the heater, the earth wire makes the heater electrically safe. Explain how this happens. (3 marks)

(b) The heater is marked '240 V 2.5 kW'. 1 kWh of electrical energy costs 8p.
For how long can the heater be run for £10? (3 marks)

(c) When the heater is working correctly the current in the live wire is equal to the current in the neutral wire. Explain this. (1 mark)

(d) A residual current device can be put into electrical circuits as a safety device to prevent people receiving shocks. It compares the currents in the live and neutral wires. If these currents are not equal then a fault is present.

The illustration below shows a residual current device connected in a circuit to a heater.

This is how it works. The live wire is wound several times round an iron core, like the primary winding in a transformer. The neutral wire is wound the same number of times round the same core. The coils are wound in such a way that if the currents in the wires are equal then the magnetic fields they create in the iron core cancel one another out.

A third coil is wound on the iron core and this is connected to a relay. The relay contacts are closed at first. When they are open, they stay open.

(i) Explain what happens in the third coil if the alternating currents in the live and neutral wires are **not** equal.
(ii) Hence explain how the circuit is made safe.
(iii) The illustration below includes a switch labelled *Test button*. Explain what happens when this switch is pressed. (2 marks)
(MEG, Jun 95, Higher Tier)

Question 5

A lamp is marked '3.0 V, 0.2 A'.

(a) Writing down the formula that you use in each part and showing your working, calculate

(i) its resistance when it is working at its normal brightness, (3 marks)
(ii) its power, (3 marks)
(iii) the charge passing through it in 10 minutes, (3 marks)

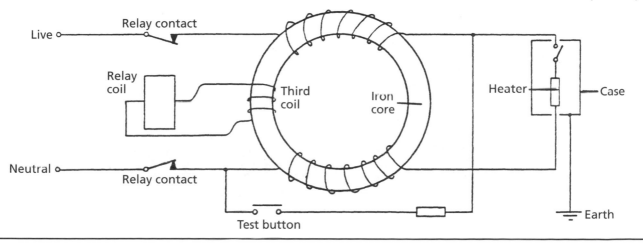

(iv) the energy it transfers in 10 minutes. (3 marks)

(b)   The only power supply available is a 12 V car battery with negligible internal resistance.

  (i) Describe and explain what would happen if the battery were connected directly to the lamp. (2 marks)

  (ii) The lamp will work at normal brightness if a resistor R is connected in series with it.

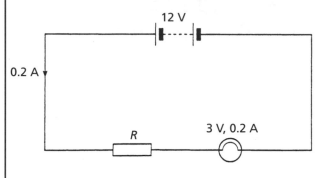

12 V

0.2 A

R 3 V, 0.2 A

  (I) Calculate the total resistance of the circuit. (1 mark)

  (II) State a suitable value for R. (2 marks)
    (WJEC, Jun 95, Higher Tier, Q9)

Question 6

(a)   Mrs Murphy has just bought three lamps for use outside in the garden. Each lamp will be at normal brightness when the voltage across its terminals is 12 V.

  Each lamp is to be wired so that it can be switched on and off independently of the others. A fourth switch, called the master control switch, will allow Mrs Murphy to switch off all the lamps at once if a fault occurs.

  (i) Draw a circuit diagram to show how the lamps are to be wired up to the switches and the power supply unit. The symbol for the power supply unit is shown for you. (7 marks)

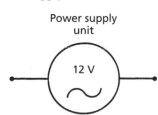

Power supply unit

12 V

  (ii) Each lamp is marked 10 W, 12 V. Calculate the current flowing in **each** lamp when it is at its normal brightness. **Show clearly how you obtain your answer.**

  (iii) Calculate the resistance of the filament of **one** of the lamps and hence find the combined resistance of **all three lamps** in the circuit drawn in (a)(i). **Show clearly how you obtain your answers.** (7 marks)

Mrs Murphy must now decide on what cable to

purchase. The shop assistant tells her that types of cable are available as shown in table 12.2.

Table 12.2

Type of cable	Maximum current in this cable in A	Cost per metre in pence
Class 1	3.0	79
Class 2	5.0	99
Class 3	13.0	149

The larger the current a cable can carry, the more expensive it is. Mrs Murphy naturally wants to **keep costs to a minimum**.

  (iv) What type of cable should Mrs Murphy buy? (2 marks)

  (v) Mrs Murphy notices that the conductor of each cable is made of copper, but the cables themselves are all of a different thickness. Which one of the three types of cable is thickest? (1 mark)

(b)   Pat carries out an investigation on the current flowing through a torch bulb when different voltages are applied across its ends. She obtains the results in table 12.3.

Table 12.3

Voltage V in volts	0.50	1.00	2.00	3.00	4.00
Current I in amperes	0.07	0.10	0.14	0.17	0.20

  (i) Plot the graph of V (y-axis) against I (x-axis) **on the graph paper below**. (3 marks)

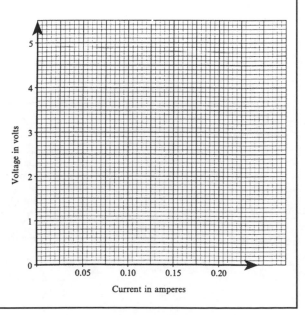

(ii) Find the resistance of the filament when the current is 0.09 A.
Show clearly how you obtain your answer.
(2 marks)

(iii) The resistance of this filament changes as the current through it changes.
This is because the temperature of the filament is not constant.
Sketch below the voltage–current graph for a metallic conductor at **constant temperature**.

(2 marks)
(NICCEA, Jun 95, Higher Tier)

Answers to examination questions

1. (a) (i) In parallel.
    (ii) $(0.4 \text{ A}) \times 5 = 2.0 \text{ A}$
    (iii) Yes since the maximum current in the circuit is less than 5 A. However, a 3 A fuse would be more suitable.
    (iv) It protects the lighting circuit by switching it off if the current gets too big. A large current could overload the wire and start a fire.
   (b) (i) Connect **Z** to **X**.
    (ii) Connect **S** to **R**.
    (iii) At the top and bottom of the stairs.

2. (a) (i) blue, live, plastic, insulator.
    (ii) **X** = earth pin **Y** = fuse.
   (b) (i) No.
    (ii) There is no way the casing can become live even if the wire were to touch since the casing is made from plastic and plastic is an electrical insulator.
    (iii) The earth wire.
   (c) (i) [For this you need to find out the current in the circuit]
    Power = current × voltage

    Rearranging gives $\text{current} = \dfrac{\text{power}}{\text{voltage}}$

    $= \dfrac{650 \text{ W}}{240 \text{ V}} = 2.7 \text{ A}$

    (ii) A 3 A fuse should be fitted.

3. (a) (i)

Table 12.1

State of boot	State of switch
Open	Closed
Closed	Open

(ii)

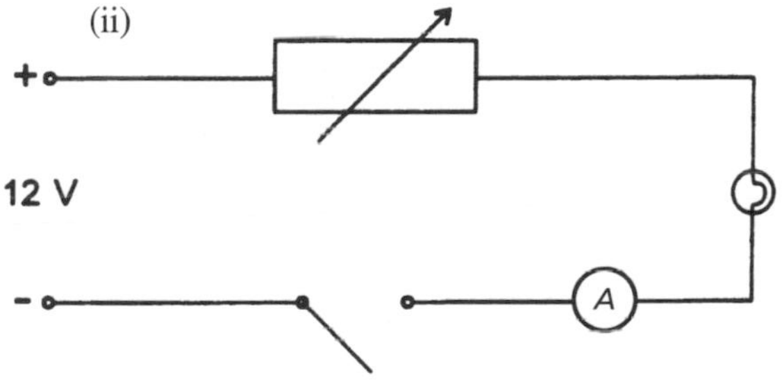

   (b) (i) Goes up.
    (ii) Decreases.
    (iii) Increases.

4. (a) The current travelling through the case passes to earth through the low resistance wire. Because the current takes the low resistance path, it rises and then blows the fuse which switches off the heater.

   (b) Number of kWh for £10 $= \dfrac{10}{0.08} = 125$

   Heater uses 5.5 kW each hour, so number of

   hours $= \dfrac{125}{2.5} = 50$ hours

   (c) The fuse and heater are in series with each other so there are no other paths for the current to follow. The current is the same all the way around a series circuit.

   (d) (i) [There is an important point to be made here. It would be easy to answer this and find that you have also answered some of the other parts of the question. To prevent this, you are always advised to read the complete question before starting your answer.] The magnetic fields due to the currents in the coils in the live and neutral circuits no longer cancel out. This residual magnetic field induces a current in the third coil causing the relay coil to be energised and thus attract the switch.
    (ii) The current in the third coil energises the relay coil and this opens the relay contacts in the live wire and this cuts off the current in the heater circuit which prevents the user from getting a shock.

(iii) When the switch is pressed, most of the current by-passes the bottom coil and the heater. The magnetic fields will not now cancel out so a magnetic field will not link the third coil and cause an induced voltage which makes a current flow. This current will cause the relay to trip (open) and switch off the heater. This enables a check to be made to see if the circuit breaker is working.

5. (a) (i) Resistance (Ω) = $\dfrac{\text{Voltage (V)}}{\text{Current (A)}}$

$$= \dfrac{3.0\ \text{V}}{0.2\ \text{A}} = 15\ \Omega$$

(ii) Power (W) = Current (A) $\times$ Voltage (V) = $3.0\ \text{V} \times 0.2\ \text{A} = 0.6\ \text{W}$

(iii) Charge (C) = Current (A) $\times$ Time (s) = $0.2\ \text{A} \times (60 \times 10\ \text{s}) = 120\ \text{C}$

(iv) Energy = Power $\times$ Time
= $0.6\ \text{W} \times (10 \times 60\ \text{s})$
= $360\ \text{J}$

(b) (i) The current would be too great and would cause the bulb to blow.

(ii) (I) Total resistance = $\dfrac{\text{Supply voltage}}{\text{Current}}$

$$= \dfrac{12\ \text{V}}{0.2\ \text{A}} = 60\ \Omega$$

(II) The total resistance is the sum of the resistance of the bulb and the resistance of resistor R.
Hence $60 = R + 15$
So, the value of R needs to be about $45\ \Omega$

6. (a) (i)

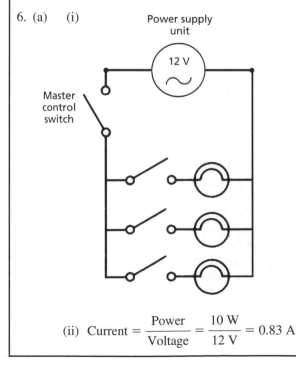

(ii) Current = $\dfrac{\text{Power}}{\text{Voltage}} = \dfrac{10\ \text{W}}{12\ \text{V}} = 0.83\ \text{A}$

(iii) Resistance of one bulb = $\dfrac{\text{Voltage}}{\text{Current}}$

$$= \dfrac{12\ \text{V}}{0.83\ \text{A}} = 14.4\ \Omega$$

Bulbs are in parallel, so the total resistance of the circuit, R is given by

$$\dfrac{1}{R} = \dfrac{1}{R_1} + \dfrac{1}{R_2} + \dfrac{1}{R_3}$$

$$= \dfrac{1}{14.4} + \dfrac{1}{14.4} + \dfrac{1}{14.4}$$

Giving $R = 4.8\ \Omega$

(iv) Largest current = $3 \times 0.83\ \text{A} = 2.49\ \text{A}$
Use Class 2

(v) Class 3

(b) (i)

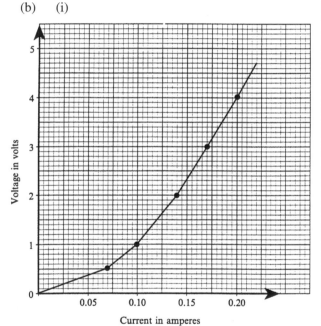

(ii) From the graph, when $I = 0.09\ \text{A}$, $V = 0.8\ \text{V}$

Resistance = $\dfrac{\text{voltage}}{\text{current}}$

$$= \dfrac{0.8\ \text{V}}{0.09\ \text{A}} = 8.9\ \Omega$$

(iii)

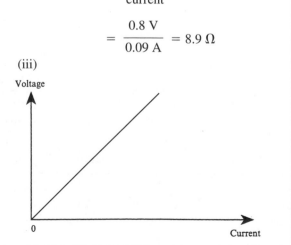

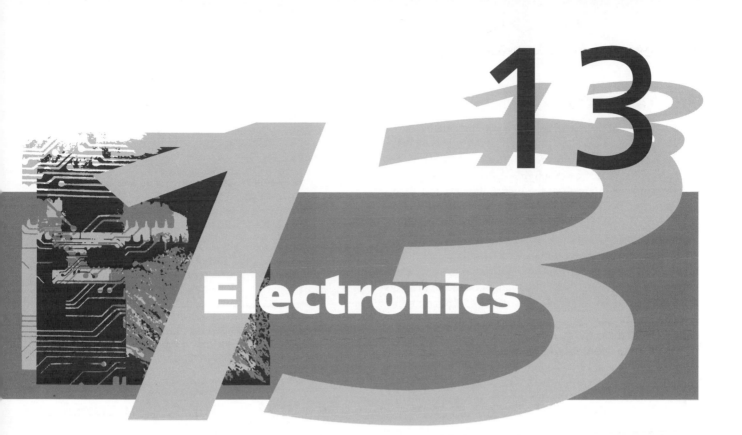

Electronics

13

WJEC & NEAB	ULEAC Syll A	ULEAC Syll B	MEG	Topic	MEG Salters'	MEG Nuffield	SEG	NICCEA
✓	✓	✓	✓	**Transducers**	✓	✓	✓	✓
✓	✓	✓	✓	**Gates**	✓	✓	✓	✓
✓		✓	✓	**The bistable**	✓	✓	✓	
✓			✓	**Capacitors**	✓			
✓			✓	**Transistors**	✓	✓	✓	
✓	✓	✓	✓	**Relays**	✓	✓	✓	✓
				Bridge rectifier circuits	✓			

1 Transducers

Light-dependent resistors (LDRs), light-emitting diodes (LEDs) and thermistors are all examples of transducers (i.e. they convert energy or information from one form to another). The resistance of a light-dependent resistor decreases as the intensity of the light falling on it increases. (LDRs are sometimes used in light meters. Selenium cells are also used in light meters. When light falls on the cell, a small p.d. results from the movement of the electrons.) LEDs are semiconductor diodes which convert electrical energy into light energy. To avoid damage by large currents, a protective resistor is usually connected in series with the LED. The diagram opposite shows a circuit with an LDR and an LED in it. When light shines on the LDR, the LED switches off.

The resistance of most thermistors decreases as their temperature increases. The next diagram shows the symbol for a thermistor.

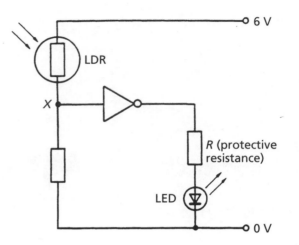

When light shines on the LDR, its resistance goes down and the voltage at X goes up. The output from the NOT gate goes from high to low and the LED switches off. The reverse happens when it gets dark.

2 Gates

The symbols and truth tables of a number of gates are shown below.

The circuit shown opposite illustrates the use of an LDR and a NOT gate to switch on an LED when it gets dark. If a thermistor replaced the LDR, then the light would come on as the temperature fell.

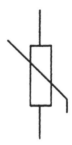

The symbol for a thermistor.

Name of gate	Symbol	Truth table			Description of gate
AND	A, B — O.P.	A	B	O.P.	The output is high if both A *AND* B are high
		0	0	0	
		0	1	0	
		1	0	0	
		1	1	1	
NAND	A, B — O.P.	0	0	1	Opposite of *AND* gate
		0	1	1	
		1	0	1	
		1	1	0	
OR	A, B — O.P.	0	0	0	The output is high if A *OR* B *OR* both are high
		0	1	1	
		1	0	1	
		1	1	1	
NOR	A, B — O.P.	0	0	1	Opposite of *OR* gate. Output high if neither A *NOR* B is high
		0	1	0	
		1	0	0	
		1	1	0	
NOT	A — O.P.	A		O.P.	An inverter
		0		1	
		1		0	

3 The bistable

A bistable has two stable states, namely Q at logic 1 and $\bar{Q}$ at logic 0, or vice versa. Two NAND gates may be combined as shown below to form a bistable. Suppose Q is

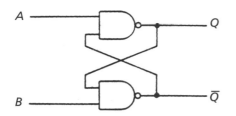

Two NAND gates connected together to make a bistable.

at logic 0 and $\bar{Q}$ is at logic 1, and A and B are both high (logic 1). If A is momentarily made low (logic 0), then Q goes high (logic 1). If B is then momentarily made low, Q resets to a low. If inputs A and B remain high, the circuit remains in one of its two stable states, depending on which of the AB inputs was last at a low voltage (see example 12). With slight modifications the bistable may be switched from one state to the other every time a pulse is fed to the input. The bistable forms the basis of memory circuits and, with additional logic, binary counters. Two NOR gates connected in the same way (i.e. cross-coupled) as shown below also form a bistable. If both inputs are 0, Q will be

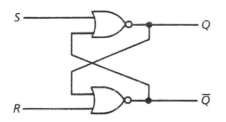

Two NOR gates connected to make a bistable.

either 0 or 1 depending upon which of the two inputs was last raised to logic 1. If S is 1 (or was last raised to 1), then $\bar{Q}$ is 1 and Q is 0.

4 Capacitors

A capacitor may be used to store charge and energy. In the circuit shown in the next diagram it is used to produce a time delay. When the switch S is closed, the output from the NOT gate is high and the lamp is off. When the switch S is opened, the capacitor charges up through the resistor R. When the voltage at A is high enough, the output from the NOT gate goes low and the lamp lights. Increasing the value of the capacitor or the resistor will increase the time delay.

5 Transistors

A transistor is a semiconductor device. It is considered to be 'off' when very little current passes through the collector circuit and 'on' when a much larger current passes through in the collector circuit. The transistor stays 'off'

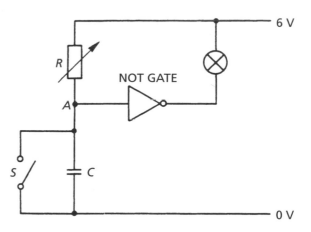

A capacitor being used to produce a time delay.

unless the base voltage and, hence, the base current, rise above a certain minimum value. When the base current rises above this minimum value, a much larger current passes through the collector circuit. The large current in the collector circuit is therefore controlled by the small base current, which can be used to switch the transistor 'on' and 'off'. See examples 5, 6 and 8.

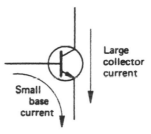

A small base current results in a larger collector current.

Because a small base current can cause a large collector current, a transistor may also be used as an amplifier.

6 Relays

A relay is a switch which is operated by an electric current. Often the relay consists of a solenoid with a switch inside it. The switch closes when a current passes through the solenoid. The figure below shows a relay being used to

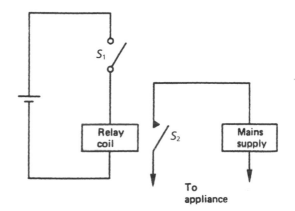

When the switch S_1 is closed, a current passes through the relay coil. The switch S_2 closes and the power to the appliance is switched on.

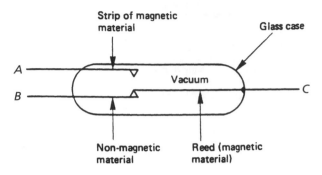

Strip of magnetic material

Glass case

A

B

Vacuum

C

Non-magnetic material

Reed (magnetic material)

A reed switch. In the presence of a magnetic field, the reed is attracted to strip A, and A and C are connected.

switch on a large current, using only the current from a battery. One common type of relay is a reed switch placed in a solenoid.

7 A bridge rectifier circuit

A semiconductor diode is a device which allows current to pass easily in one direction only. The four diodes shown below are arranged to form a 'bridge rectifier'. They 'steer' the current in the desired direction. The current can only pass in the direction shown by the arrows. It will always pass through the load from *A* to *B*.

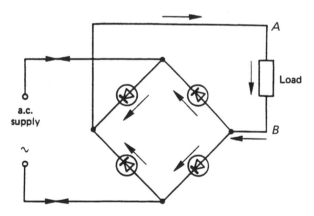

A bridge rectifier circuit. The current always flows in the same direction through the load.

The next illustration (a) shows the current through the load. The output from the bridge rectifier is said to be full-wave rectified.

If a capacitor is placed across the load (i.e. across the output from the bridge rectifier) the variations in the current are reduced. The capacitor charges up while the output voltage is rising, and discharges when the output voltage is falling. The smaller variations in the current which remain are known as the ripple current (part b).

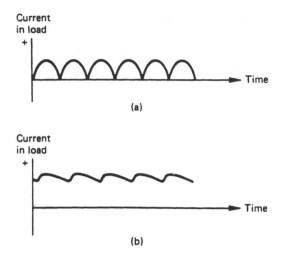

Current in load

+

Time

(a)

Current in load

+

Time

(b)

(a) Unsmoothed d.c. (b) Smoothed d.c.

Worked examples

Example 1

Which of the diagrams below shows the possible logic states of an AND gate?

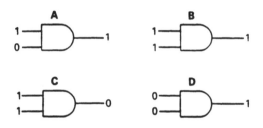

Solution

[See section 2.]

Answer **B**

Examples 2 and 3

The circuit below is designed to switch on a buzzer. The variable resistor is adjusted so that the buzzer is just off.

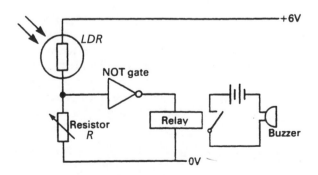

2. The buzzer will be switched on if
   A the room gets darker
   B the room gets lighter

C the room temperature rises
D the room temperature falls

3. The purpose of the resistor *R* is
A to prevent too much current flowing through the LDR
B to prevent too much current flowing through the NOT gate
C to adjust the temperature at which the buzzer is switched on
D to adjust how dark/light the room must get before the buzzer switches on

Solutions

2. [When the room gets darker the resistance of the LDR goes up. The voltage drop across it goes up, so the input voltage to the NOT gate goes down, i.e. the logic changes from 1 to 0. The output from the NOT gate goes to 1 and the buzzer is switched on.]

Answer **A**

3. Answer **D**

Example 4

The graphs below show how the two inputs, *X* and *Y*, of an AND gate vary with time.

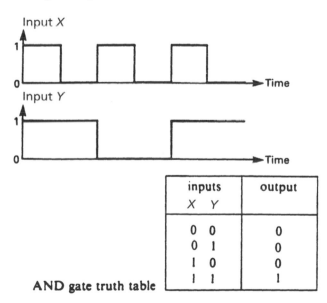

inputs X Y	output
0 0	0
0 1	0
1 0	0
1 1	1

AND gate truth table

Which of the graphs shown below shows how the output varies with time?

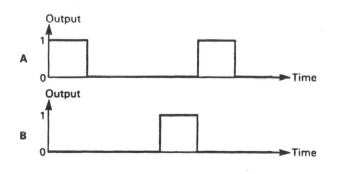

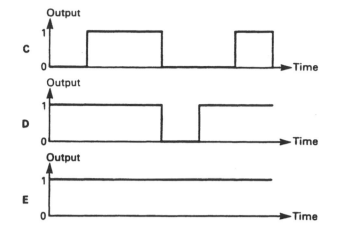

Solution

[The output is 1 when both inputs are 1. **A** correctly shows the time when both inputs are 1.]

Answer **A**

Examples 5 and 6

The diagram shows a circuit for a transistor switch.

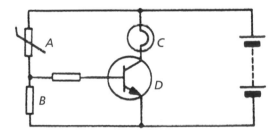

5. Which of the lettered symbols on the diagram represents the transistor?
6. In this circuit the transistor is switched on by a change in
A temperature
B light intensity
C pressure
D moisture level

(ULEAC, Intermediate Tier)

Solutions

5. Answer **D**
6. [**A** is a thermistor. The resistance of a thermistor changes when the temperature changes. This change in resistance changes the voltage of the base of the transistor and will cause the transistor to switch on (see section 5).]

Answer **A**

Example 7

In the circuit shown below the inputs to the first logic gate

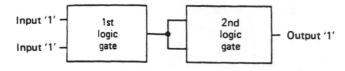

are both at logic level '1'. The output from the second gate is also at logic level '1'.

The two gates could be

A the first an AND gate and the second a NAND gate
B the first a NAND gate and the second an AND gate
C the first an AND gate and the second a NOR gate
D both NAND gates

Solution

[Use the logic tables in section 2.]

Answer **D**

Example 8

In only one of the circuits below the lamp lights. Which circuit is it?

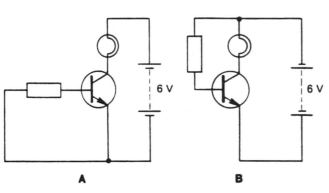

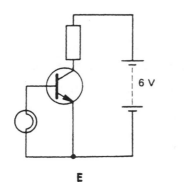

Solution

[When a small current passes from the base to the emitter, this switches 'on' the transistor, causing a large current to pass from the collector to the emitter. In **B** the battery is the wrong way round.]

Answer **D**

Example 9

In the circuit below a voltmeter V is connected across lamp L.

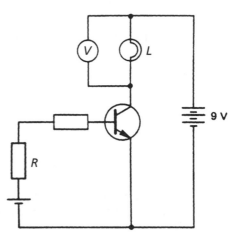

Which of the changes in table 13.1 could occur at lamp L and voltmeter V, if the resistor R is reduced in value?

Table 13.1

	Lamp L	Reading on voltmeter V
A	goes out	decreases
B	goes out	increases
C	no change	no change
D	lights up	decreases
E	lights up	increases

Solution

[If the current in the resistor R is increased the current in the base of the transistor will increase. This could switch the transistor on and the lamp could light. If this happened a current would pass through the lamp and the voltage across it would increase.]

Answer **E**

Example 10

A boy designs a model railway signal to operate from the circuit shown below.

The switch marked S is a reed switch which closes when a magnet is brought close to it. The light-emitting diodes are the lamps for the signal.

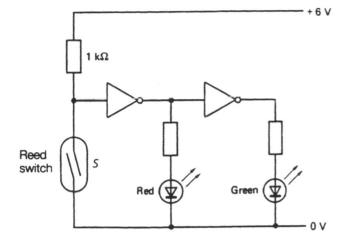

(i) When the switch S is open, which of the two lights are lit? Explain your answer. (4 marks)

(ii) The boy fixes a magnet underneath the train and fixes the red switch on the track so that the train passes over it. What happens to the signal lights as the train passes over the reed switch? (4 marks)

Solution

(i) Green. When S is open, the input to the first NOT gate is high. The output is low, so the red light is off. The output from the second gate is high, and the green light is on.

(ii) When the train passes over the red switch, it closes, and the input to the first gate goes low. Its output goes high, turning on the red light. The output from the second gate goes low and the green light goes off.

Example 11

In the circuit shown below the switch S is closed when the relay coil is energised.

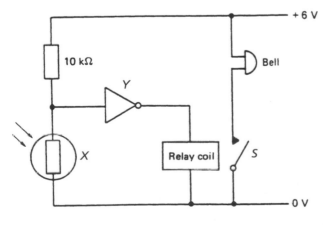

(i) Name the component labelled X. (1 mark)

(ii) Complete table 2, the truth table for the gate labelled Y.

Table 13.2

Input	Output
0	
1	

What is the gate called which is represented by the above truth table? (2 marks)

(iii) What happens to the input voltage of the logic gate when light shines on the component X? Explain your answer. (3 marks)

(iv) Explain what happens in the rest of the circuit as a result of light shining on the component X. (3 marks)

(v) What is the value of the input voltage at the logic gate if the resistance of the component X is 20 kΩ? (3 marks)

Solution

(i) Light-dependent resistor.

(ii)

Table 13.2

Input	Output
0	1
1	0

A NOT gate.

(iii) It goes low. The resistance of the LDR goes down and the voltage input to the gate goes low.

(iv) The output from the gate goes high and a current passes through the relay coil. This closes the switch S, allowing a current to flow through the bell, and the bell rings.

(v) The voltage across X is twice the voltage across the 10 kOhm resistor, so the voltage across the 10 kOhm resistor is 2 V. The input voltage to the gate is therefore 6 V − 2 V = 4 V.

Example 12

(a) Draw a truth table for the two-input NAND gate shown in Fig. A. (2 marks)

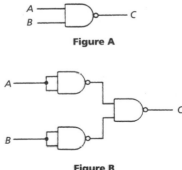

Figure A

Figure B

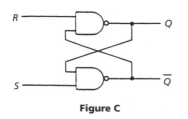

Figure C

(b) Three NAND gates are connected as shown in Fig. B. Draw the truth table for the arrangement. What is the resulting gate called? (3 marks)

(c) Two NAND gates are connected as shown in Fig. C. Complete the truth table shown below. The initial condition is shown on the first line of the sequence and you must make each change in the sequence indicated, moving one line at a time down the table. (4 marks)

Sequence		R	S	Q	$\bar{Q}$
	1	1	1	0	1
	2	0	1		
	3	1	1		
	4	1	0		
	5	1	1		

(d) The above logic is that of the burglar system shown in Fig. D. The alarm rings when it receives a logic 1 from the output of the NAND gate to which it is connected. A switch will close when a window in the house is opened and open again when the window is closed. Explain why
  (i) the alarm rings when one of the switches is closed, (3 marks)
  (ii) the alarm does not stop ringing if the burglar closes the window. (3 marks)

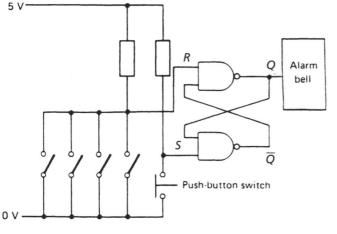

Figure D

(e) How can the alarm be switched off and reset? (2 marks)

(f) Why are two NAND gates connected as shown in Fig. B referred to as (i) a bistable, (ii) a flip-flop? (3 marks)

Solution

(a)

A	B	C
0	0	1
0	1	1
1	0	1
1	1	0

(b)

A	B	C
0	0	0
0	1	1
1	0	1
1	1	1

An OR gate
[The first two NAND gates which have their two inputs joined together are NOT gates and invert the input.]

(c)

	R	S	Q	$\bar{Q}$
1	1	1	0	1
2	0	1	1	0
3	1	1	1	0
4	1	0	0	1
5	1	1	0	1

(d) (i) When one of the switches is closed the input R goes low (logic 0). As shown in the table above, when R goes to logic 0, Q goes to logic 1 and the alarm will ring.
  (ii) Closing the window causes the logic of R to change from logic 0 to logic 1 and, as shown in the table above, the output Q remains at logic 1 and the alarm continues ringing. [The alarm is said to be latched.]

(e) Close all the windows, then push and release the push-button switch.

(f) It is called a bistable because it has two stable states. The outputs are either logic 0 and logic 1, or logic 1 and logic 0. It is called a flip-flop because it 'flips' from one state to the other when R goes from logic 1 to logic 0 and 'flops' back again when S goes from logic 1 to 0.

Example 13

You are required to design a circuit so that an alarm bell rings when a push-button switch is closed (i.e. is taken to logic 1), and at the same time it is also daylight or raining or both. In order to do this, you are provided with a light sensor and a water sensor. The output from the light sensor is high (logic 1) when it is daylight and the output from the water sensor is high (logic 1) when it is raining. The table below summarises the requirements.

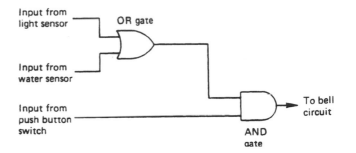

Push-button switch	Light sensor	Water sensor	Output
0	0	0	0
0	0	1	0
0	1	0	
0			
1			
1			
1			
1			

(a) Complete the table. (5 marks)
(b) The desired result may be achieved by using two 2-input logic gates. Draw a circuit you could use showing the two logic gates. The details of the bell circuit are not required. (5 marks)

Solution

(a)

Push-button switch	Light sensor	Water sensor	Output
0	0	0	0
0	0	1	0
0	1	0	0
0	1	1	0
1	0	0	0
1	0	1	1
1	1	0	1
1	1	1	1

Example 14

The circuit drawn below is that used in a tomato waterer. The contacts labelled A are buried in the soil.

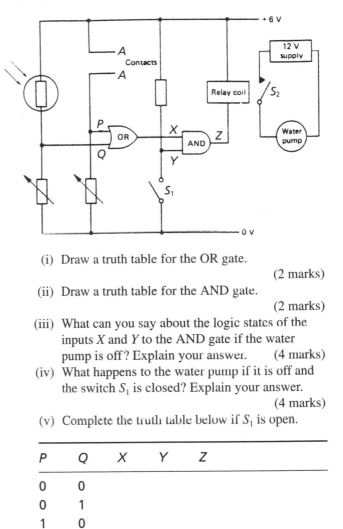

(i)   Draw a truth table for the OR gate. (2 marks)
(ii) Draw a truth table for the AND gate. (2 marks)
(iii) What can you say about the logic states of the inputs X and Y to the AND gate if the water pump is off? Explain your answer. (4 marks)
(iv) What happens to the water pump if it is off and the switch S_1 is closed? Explain your answer. (4 marks)
(v)   Complete the truth table below if S_1 is open.

P	Q	X	Y	Z
0	0			
0	1			
1	0			
1	1			

(vi) When the soil is very dry, it is dark and the switch S_1, is open, is the water pump on or off? Explain your answer. (4 marks)

Solution

(i)

P	Q	X
0	0	0
0	1	1
1	0	1
1	1	1

(ii)

X	Y	Z
0	0	0
0	1	0
1	0	0
1	1	1

(iii) They must both be high. If X and Y are high, then Z is high, so there is no voltage across the relay coil. No current passes through the coil and S_2 is open.

(iv) The pump starts. When S_1 is closed, the voltage input at Y is low and Z goes low. There is a voltage across the relay coil, a current passes and the switch S_2 closes, starting the pump.

(v)

P	Q	X	Y	Z
0	0	0	1	0
0	1	1	1	1
1	0	1	1	1
1	1	1	1	1

(vi) On. When the soil is dry and it is dark, inputs P and Q are both low. As the truth table above shows, in this situation Z is low. There is a voltage across the relay coil, a current passes through it, and the switch S_2 closes, turning the pump on.

Example 15

In the circuit shown below the lamp L is lit when the light falling on the LDR falls below a certain intensity.

(i) What do the letters 'LDR' stand for? (2 marks)

(ii) What do the letters 'NO' stand for? (2 marks)

(iii) The resistance of the LDR varies between 300 Ohm and 10 kOhm. What is the voltage at the point X when its resistance is

(a) 300 Ohm?

(b) 10 kOhm? (6 marks)

(iv) What change occurs in the current flowing from the collector to the emitter of the transistor as the

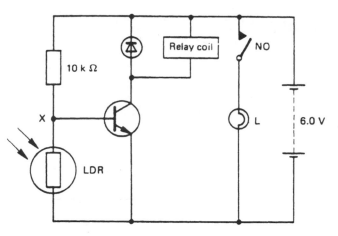

light falling on the LDR suddenly decreases in intensity so that the resistance of the LDR changes from 300 Ohm to 10 kOhm? (2 marks)

(v) Explain why this change in current switches the lamp on. (3 marks)

Solution

(i) Light-dependent resistor.

(ii) Normally open.

(ii) Let I be current through the 10 kOhm resistor and then

(a) $V = IR$

$$6\,V = I(10\,000 + 300)\,\text{Ohm}$$

$$\Rightarrow I = \frac{6}{10\,300}\,\text{A}$$

Voltage across LDR $= \left(\dfrac{6}{10\,300} \times 300\right)$

$$= 0.17\,\text{V}$$

$$Vx = 0.17\,\text{V}$$

(b) If the resistance of the LDR is equal to the resistance of the resistor then the voltage across each will be the same, namely 3.0 V. Hence $Vx = 3.0$ V.

(iv) The voltage at X changes from 0.18 V to 3 V and the transistor switches on. A current passes from the collector to the emitter.

(v) The current passing from the collector to the emitter flows through the relay. The switch closes and the lamp lights.

Example 16

Harriet has built the circuit shown below. She wants to use it as an amplifier.

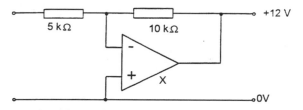

(a) (i) What is the name of component X? (1 mark)

(ii) Label the input and output of the amplifier.

(2 marks)

(iii) Use the rule $gain = -\dfrac{R_F}{R_I}$ to calculate the

gain of the amplifier. (2 marks)

(b) Harriet intends to apply a test signal to her amplifier. She looks at the test signal with an oscilloscope. The diagram shows what she sees on the oscilloscope screen.

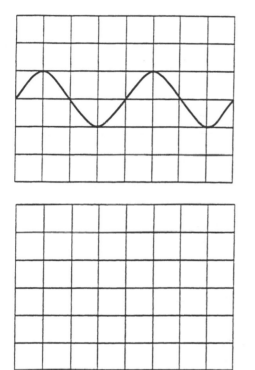

The timebase of the oscilloscope is sct at 2 ms/division. Calculate the frequency of the test signal. (3 marks)

(c) Harriet applies the test signal to her amplifier. She connects the oscilloscope to the output, without changing any of its settings. Draw on the diagram to show what she will see on the screen. (3 marks)

(MEG Salters', Jun 95, Higher Tier)

Solution

(a) (i) Amplifier

(ii)

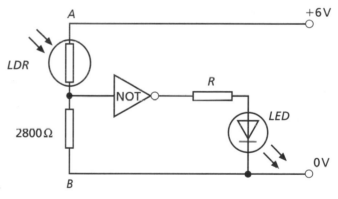

(iii) Gain $= -\dfrac{R_F}{R_I} = \dfrac{-10\ k\Omega}{5\ k\Omega} = -2$

(iv) All positive values are made negative and vice versa.

(b) 1 wavelength = 4 divisions
To travel 4 divisions takes 4×2 ms = 8 ms

Hence period = 0.008 s so frequency

$$= \frac{1}{period} = \frac{1}{0.008} = 125\ \text{Hz}$$

(c) The trace will be inverted and have twice the amplitude of the input signal. [This is because the gain is -2.]

Example 17

In this question you may find these equations useful:

$$current = \frac{p.d.}{resistance}$$

$$p.d. = current \times resistance$$

A photographic 'darkroom' must be kept dark while films are developed. The circuit shows a device which warns people not to enter the room when it is dark inside. The LDR is inside the darkroom. The LED is outside the room above the door.

(a) The light in the darkroom is on. The resistance of the LDR is 200 Ω.

(i) Calculate the total resistance between A and B. (1 mark)

(ii) Calculate the current in the LDR. (3 marks)

(iii) Calculate the p.d. across the 2800 Ω resistor.

(2 marks)

The input to the NOT gate is logic 1.

(iv) What is the logic output of the NOT gate?

(1 mark)

(v) Is the LED on? (1 mark)

(b) The darkroom light is now switched off.

(i) What happens to the resistance of the LDR?

(1 mark)

(ii) What happens to the LED? (1 mark)

(c) What is the purpose of resistor R? (2 marks)

Solution

(a) (i) Total resistance = (2800 + 200) = 3000 Ω

(ii) Current = $\dfrac{\text{p.d.}}{\text{resistance}}$

$$= \frac{6\text{ V}}{3000\text{ }\Omega} = 0.002\text{ A}$$

(iii) p.d. = current × resistance
= 0.002 A × 2800 Ω = 5.6 V

(iv) 0

(v) No.

(b) (i) It increases.

(ii) It comes on.

(iii) To limit the size of the current passing through the NOT gate and the LED since too much current can damage them.

Examination questions

(Numerical answers and hints on solutions will be found at the end of the chapter.)

Question 1

The diagram below shows a sprinkler system used in a large shop.

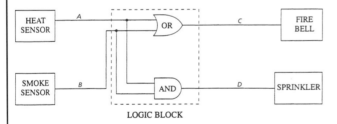

LOGIC BLOCK

The smoke sensor gives logic 1 when it detects smoke. The heat sensor gives logic 1 when it detects heat.

(a) Complete the truth table below for the logic block.

(2 marks)

A	B	C	D
0	0		
0	1		
1	0		
1	1		

(b) State what happens if smoke is detected but not heat. (1 mark)

(WJEC, Jun 95, Intermediate Tier)

Question 2

(a) The diagram below shows one way of protecting a valuable vase in a museum. The idea is that an alarm bell rings if a burglar lifts up the vase.

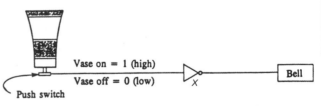

(i) Write down the name of the logic gate which is used at X. (1 mark)

(ii) Explain how this logic gate will cause the intended effect. (2 marks)

(iii) Suggest another output transducer which could be used instead of the alarm bell.

(1 mark)

(b) In order to improve the security system, another sensor is added which detects a light being switched on at night. The improved system is shown below.

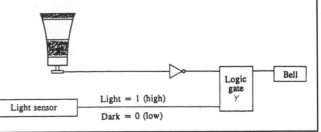

(i) Write down the name of the logic gate Y.
(1 mark)

(ii) Besides the two sensors already used, suggest one other sensor which could be used to detect the presence of a burglar. (1 mark)
(SEG, Spec, Intermediate Tier)

Question 3

You have been asked to design a circuit which will automatically switch on a car seat heater only if the weather is cold and the driver is sitting in the car. The components available are

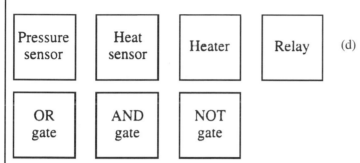

The pressure sensor gives logic 1 (high) when the pressure is high.
The heat sensor gives logic 1 (high) when the temperature is high.

In the space below draw the components you would use, correctly joined together.

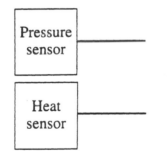

(WJEC, Jun 95, Intermediate Tier)

Question 4

When a charged capacitor is connected across a resistor the voltage across it decreases. The diagram shows a way of charging a capacitor from a battery. The capacitor can then be discharged through a resistor.

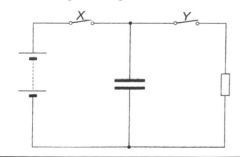

(a) Explain how switches X and Y should be used to charge the capacitor up to the battery voltage.
(1 mark)

(b) Explain how switches X and Y should be used to discharge the capacitor through the resistor.
(1 mark)

(c) The time taken for the capacitor to discharge to half its initial value depends on the values of the capacitor and resistor. How would this time be affected in the circuit shown above if:
(i) a capacitor with a bigger capacitance than that shown was used (together with the same resistor)?
(ii) a resistor with a bigger resistance than that shown used (together with the same capacitor)? (2 marks)

(d) The next diagram shows a transistor switch connected to a lamp. The input sensor is the charging/discharging capacitor circuit referred to above. The output device is a lamp.

The transistor switch comes on when the transistor input is high.

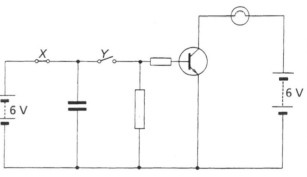

In the circuit shown, switch X is closed and switch Y is open.
Switch X is now opened.
(i) Describe what happens to the lamp when switch Y is closed and kept closed.
(ii) Suggest a use for such a circuit. (3 marks)
(NEAB, Jun 95, Intermediate Tier)

Question 5

(a) The diagram shows two resistors connected in series. The p.d. between A and C is 9 V.

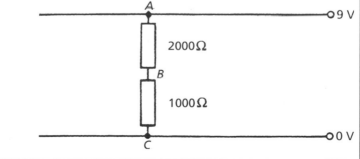

(i) Calculate
1. the total resistance between *A* and *C*
2. the current flowing in the resistors
3. the p.d. across the 1000 Ω resistor.

(ii) Which one of the following is shown in the diagram?
Tick the box next to the correct answer.
a rheostat ☐
a potential divider ☐
a variable resistor ☐

(7 marks)

(b) A transistor can be used to switch on a lamp. The circuit is shown below.

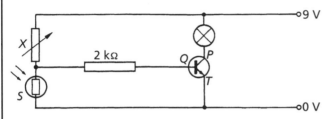

(i) Name the following parts in the diagram.
1. *S* is ....
2. On the transistor,
   P is the ....
   Q is the ....
   T is the ....

(ii) Component *S* has a large resistance when it is dark and its resistance decreases as it gets lighter.
In the light,
1. is the p.d. across *S* high or low?
2. is the transistor switched ON or OFF?
3. is the lamp switched ON or OFF?

(iii) Suggest a practical use for this circuit.

(iv) What is the purpose of the component *X*?

(12 marks)
(NEAB, Jun 94, Intermediate Tier)

Question 6

(a) The alarm circuit shown below includes a bistable latch. The purpose of the circuit is to sound a buzzer if a person enters a room.

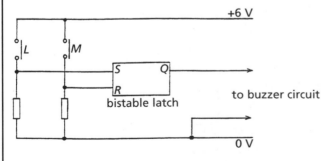

(i) *L* is a switch which closes when the door opens.
What is the purpose of switch *M*? (2 marks)

(ii) The maker of the circuit tests it by closing switches *L* and *M* for 2 seconds at a time in the sequence shown by these timing diagrams.

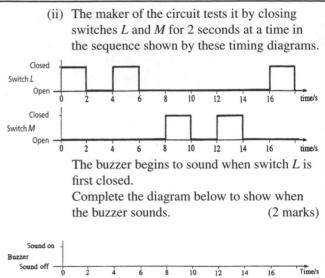

The buzzer begins to sound when switch *L* is first closed.
Complete the diagram below to show when the buzzer sounds. (2 marks)

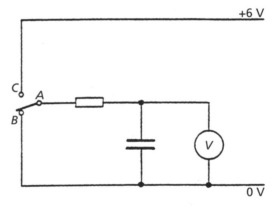

(b) A switch, resistor and capacitor are connected in series to form part of the circuit shown below.

(i) The switch contact joins *A* to *B*. The voltmeter reads zero.
What does this tell us about the capacitor?
(1 mark)

(ii) The switch contact is moved so that it joins *A* to *C*. The voltmeter reading rises slowly so that after about 30 s the reading has reached 3 V and is still rising.
Explain why this happens. (3 marks)

(c) The resistor and capacitor are added to the latch circuit of part (a) so that the output, *Q*, of the latch is connected to the resistor. The connections to the buzzer circuit are now across the capacitor.
The circuit is as shown below.

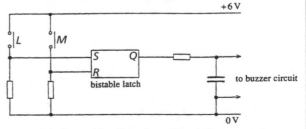

Explain how this alteration affects the room alarm system. (2 marks)
(MEG, Jun 95, Higher Tier)

Answers to examination questions

1. (a)

A	B	C	D
0	0	0	0
0	1	1	0
1	0	1	0
1	1	1	1

   (b) The fire bell sounds but the sprinkler will not work.

2. (a) (i) NOT gate.
      (ii) When the vase is removed the switch will give a signal of 0 which is turned into a 1 by the NOT gate X which starts the bell ringing.
      (iii) A flashing light.
   (b) (i) AND gate.
      (ii) A heat (infra-red) sensor to sense body heat.

3. *See the illustration below.*
   [The heat sensor will produce a 1 when the temperature is high. To produce an output of 1 when the temperature is low, the signal needs to be passed through a NOT gate.]

4. (a) Switch X should be closed. [It does not matter whether switch Y is open or closed.]
   (b) Switch X is opened and switch Y is closed.
   (c) (i) Time would increase.
      (ii) Time would increase.
   (d) (i) The bulb lights for and stays on for a short period and then goes off. [You would be wasting your time here if you went into how the circuit works because this is not asked for.]
      (ii) Lights in a corridor of a hotel. [They allow the guest to reach their room before the light is switched off.]

5. (a) (i) 1. $R = (2000 + 1000) = 3000\ \Omega$

2. $I = \dfrac{V}{R} = \dfrac{9\ \text{V}}{3000\ \Omega} = 0.003$ A (3 mA)

3. $V = \dfrac{9 \times 1000}{(2000 + 1000)}$ [This is the potential divider equation.]
     $= 3$ V

    (ii) The only one which should be ticked is a potential divider.

   (b) (i) 1. *S* is a light dependent resistor (LDR)
        2. *P* is the collector
          Q is the base
          T is the emitter.
    (ii) In the light
      1. low
      2. OFF [If the resistor has a low resistance then the p.d. across it will be low.]
      3. OFF
    (iii) A night light in a corridor.
    (iv) Provides adjustment so you can determine the light level which switches the light on/off.

6. (a) (i) [*S* is the SET and *R* is the RESET of the bistable. To change the output of the bistable from OFF to ON the SET input has to be changed. The output now stays the same even if the SET input is changed back again. The only way to change the bistable from ON to OFF is to change the RESET input. The output will stay OFF even if the RESET is changed back again.]
     Switch *M* is the RESET switch which is used to switch the alarm off since closing the door will not switch the alarm off.
    (ii) *See the illustration overleaf.*
    [Once switch *L* is closed, the buzzer will sound no matter how many times *L* is subsequently open or closed. The buzzer therefore stays on until switch *M*, the

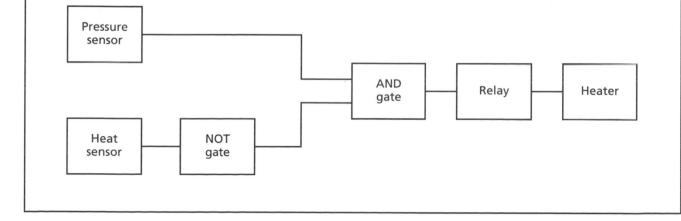

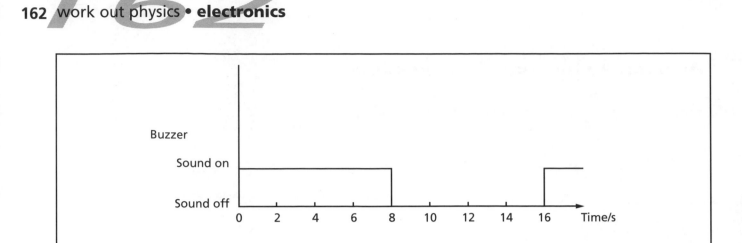

RESET is closed. This turns the buzzer off and the buzzer will stay off despite changes in switch M. The only way to turn it on again is to close switch L so the buzzer will be on at $T = 16$ s.]

(b) (i) The capacitor is uncharged.
    (ii) Current travels through the resistor and charge starts to build up on the capacitor. Because the resistor has a high resistance, the charging process takes some time. The charging process will stop and the voltmeter will read 6 V when the capacitor is fully charged.

(c) The buzzer will start off with a low volume and then get louder as the capacitor becomes fully charged. When the alarm is reset the buzzer will get softer until it stops.

14

Magnetism and Electromagnetism

WJEC & NEAB	ULEAC Syll A	ULEAC Syll B	MEG	Topic	MEG Salters'	MEG Nuffield	SEG	NICCEA
✓	✓	✓	✓	**Magnetism**	✓	✓	✓	✓
✓	✓	✓	✓	**Electromagnetism**	✓	✓	✓	✓
✓	✓	✓	✓	**The electric motor and dynamo**	✓	✓	✓	✓
✓	✓	✓	✓	**Faraday's law of electromagnetic induction**	✓	✓	✓	✓
✓	✓	✓	✓	**Transformers**	✓		✓	✓
✓			✓	**Left-hand and right-hand rules**	✓		✓	✓

1 Magnetism

Certain substances, including iron, steel, cobalt and nickel, are magnetic.

The region of space around a magnet where its magnetic influence may be detected is known as a magnetic field, and contains something we call magnetic flux. The flux patterns may be shown by sprinkling iron filings around a magnet and gently tapping the surface, or by using plotting compasses. The flux patterns resulting from a bar magnet and a U-shaped magnet are shown in the diagram below.

Magnets attract unmagnetised pieces of iron and steel. To discover whether a bar of iron or steel is magnetised, bring each end up to the N-pole of a suspended magnet. If one end repels the N-pole, the bar is magnetised (like magnetic poles repel each other). If both ends attract the N-pole, the bar is not magnetised.

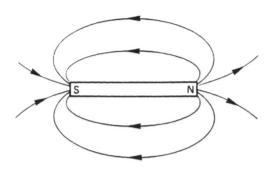

(a)

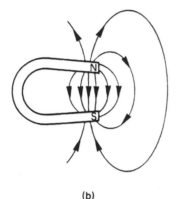

(b)

The lines of flux (a) due to a bar magnet and (b) due to a U-shaped magnet.

A bar of magnetic material may be magnetised by putting it in a solenoid carrying a current. It may be demagnetised by withdrawing it slowly from a solenoid in which an alternating current is passing. Recording heads in a tape recorder use solenoids to magnetise the tape.

Materials such as steel and alcomax (a steel-like alloy) are difficult to magnetise and are said to be hard. Iron and mumetal (a nickel alloy) are relatively easy to magnetise, but they do not retain their magnetism, and are said to be soft.

2 Electromagnetism

When a conductor carries a current it produces a magnetic field. The flux patterns resulting from a current passing through a straight wire and a solenoid are shown below.

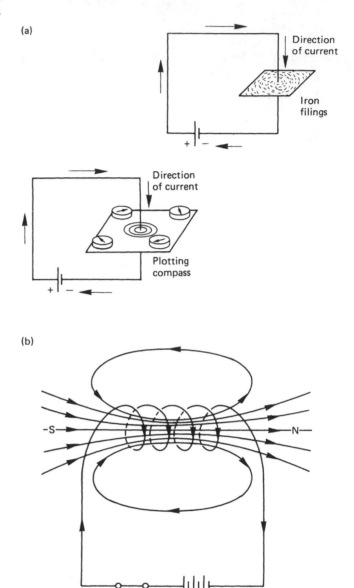

(a)

(b)

The flux patterns caused by (a) a current passing through a straight wire and (b) a current passing through a solenoid.

The direction of the field may be found by use of the 'screw rule', which states that if a right-handed screw is turned so that it moves forward in the same direction as the current, its direction of rotation gives the direction of the magnetic field.

The effect is used in electromagnets, which are usually solenoids wound round iron or mumetal. Iron and mumetal are soft magnetic materials that quickly lose their magnetism when the current in the solenoid is switched off.

Electromagnetic relays use the magnetic effect of a current passing through a solenoid to operate a switch. The 'pull on' current is the value of the current in the solenoid needed to close the switch, and the 'switch off' current is the value of the current when the switch opens.

3 The electric motor and the dynamo

(See examples 8 and 11.) Electric motors and dynamos basically consist of a coil which can rotate in a magnetic field. In the motor a current passes through the coil, and the resulting forces on the coil cause it to rotate. A split-ring commutator reverses the direction of the current every half-revolution, thus ensuring continuous rotation. The turning effect may be increased by increasing the current, increasing the number of turns on the coil or increasing the strength of the magnetic field. In the dynamo the coil is rotated, and as the wires cut the lines of flux, an e.m.f. is induced across the ends of the coil. In an a.c. dynamo the ends of the coil are connected to slip rings. If a commutator replaces the slip rings, the output is d.c.

4 Faraday's law of electromagnetic induction

Whenever an e.m.f. is induced in a conductor due to the relative motion of the conductor and a magnetic field, the size of the induced e.m.f. is proportional to the speed of the relative motion.

Alternatively, the law may be stated:
Whenever there is a change of magnetic flux linked with a circuit an e.m.f. is induced. The e.m.f. is proportional to the rate of change of flux linked with the circuit.

If the circuit is complete, the induced e.m.f. produces a current. This effect is used in dynamos, transformers and playback heads of tape recorders.

5 Transformers

A primary and secondary coil are wound on a continuous core made of a soft magnetic alloy (cores of modern alloys have largely replaced iron cores). The alternating current passing through the primary causes a changing flux in the secondary and hence an induced e.m.f. across the secondary coil (see example 9). Most modern transformers are made so that there is negligible magnetic flux loss and

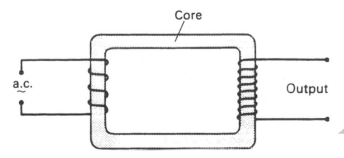

A step-up transformer.

$$\frac{\text{Voltage across secondary}}{\text{Voltage across primary}} =$$

$$\frac{\text{number of turns on secondary}}{\text{number of turns on primary}}$$

$$\text{Efficiency} = \frac{\text{power out}}{\text{power in}} = \frac{(VI)_{\text{secondary}}}{(VI)_{\text{primary}}}$$

Transformers have high efficiencies, and in many cases a good approximation is

$$(VI)_{\text{secondary}} = (VI)_{\text{primary}}$$

Transformers will not work with direct current since there is no continually changing flux.

As in all appliances, the switch and fuse must be in the live wire.

Electricity is transmitted across the British countryside on the National Grid. High voltages are used, as low-voltage power lines are wasteful and inefficient, because of the large currents needed. The energy lost as heat in the wires increases rapidly as the current increases (energy lost as heat $= I^2 Rt$: see page 135, section 6). Step-up transformers are used at the power station end of the transmission line and step-down transformers at the consumer end. Since

$$\text{Power} = \text{Voltage} \times \text{Current}$$

a given power can be transmitted at a lower current if a high voltage is used.

6 Left-hand and right-hand rules

Left-hand rule for the motor effect

Hold the first finger, second finger and thumb of the left hand mutually at right angles, so that the First finger points in the direction of the magnetic Field, the seCond finger in the direction of the Current, then the THuMb points in the direction of the THrust or Motion.

Right-hand rule for the dynamo effect

Hold the first finger, second finger and thumb of the right hand mutually at right angles, so that the First finger points in the direction of the Field, the thumb in the direction of the Motion, then the second finger points in the direction of the induced Current.

Worked examples
Example 1

Jamie holds a wire between two strong magnets as shown in the diagram. The current in the wire is 2 A.
There is an upward force on the wire.

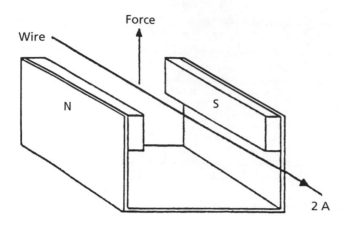

Force

Wire

N

S

2 A

When Jamie increases the current to 4 A the upward force will

A   increase
B   decrease
C   stay the same
D   reverse.

Solution

[The force on the wire depends on the strength of the magnetic field and on the magnitude of the current. The greater the current, the greater the force. The direction of the force will change if the direction of the current changes.]

Answer **A**

Example 2

Which of these circuits (**A**, **B**, **C** or **D**) shows the safest way to connect a transformer to the mains?

Solution

[For safety a switch and fuse should be in the live wire. The core should be earthed.]

Answer **C**

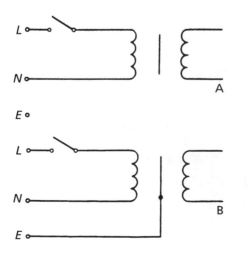

Example 3

The diagram shows a simple electric motor.

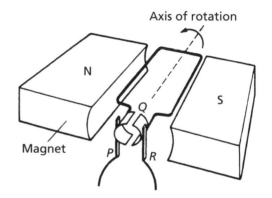

Axis of rotation

N

S

Magnet

P   Q   R

The combined function of the parts labelled *P*, *Q* and *R* is to

A   reverse the current in the coil every half revolution
B   reverse the current in the coil every quarter revolution
C   reverse the polarity of the field produced by the magnet
D   control the magnitude of the current
E   control the magnitude of the magnetic field

Solution

[The commutator and brushes reverse the direction of the current every half revolution. See section 3 and example 8.]

Answer **A**

Example 4

Which of the following diagrams shows a transformer with the **correct** number of turns on the coils to change 240 V a.c. into 12 V a.c.?

Solution

[The voltage changes from 240 V to 12 V. The voltage

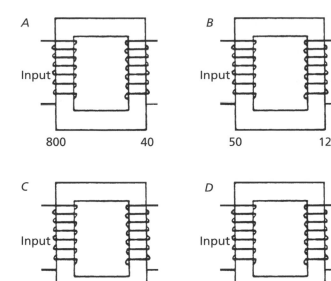

A	B
Input	Input
800 40	50 12

C	D
Input	Input
240 100	500 50

drops by a factor of 20. Therefore the turns must drop by a factor of 20. Or, as the equation in section 5 states, the ratio of the voltages is the ratio of the turns.

$$\frac{240\ \text{V}}{12\ \text{V}} = \frac{800\ \text{turns}}{40\ \text{turns}}$$

This is only true for A.]

Answer **A**

Example 5

The diagram shows the transmission system between a power station and a home.

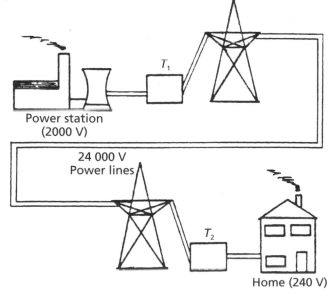

Power station (2000 V)

24 000 V Power lines

Home (240 V)

Which of the following correctly describes the transformers T_1 and T_2?

	T_1	T_2
A	step-up	step-up
B	step-up	step-down
C	step-down	step-up
D	step-down	step-down

Solution

[The power is transmitted at a high voltage as this reduces the energy loss in transmission. The voltage is stepped up to transmit and stepped down for the consumer. T_1 is a voltage step-up transformer and T_2 is a voltage step-down transformer. See section 5.]

Answer **B**

Example 6

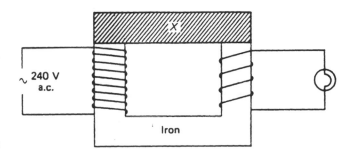

The circuit shown in the diagram was set up in order to demonstrate a step-down transformer. The lamp glowed dimly. The lamp would glow more brightly if

A the number of turns on the primary coil were reduced

B the iron were replaced by copper

C the shaded section of iron, X, were removed

D the number of turns on the secondary coil were reduced

Solution

[The voltage across the secondary coil depends on the turns ratio (section 5). Reducing the number of turns on the primary coil increases the turns ratio and, hence, increases the voltage across the secondary coil. Reducing the number of turns on the secondary coil reduces the turns ratio and, hence, reduces the voltage across the secondary coil. Copper is non-magnetic and so is not suitable for the core. The flux through the secondary coil will be reduced if the magnetic circuit is not complete, so removing X will decrease the voltage across the secondary coil.]

Answer **A**

Example 7

(a) (i) Explain why iron rather than hard steel is used in electromagnets.

(ii) State two ways of increasing the strength of an electromagnet. (4 marks)

(b) Two pieces of steel, A and B, are lying on a bench. The N-pole of a magnet when brought up to one end of A attracts it, and when brought up to one end of B repels it. What can you deduce about the state of magnetism of A and B? (3 marks)

Solution

(a)   (i)   Iron is more easily magnetised than hard steel and loses its magnetism more easily. If the magnetising current were switched off, the iron would lose its magnetism but the steel would retain some magnetism.

(ii)   Increase the number of turns. Increase the current.

(b)   *B* is magnetised. *A* may be magnetised. The end of *B* to which the magnet was brought up is a N-pole. [Repulsion is a test for magnetism. If the N-pole of the magnet were to be brought up to the S-pole of *B*, then attraction would occur. If *B* were not magnetised, it would still be attracted by the magnet, because the N-pole of the magnet would induce an S-pole on *B*. The N-pole and the S-pole would attract each other. Attraction is not a test for magnetism, so *A* may or may not be magnetised.]

Example 8

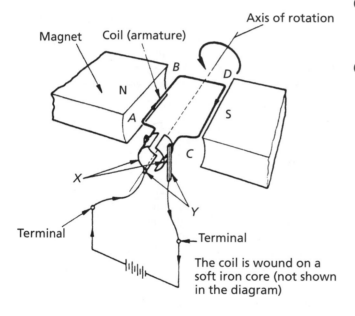

The diagram shows a d.c. electric motor.

(a)   What is the part *X* called?   (1 mark)

(b)   What is the part *Y* called?   (1 mark)

(c)   Explain why the coil of the motor rotates continuously when the motor is connected to a d.c. supply.   (7 marks)

(d)   Suggest three ways in which the motor could be made to go faster.   (3 marks)

(e)   What would happen to the rotation if, when the direction of the current in the coil were reversed, at the same time the direction of the field were reversed?   (2 marks)

(f)   If the motor is 3 W, 6 V and its efficiency is 20%, what is
        (i)   the current it takes?
        (ii) the energy it can supply in 10 s?   (6 marks)

Solution

(a)   Split-ring commutator.

(b)   Brushes.

(c)   The wires *AB* and *CD* are current-carrying conductors and they are in a magnetic field which is perpendicular to the wire. Each side of the coil has a force acting on it. The force on *AB* is downwards and the force on *CD* is upwards. This couple [two equal and opposite parallel forces] would cause the coil to rotate through 90° from the position shown. As the coil passes through the vertical position, the two sections of the split-ring commutator are connected, via the brushes, to the opposite terminals of the battery. The current in the coil is reversed and the forces acting on the sides of the coil are reversed in direction. The forces rotate the coil through 180° until it is next in the vertical position, when the current and the forces are again reversed. The coil thus continues to rotate.

(d)   Increase the current. Increase the strength of the magnetic field. Increase the number of turns on the armature coil.

(e)   If both the current and the field were reversed at the same time, the forces would continue to act in the same direction and the coil would not rotate continuously.

(f)   (i)   Power (watts) = potential difference (volts) × current (amperes) [page 135, section 6]

3 W = 6 V × current

$$\Rightarrow \text{Current} = \frac{3\text{ W}}{6\text{ V}} = 0.5\text{ A}$$

(ii)   The 3 W is the power taken in. We need to know the power got out, so we use the equation on page 135, section 6.

$$\text{Efficiency} = \frac{\text{power out}}{\text{power in}}$$

[Remember to put the efficiency as a fraction or decimal, NOT as a percentage.]

$$0.2 = \frac{\text{power out}}{3\text{ W}}$$

Power out = 0.2 × 3 W = 0.6 W = 0.6 J/s
So, in 10 s the energy supplied   = 0.6 J/s
                                        × 10 s
                                        = 6 J.

Example 9

The diagram represents a transformer with a primary coil of 400 turns and a secondary coil of 200 turns.

(a)   If the primary coil is connected to the 240 V a.c. mains what will be the secondary voltage?   (1 mark)

(b)   Explain carefully how the transformer works.
                                                        (4 marks)

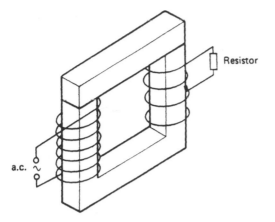

(c) Calculate the efficiency of the transformer if the primary current is 3 A and the secondary current 5 A.
(4 marks)

(d) Give reasons why you would expect this efficiency to be less than 100%. (3 marks)

(e)

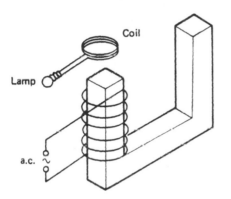

The secondary coil is removed and a small coil connected to a low voltage lamp is placed as shown. Explain the following observations:
(i) the lamp lights; (2 marks)
(ii) if the coil is moved upwards, the lamp gets dimmer; (2 marks)
(iii) if an iron rod is now placed through the coil, the lamp brightens again; (2 marks)
(iv) the lamp will not light if a d.c. supply is used instead of an a.c. one. (2 marks)

Solution

(a) 120 V [This assumes that there are no flux losses.]
(b) When an a.c. is connected to the primary coil there is a constantly changing magnetic flux in the iron core. This continuously changing flux passes through the secondary coil, and hence there is an induced e.m.f. in the secondary coil. Since the same e.m.f. is induced in each turn of the secondary coil and all the turns are in series, halving the number of turns of the secondary coil halves the induced e.m.f.
(c) Power input $= (VI)_{\text{primary}} = 240 \text{ V} \times 3 \text{ A}$
$= 720 \text{ W}$ [see page 135, section 6]
Power output $= (VI)_{\text{secondary}} = 120 \text{ V} \times 5 \text{ A}$
$= 600 \text{ W}$

$$\text{Efficiency} = \frac{\text{power out}}{\text{power in}} \quad [\text{see section 5}]$$
$$= \frac{600 \text{ W}}{720 \text{ W}} = 0.83 = 83\%$$

(d) The efficiency is less than 100% because of
(i) copper losses, that is the heat produced in the wires
(ii) eddy currents flowing in the iron core produce heat
(iii) magnetic leakage; not all the flux in the iron on the primary side reaches the secondary side
(iv) the work done in continually magnetising and demagnetising the iron core (this is called the hysteresis loss).

(e) (i) Some of the magnetic flux around the primary passes through the coil. This changing flux means that an e.m.f. is induced in the coil. This e.m.f. produces a current in the coil which lights the lamp.
(ii) When the coil is moved further away from the primary, the flux passing through it decreases. The rate of change of flux is less and the magnitude of the induced e.m.f. is reduced.
(iii) The iron rod will become magnetised and will increase the flux passing through the coil. The rate of change of flux is increased and hence the e.m.f. is increased.
(iv) When a d.c. supply is used the flux through the coil is constant. If there is no change of flux there is no induced e.m.f.

Example 10

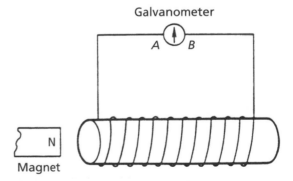

(a) The diagram shows a solenoid connected to a galvanometer. Explain why
(i) if the magnet is held stationary at the end of the coil, there is no deflection of the galvanometer pointer
(ii) if the magnet is moved towards the solenoid there is a deflection of the galvanometer pointer
(iii) the faster the magnet moves towards the solenoid the greater is the deflection of the galvanometer pointer. (6 marks)

(b) A transformer has 400 turns in the primary winding and 10 turns in the secondary winding. The primary e.m.f. is 250 V and the primary current is 2.0 A.

Calculate
   (i) the secondary voltage, and
   (ii) the secondary current, assuming 100% efficiency. (6 marks)
Transformers are usually designed so that their efficiency is as close to 100% as possible. Why is this?
Describe *two* features in transformer design which help to achieve high efficiency. (4 marks)

Solution

(a)　(i) If the magnet is stationary the flux through the coil is not changing. An e.m.f. will only be induced in the coil if the flux through it is changing.

(ii) When the magnet is moved towards the solenoid the flux through the coil is changing and hence there is an induced e.m.f.

(iii) The magnitude of the induced e.m.f. depends on the rate of change of flux. If the magnet is moved faster the rate of change of flux is greater and hence the induced e.m.f. increases and the resulting current increases.

(b)　(i)
$$\frac{\text{Voltage across secondary}}{\text{Voltage across primary}}$$
$$= \frac{\text{number of turns on secondary}}{\text{number of turns on primary}}$$

[see section 5]

$$\frac{\text{Voltage across secondary}}{250 \text{ V}} = \frac{10 \text{ turns}}{400 \text{ turns}}$$

Voltage across secondary 6.25 V

(ii) $(VI)_{\text{primary}} = (VI)_{\text{secondary}}$ [see section 5]
250 V × 2.0 A = 6.25 V × $I_{\text{secondary}}$
$I_{\text{secondary}} = 80$ A
The greater the losses the greater is the cost of running the transformer and the hotter the transformer becomes when being used. Large transformers have to be cooled and this can be expensive. The core is laminated, that is it is made up of strips insulated from each other. This reduces the eddy currents that flow in the core because of the changing magnetic flux. The coils are wound using low resistance material so that the heating effect in the coils is reduced to a minimum.

Example 11

The diagram shows a simple form of a.c. dynamo.
(a)　(i) What are the names of the parts labelled *A* and *B*?

(ii) What would be the effect of doubling the number of turns on the coil if the speed of rotation remained unchanged?
(iii) Which of the output terminals is positive if the coil is rotating in the direction shown in the diagram (anticlockwise)? (8 marks)

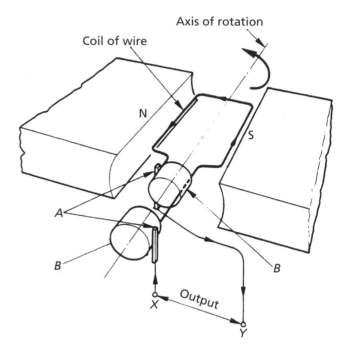

(b) The rotating dynamo coil has its ends connected to an oscilloscope which measures the voltage across the coil. As the coil passes through the position shown, the oscilloscope reads 12 mV. What will the oscilloscope read when the coil is in a vertical plane? Explain your answer. (3 marks)

(c) Sketch a graph showing how the p.d. across the ends of the rotating coil varies with time for an a.c. dynamo. On the same sheet of paper, vertically below the first graph and using the same time scale, sketch graphs to show the effect of (i) doubling the speed of rotation and at the same time keeping the field and the number of turns constant; (ii) doubling the number of turns on the coil and at the same time doubling the speed of rotation of the coil, keeping the field constant. (9 marks)

Solution

(a)　(i) The brushes are labelled *A* and the slip rings labelled *B*.
(ii) The induced e.m.f. would double.
(iii) *Y* is the positive terminal.

(b) 0 V. The p.d. across the ends of the coil is zero when the coil is in a vertical plane, because in this position the sides of the coil are moving parallel to the lines of flux and the flux through the coil is not changing; hence, there is no induced e.m.f.

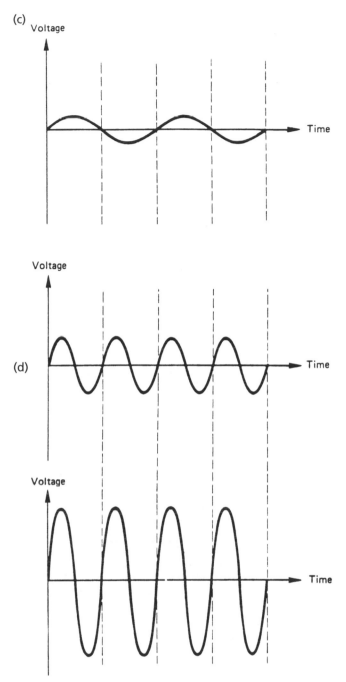

(c)

(d)

[In (ii) the voltage goes up by a factor of 4 because doubling the number of turns doubles the voltage, and doubling the speed of rotation also doubles the voltage. Doubling the speed of rotation also *doubles the frequency*.]

Example 12

(a)

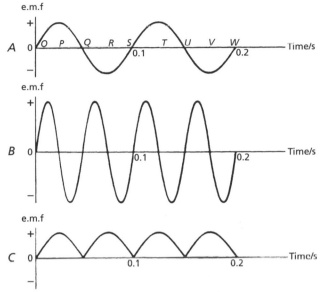

Graph *A* above shows how the e.m.f. produced by a simple dynamo varies with time.

Graphs *B* and *C* show how the e.m.f. produced by the same dynamo varies with time after certain alterations and modifications have been made.

  (i) How many revolutions has the coil of the dynamo made in the time interval *OT* on graph *A*?   (1 mark)

 (ii) What is the frequency of the alternating e.m.f. as shown by graph *A*?   (2 marks)

(iii) Which letters on graph *A* correspond to the plane of the coil of the dynamo being parallel to the magnetic field?   (2 marks)

(iv) Explain why the e.m.f. at *Q* is zero.   (3 marks)

 (v) What alteration has been made for the dynamo to produce the e.m.f. represented by graph *B*?   (2 marks)

(vi) What modification has been made to the dynamo for it to produce the e.m.f. represented by graph *C*? Illustrate your answer with sketches showing the original and the modified arrangements.   (4 marks)

(b) A dynamo is driven by a 5 kg mass which falls at a steady speed of 0.8 m/s. The current produced is supplied to a 12 W lamp which glows with normal brightness.

Calculate the efficiency of this arrangement.

  (6 marks)

Solution

(a)   (i) 1.25.

      (ii) 1 cycle takes 0.1 s. Frequency = 10 Hz.

    (iii) *P*; *R*; *T*; *V*. The e.m.f. is a maximum because

the wire on the sides of the coil is cutting the field at the greatest rate.

(iv) At Q the plane of the coil is perpendicular to the magnetic field. The sides of the coil are moving parallel to the field, so they are not cutting the lines of flux and the rate of change of flux is zero.

(v) The speed of rotation has been doubled.

(vi) Slip rings have been replaced by split rings (see diagram).

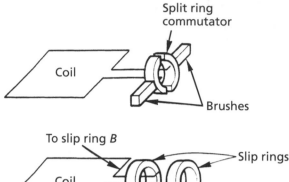

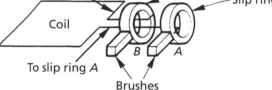

(b) Work done in 1 s by mass falling = force × distance travelled in 1 s = 50 N × 0.8 m/s = 40 W [The force on the 5 kg mass is 50 N.]
Work got out = 12 W

$$\text{Efficiency} = \frac{\text{work out}}{\text{work in}} = \frac{12}{40} = 0.3 \text{ or } 30\%$$

Example 13

(i) The diagram below shows a moving coil loudspeaker. Make a sketch of the magnet and show on your sketch the nature and the position of the magnetic poles.

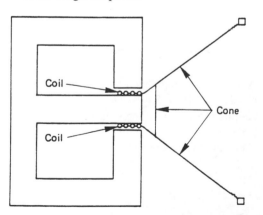

(ii) Explain what would be heard if a 1 V, 50 Hz supply were connected across the terminals of the loudspeaker. What difference would be heard if the supply were changed to a 2 V, 100 Hz one?
(8 marks)

Solution

(i)

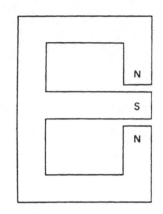

(ii) The wire of the coil is in a magnetic field. When a current flows in the wire of the coil a force acts on the wire. This force is perpendicular to both the field and the current. When an alternating current is flowing in the coil, the coil will move to the left and then to the right each time the current changes direction. The cone will move in and out 50 times every second and a note of 50 Hz will be heard.

If the supply were changed to 2 V, 100 Hz, the amplitude of the vibration would be greater, so a louder note would be heard. The cone would complete 100 cycles every second and a note of 100 Hz would be heard.

Example 14

(a) Explain why on the National Grid system
  (i) very high voltages are used
  (ii) alternating current is used.   (5 marks)

(b) A town receives electricity via the National Grid system at 100 kV at a rate of 40 MW. The cable connecting the town to the power station has a total resistance of 4 Ohm. What is
  (i) the current passing through the cable?
  (ii) the power loss as a result of heating in the cable?   (5 marks)

Solution

(a) (i) A certain amount of energy may be transmitted by using a high voltage and a low current, or a low voltage and a high current. The former is chosen because the heating effect depends on the square of the current and the loss of heat energy is reduced markedly by using a high voltage. Low current also means that the wires do not have to be so thick and the cost of supporting the wires is reduced.

  (ii) Using alternating current means that transformers can be used to step up the voltage before transmission and also they reduce it again at the end of the power line.

(b) (i) Power = (p.d.) × (current) [see page 135, section 6]
$$40 \times 10^6 \text{ W} = 100 \times 10^3 \text{ V} \times \text{current}$$
current = 400 A

(ii) Power loss = $I_2 R$ watt [see page 135, section 6]
$- (400^2 \times 4) \text{ W} = 6.4 \times 10^5 \text{ W}$

Example 15

Calculators may be powered by the mains using an adapter or by dry batteries.

(i) Why is an adapter necessary if the mains is used?
(2 marks)

(ii) Outline the principle by which the adapter works (no details of the electrical circuit or a diagram are necessary).
(4 marks)

Solution

(i) Calculators run off about 6 V and the mains is 240 V. The adapter reduces the mains voltage before supplying the calculator.

(ii) The adapter has a transformer which consists of a primary coil and a secondary coil wound on a soft magnetic alloy. The 240 V is connected across the primary coil. This alternating voltage causes a changing flux in the soft magnetic alloy, and this changing flux passes through the secondary coil. The changing flux in the secondary coil results in e.m.f. being induced in it. The secondary coil has only one-fortieth of the turns the primary coil has on it and hence the voltage is reduced by one-fortieth (if 6 V is required). A rectifier converts the a.c. voltage to a d.c. voltage.

Examination questions

(Numerical answers and hints on solutions will be found at the end of the chapter.)

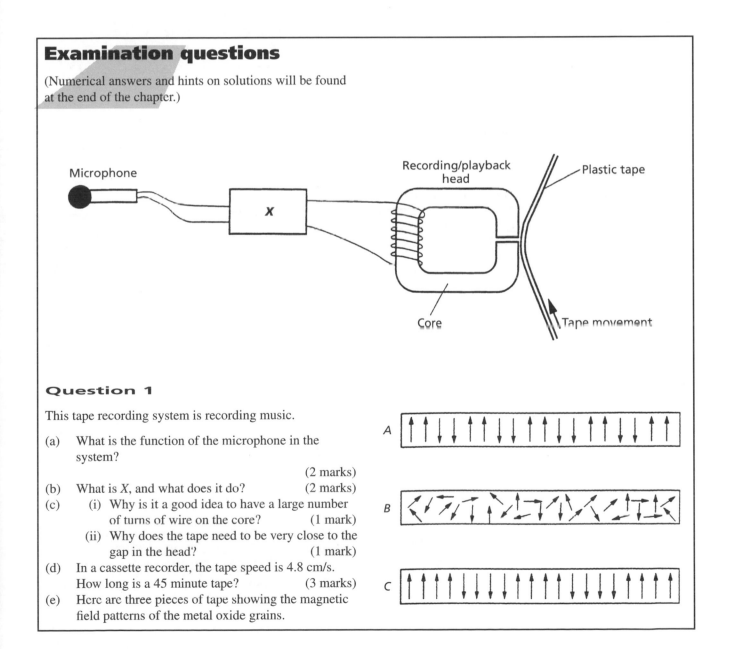

Question 1

This tape recording system is recording music.

(a) What is the function of the microphone in the system?
(2 marks)

(b) What is X, and what does it do? (2 marks)

(c) (i) Why is it a good idea to have a large number of turns of wire on the core? (1 mark)

(ii) Why does the tape need to be very close to the gap in the head? (1 mark)

(d) In a cassette recorder, the tape speed is 4.8 cm/s. How long is a 45 minute tape? (3 marks)

(e) Here are three pieces of tape showing the magnetic field patterns of the metal oxide grains.

(i) One of the tapes is unrecorded. Explain which one it is. (1 mark)

(ii) One of the tapes is recorded with a high frequency sound. Explain which one it is.
(MEG Nuffield, Jun 95, Intermediate Tier, Q11)

Question 2

Susan has bought a large relay for use in a project. It is shown in the diagram. She would like to know what it is made of and how it works.

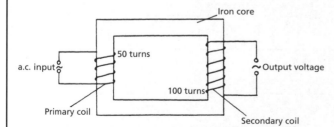

(a) (i) Draw lines on the diagram to show the magnetic field around the coil when there is a current in it. (2 marks)

(ii) Susan has been told that the coil is wound around a soft iron core. Explain why the core is made of **soft iron**. (2 marks)

(b) Susan thinks that the armature is made of steel.

(i) Describe and explain what would happen to a steel armature after the supply to the coil has been switched off. (3 marks)

(ii) Suggest a better material for the armature. (1 mark)

(c) Does it matter which way round the coil is connected to a power supply? Explain your answer. (2 marks)

(MEG Salters', Intermediate Tier, Q1)

Question 3

(a) What is the difference between alternating current and direct current? (2 marks)

The diagram below shows a simple transformer.

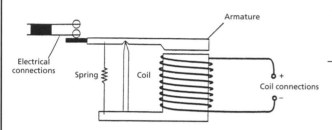

The graph shows how the output varies with time.

The number of turns on the secondary coil is changed from 100 to 150.

(b) On the graph show how the output voltage now varies with time. (2 marks)

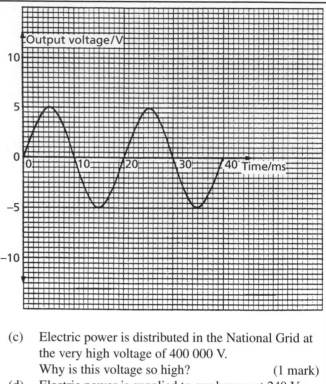

(c) Electric power is distributed in the National Grid at the very high voltage of 400 000 V. Why is this voltage so high? (1 mark)

(d) Electric power is supplied to our homes at 240 V. How is the voltage reduced from 400 000 V? (2 marks)

(e) Why is it necessary to reduce the voltage to 240 V? (1 mark)

(MEG, Jun 95, Intermediate Tier)

Question 4

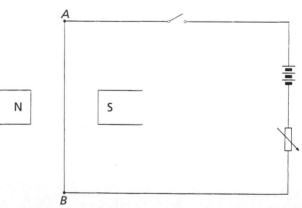

The diagram shows a flexible wire AB hanging between the poles of a powerful magnet. The switch is closed so that a current flows and the wire experiences a force at right angles to, and out from, the paper.

(a) Mark on the diagram the direction of the current through the wire AB when the switch is closed. (1 mark)

(b) State what effect (if any) the following alterations would have on the force on the wire AB:

(i) increasing the size of the current; (1 mark)

(ii) reversing the battery; (1 mark)

(iii) using weaker magnetic poles. (1 mark)

(WJEC, Jun 95, Intermediate Tier, Q6)

Question 5

The diagram shows apparatus used to investigate induced currents.

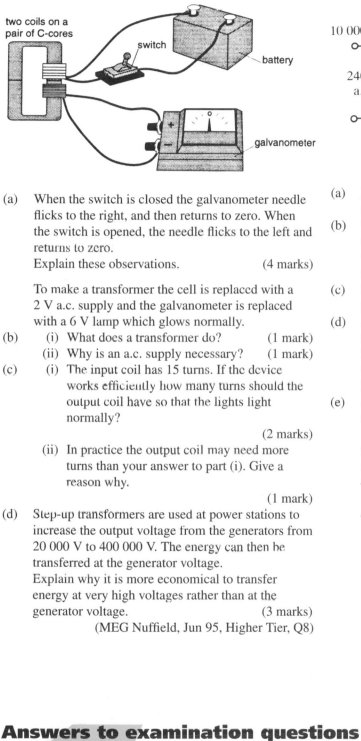

(a) When the switch is closed the galvanometer needle flicks to the right, and then returns to zero. When the switch is opened, the needle flicks to the left and returns to zero.
Explain these observations. (4 marks)

To make a transformer the cell is replaced with a 2 V a.c. supply and the galvanometer is replaced with a 6 V lamp which glows normally.

(b) (i) What does a transformer do? (1 mark)
(ii) Why is an a.c. supply necessary? (1 mark)

(c) (i) The input coil has 15 turns. If the device works efficiently how many turns should the output coil have so that the lights light normally?
(2 marks)
(ii) In practice the output coil may need more turns than your answer to part (i). Give a reason why.
(1 mark)

(d) Step-up transformers are used at power stations to increase the output voltage from the generators from 20 000 V to 400 000 V. The energy can then be transferred at the generator voltage.
Explain why it is more economical to transfer energy at very high voltages rather than at the generator voltage. (3 marks)
(MEG Nuffield, Jun 95, Higher Tier, Q8)

Question 6

The diagram below shows a transformer used to reduce the value of a voltage applied to its input leads.

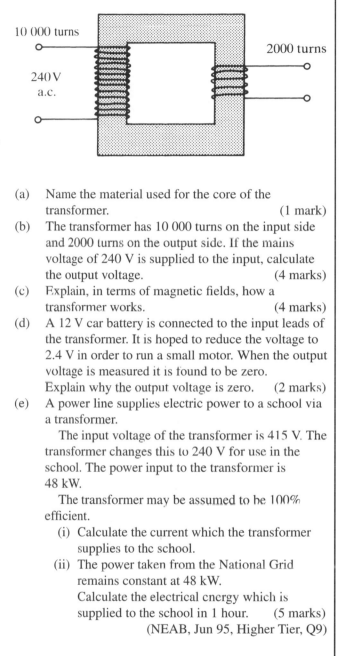

10 000 turns

240 V
a.c.

2000 turns

(a) Name the material used for the core of the transformer. (1 mark)

(b) The transformer has 10 000 turns on the input side and 2000 turns on the output side. If the mains voltage of 240 V is supplied to the input, calculate the output voltage. (4 marks)

(c) Explain, in terms of magnetic fields, how a transformer works. (4 marks)

(d) A 12 V car battery is connected to the input leads of the transformer. It is hoped to reduce the voltage to 2.4 V in order to run a small motor. When the output voltage is measured it is found to be zero.
Explain why the output voltage is zero. (2 marks)

(e) A power line supplies electric power to a school via a transformer.
The input voltage of the transformer is 415 V. The transformer changes this to 240 V for use in the school. The power input to the transformer is 48 kW.
The transformer may be assumed to be 100% efficient.
(i) Calculate the current which the transformer supplies to the school.
(ii) The power taken from the National Grid remains constant at 48 kW.
Calculate the electrical energy which is supplied to the school in 1 hour. (5 marks)
(NEAB, Jun 95, Higher Tier, Q9)

Answers to examination questions

1. (a) It is a transducer which converts sound energy into electrical energy.
(b) X is an amplifier. It converts the small varying electrical signal into a similar varying signal but having a larger amplitude.
(c) (i) A stronger magnetic field is produced.
(ii) The strength of the magnetic field decreases

with distance so the tape needs to be as close so that the field can be picked up by the head.

(d) 45 minutes = 45×60 s = 2700 s
Distance = speed × time = (4.8 cm/s) × (2700 s) = 12 960 cm = 129.6 m

(e) (i) B
Because there is no magnetic pattern on the

tape. [The arrows showing the field are arranged randomly.]

(ii) *A*

The arrows showing the stored field changes direction in a smaller distance so the note changes more frequently.

2. (a) (i)

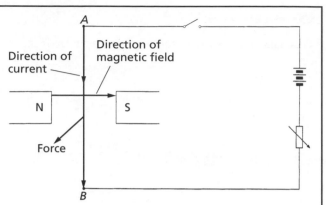

(ii) The soft iron core concentrates the magnetic field therefore making the coil a stronger electromagnet.

(b) (i) The armature would become permanently magnetised and it would move down and stick to the coil thus keeping the electrical connections closed.

(ii) Soft iron [soft iron will lose its magnetism when the coil is switched off].

(c) Turning the coil around will reverse the polarity of the ends of the electromagnet. Since the armature will be attracted to a north and a south pole, the arrangement will still work in the same way.

3. (a) Direct current travels in a single direction around a circuit depending on the way round the battery is connected (current travels from the positive to the negative terminal). Alternating current changes direction in the circuit many times a second.

(b) A similar graph should be drawn but having an amplitude of 7.5 V. [The original input was 2.5 V since this was doubled because there are double the turns on the secondary compared with the primary. with 150 turns, the input will be trebled so the output voltage will have an amplitude of 7.5 V.]

(c) So the current can be kept small thus minimising energy losses along the wires.

(d) A step-down transformer is used having many times fewer turns on the secondary coil compared to the primary.

(e) This is the voltage of the mains domestic supply.

4. (a) [Flemming's left-hand rule is used to predict the direction of the force on the wire. The magnetic field always points from north to south (from left to right on the diagram) and the current will flow from *A* to *B*.] Applying the law gives the force acting perpendicularly out of the paper.

(b) (i) The force would be bigger.

(ii) The force would be opposite in direction (into the page in this case).

(iii) The force would be weaker.

5. (a) When the switch is closed the current passes through the coil creating a magnetic field. This magnetic field does not reach its full strength right away and when the switch is opened it does not die down to zero straightaway. Instead, the field changes over a period of time and this changing magnetic field links the coil connected to the galvanometer and induces an e.m.f. across its terminals. The e.m.f. is only produced while the field changes and will be zero at all other times.

(b) (i) It changes voltages either up or down.

(ii) There has to be a changing current in order to produce a changing magnetic field. Alternating current changes size and direction many times a second.

(c) (i) [To go from 2 V input to 6 V output there has to be three times the number of turns on the secondary compared with the primary.] Since the primary coil has 15 turns the secondary coil would need $3 \times 15 = 45$ turns.

(ii) The transformer will not be 100% efficient. [Some of the energy associated with the magnetic field will be transferred to heat in the core.]

(d) By keeping the voltage high, it is possible to reduce the current and still keep the power transmitted the same [$P = VI$]. Since the heating effect of an electric current is given by $H = I^2Rt$ by keeping the current, I, small, the heat lost is kept small.

6. (a) Soft iron

(b) $$\frac{\text{Voltage across primary}}{\text{Voltage across secondary}}$$

$$= \frac{\text{Number of turns on primary}}{\text{Number of turns on secondary}}$$

$$\frac{240}{Vs} = \frac{10\ 000}{2000}$$

Hence, output voltage $Vs = 48$ V

(c) Because there is alternating current in the primary

coil, a changing magnetic field is produced around this coil and in the core. The field cuts the coils in the secondary and induces an e.m.f. in it. The size of the alternating voltage produced in the secondary coil depends on the size of the original voltage along with the number of turns in each coil.

(d) Transformers only work when a.c. is used as the input. With d.c. supplies there is no changing field produced and therefore no electromagnetic induction.

(e) (i) Power = Current × Voltage
48 000 W = Current × 240 V
Current = 200 A

(ii) 48 kW means that 48 000 Joules are supplied each second.
In one hour there are 60 × 60 = 3600 s
So, in one hour energy = 48 000 × 3600
= 170 000 000 J = 170 MJ

15

Radioactivity

WJEC & NEAB	ULEAC Syll A	ULEAC Syll B	MEG	Topic	MEG Salters'	MEG Nuffield	SEG	NICCEA
✓	✓	✓	✓	**Radioactivity and its detection**	✓	✓	✓	✓
✓	✓	✓	✓	**Different kinds of radiation**	✓	✓	✓	✓
✓	✓	✓	✓	**The structure of the atom**	✓	✓	✓	✓
✓	✓	✓	✓	**Radioactive decay and half-life**	✓	✓	✓	✓
✓	✓			**Equations for radioactive decay**				✓
✓	✓	✓	✓	**Uses of radioactivity**	✓	✓	✓	✓
✓	✓	✓	✓	**Safety**	✓	✓	✓	✓
✓	✓			**Nuclear power**	✓		✓	✓

1 Radioactivity

Radioactivity and its detection

Radioactivity originates in the nucleus of an atom and is a random process. Radioactive substances produce radiation which has the ability to ionise the air surrounding them and it is this property which is used to detect them. In a cloud chamber, alcohol vapour condenses along the path of the ionising radiation. In a Geiger–Müller tube, a pulse of current, produced by the ionising particle entering the tube, flows between the central electrode and the metal tube surrounding it. These pulses may be counted by a ratemeter (in counts per second) or by a scalar which gives the total counts.

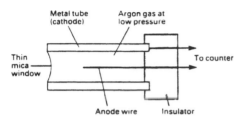

A Geiger–Müller tube.

When measuring radioactivity you need to measure the background radiation which is the radiation present when no source is near and subtract it from the reading with the source present.

Different kinds of radiation

α radiation

This is the emission of positively charged helium nuclei from the nucleus of an atom. It is stopped by a sheet of paper. It can be deflected by electric and magnetic fields, but the experiment cannot be conducted in air because the path of an α particle in air is only about 5 cm. α radiation is the most highly ionising of the three radiations. Alpha particles consist of two neutrons and two protons and have a +2 charge.

β radiation

This is the emission of electrons from the nucleus of the atom. It is absorbed by about 3 mm of aluminium foil. It is easily deflected by electric and magnetic fields and is less ionising than α particles. A beta particle is emitted when a proton changes into a neutron.

γ radiation

This is the emission of electromagnetic waves from the nucleus of an atom. It is unaffected by electric and magnetic fields. It is very penetrating but about 4 cm of lead will absorb most of it. It is only weakly ionising.

2 The structure of the atom

The positive charge and most of its mass is concentrated into a very small volume within the centre of the atom, called the nucleus. Electrons have a very small mass, a negative charge and orbit the nucleus. Inside the nucleus are two types of particles: protons and neutrons. They both have the same relative mass but the proton have a positive charge whereas neutrons have no charge. If the atom is electrically neutral, then the number of protons must be equal to the number of orbiting electrons. These details are summarised in table 1.

Table 1

	Mass	Charge
proton	1	+1
neutron	1	0
electron	negligible (1/2000)	−1

Proton number and nucleon number

Proton number (sometimes called atomic number) = number of protons in nucleus.

Nucleon number (sometimes called mass number) = number of protons in the nucleus + number of neutrons in the nucleus.

All atoms of the same element have the same number of protons in the nucleus and hence contain the same number of orbiting electrons.

Isotopes

The illustration below shows the three isotopes of hydrogen.

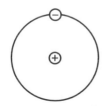

Normal hydrogen $_1^1H$
1 proton + 1 electron

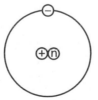

Deuterium $_1^2H$
1 proton, 1 neutron + 1 electron

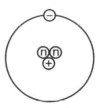

Tritium $_1^3H$
1 proton, 2 neutrons + 1 electron

They all contain the same numbers of protons and electrons but they differ in the number of neutrons in the nucleus. They are called isotopes of hydrogen. Isotopes have the same proton numbers but different nucleon numbers. An isotope of an element which is radioactive, is called a radioisotope.

3 Radioactive decay and half-life

Radioactive decay

Radioactive decay occurs when the nucleus of an atom is unstable and is able to regain its stability by either emitting particles (alpha and beta) or by emitting excess energy in the form of gamma radiation. If the nucleus splits up then it will emit radiation and a different atom with a different number of protons will be formed.

Half-life

The half-life of a radioactive source is the time taken for half the number of radioactive atoms in a sample to decay. This will also be the time taken for the radioactivity of the sample to fall to half of its original value.

Half-life may be measured by having a Geiger–Müller tube connected to a ratemeter positioned at a suitable distance from the source. The count rate is plotted against time (to get the correct count rate, the background rate must be subtracted from the reading on the ratemeter). The half-life is calculated from the graph.

Equations for radioactive decay

$^{238}_{92}$U (uranium-238) decays by α ($^4_2$He) emission into thorium. The equation representing the decay is

$$^{238}_{92}\text{U} \xrightarrow[\text{emission}]{\alpha} {}^{234}_{90}\text{Th} + {}^4_2\text{He}$$

(The sum of the nucleon (mass) numbers on the right-hand side of the equation is equal to 238, and the sum of the proton (atomic) numbers is 92.)

$^{239}_{92}$U (uranium-239) decays by β ($^{\ 0}_{-1}$e) emission into neptunium. The equation representing the decay is

$$^{239}_{92}\text{U} \xrightarrow[\text{emission}]{\beta} {}^{239}_{92}\text{Np} + {}^{\ 0}_{-1}\text{e}$$

4 Uses of radioactivity

Radiotherapy

If radiation causes ionisation in living cells it can cause damage and possibly cause cancer. If a larger dose of radiation is used then it is possible to kill cells completely.

Such radiation can be used to kill cancerous cells and stop them from spreading.

Sterilisation of blood and medical instruments

Blood and medical instruments are sterilised to rid them of harmful micro-organisms.

Detection of flaws

If a large gamma source is positioned inside a steel pipe and a photographic plate positioned outside the pipe, the resulting photograph will expose any flaws in the pipe (e.g. bubbles or cracks).

Thickness control

In paper mills the thickness of paper can be controlled by measuring how much β radiation is absorbed by the paper. The source is placed on one side of the paper and the detector on the other side.

Tracers

A small quantity of a radioactive isotope is mixed with a non-radioactive isotope, enabling the path of the element to be followed in a plant or animal or human.

5 Safety

Because of the dangers of radioactivity there are certain precautions which should be taken when dealing with radioactive sources the main ones being:

* Always handle sources with tongs and not with your fingers.
* Never point sources towards yourself or other people.
* Never eat or drink in a room near the sources.
* Cover up any cuts or sores.
* When not in use, keep sources in lead-lined boxes.
* Ensure that the exposure time for any radiation is kept to minimum.

6 Nuclear power

Nuclear fission

In a nuclear power station nuclei of ^{235}U (or sometimes ^{239}Pu) absorb a neutron and split into two parts. This is an example of nuclear *fission*. The splitting produces energy and two or three more neutrons. These neutrons may penetrate other ^{235}U nuclei, thus releasing more neutrons and starting a chain reaction. The next diagram shows this process.

Slow-moving neutrons are more easily absorbed than fast-moving ones, and moderators (usually water or

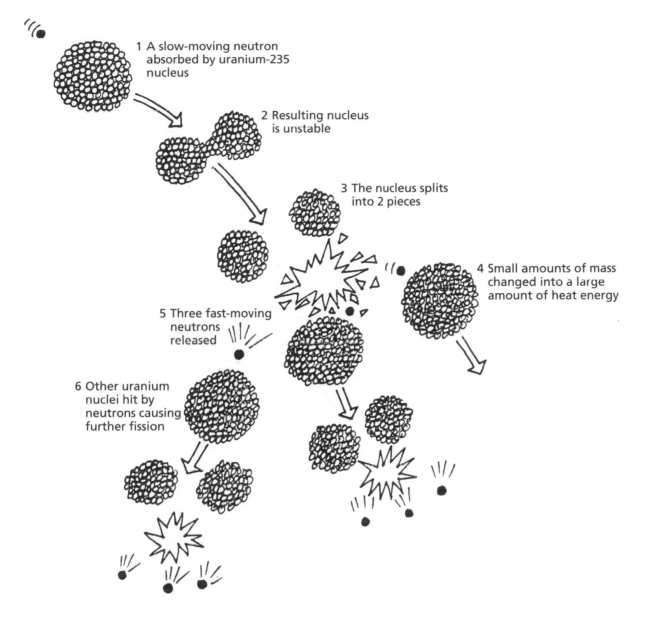

1 A slow-moving neutron absorbed by uranium-235 nucleus

2 Resulting nucleus is unstable

3 The nucleus splits into 2 pieces

4 Small amounts of mass changed into a large amount of heat energy

5 Three fast-moving neutrons released

6 Other uranium nuclei hit by neutrons causing further fission

A chain reaction showing how nuclear fission occurs.

graphite) are used to slow the neutrons down. The heat is extracted by means of a 'heat extractor fluid' which flows through the reactor and absorbs heat. A heat exchanger converts the heat into steam and the steam is used to drive the turbine generators.

The speed of the chain reaction is controlled by the use of control rods which absorb neutrons. When these rods are lowered into the reactor, the speed of the reaction is slowed down.

The energy produced comes from the conversion of a small amount of matter into energy. An atomic bomb is an uncontrolled chain reaction, as a result of nuclear fission.

Safety precautions in British nuclear power stations include:

(i) Regular monitoring of all personnel to check their exposure to radioactivity.

(ii) Careful disposal of all radioactive waste products (especially those with a long half-life).

(iii) An automatic shutdown of the reactor should the temperature of the reactor go above a certain safe limit.

(iv) Monitors throughout the whole plant to ensure that it is functioning properly and that there is no release of radioactive materials. This includes the monitoring of any water or gases which are discharged into the outside world (they are filtered before discharge).

(v) Barriers throughout the plant to prevent the release of radioactive material.

(vi) The wearing of protective clothing.

(vii) A complete shutdown every two years for a complete overhaul.

Nuclear fusion

The Sun's power comes from fusion. One example of fusion is the fusion of hydrogen-2 (deuterium) with hydrogen-3 (tritium). This results in the release of energy. The hydrogen bomb is an uncontrolled fusion reaction. The technical problems of producing energy by a controlled fusion reaction have not yet been solved. One problem is that all ordinary containers melt at the temperature of the reaction. For more details on fusion, refer to chapter 19.

Worked examples

Example 1

Radiation although dangerous, can be used in the treatment of cancer. This is because
A   radiation kills cells in the body
B   radiation only kills the cancerous cells
C   radiation can be easily detected
D   radiation can be used to sterilise medical instruments

Solution

[Radiation kills all cells whether cancerous or not. By killing the cells you can stop them from spreading throughout the body.]

Answer **A**

Example 2

The isotope $^{14}_{6}C$ contains
A   14 protons
B   6 neutrons
C   6 electrons
D   8 protons

Solution

[C contains 6 protons (and hence 6 electrons) and $14 - 6 = 8$ neutrons.]

Answer **C**

Example 3

Which one of the following particles has the greatest mass?
A   protons
B   electrons
C   neutrons
D   alpha particles

Solution

[Alpha particles are the nuclei of helium atoms and have 2

protons and 2 neutrons giving a mass number of 4. The mass of an electron is very small.]

Answer **D**

Example 4

The device shown is to check the thickness of polythene.

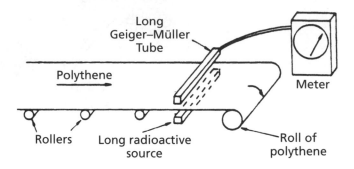

If a source which emits α, β and γ radiation is used, which are likely to penetrate the polythene to the detector?
A   both α and γ
B   both α and β
C   both β and γ
D   all of α, β and γ

Solution

[See section 1, 'Different kinds of radiation'. Remember that α radiation is absorbed by a few centimetres of air and a sheet of paper. It will not penetrate the air and polythene between the source and the Geiger–Müller tube.]

Answer **C**

Examples 5 and 6

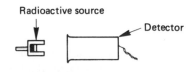

A detector of radioactivity is placed opposite a radioactive source as shown in the diagram. It is found that radiation from the source is detected only if the distance between the source and the detector is less than 5 cm.
5.   The source is probably emitting
    A   alpha radiation only
    B   beta radiation only
    C   gamma radiation only
    D   alpha and beta radiation only
    E   alpha, beta and gamma radiation
6.   The source is changed. A sheet of paper placed between the new source and the detector is found to reduce the count rate. A sheet of aluminium is found to stop it completely. The source is emitting

A gamma radiation only
B beta and gamma radiation only
C alpha and gamma radiation only
D alpha and beta radiation only
E alpha, beta and gamma radiation

Solutions

5. [β and γ radiation travel much more than 5 cm in air.]

Answer **A**

6. [α particles would be stopped by a sheet of paper and β particles by the sheet of aluminium. γ rays would penetrate the aluminium.]

Answer **D**

Example 7

A spent fuel rod containing radioactive material from a nuclear reactor has an activity of 40 000 Bq. It must be stored until the activity is 20 000 Bq. The half-life of the radioactive material is 1 month. (1 Bq = 1 count per second.)

For how long must the fuel rod be stored?
A 2 weeks
B 1 month
C 2 months
D 4 months

Solution

[The activity halves in 1 month. See section 3, 'Half-life'.]

Answer **B**

Example 8

The half-life of a radioactive substance is 10 hours. The original activity of a sample is measured and found to be 1200 counts per minute. Which of the following statements is correct?
A After 10 hours the count rate will be 120 counts per minute
B After 20 hours the count rate will be 300 counts per minute
C After 40 hours the count rate will be 150 counts per minute
D After 50 hours the count rate will be 240 counts per minute

Solution

[The activity halves every 10 hours. So
after 10 hours the activity will be 600 counts per minute
after 20 hours the activity will be 300 counts per minute

after 30 hours the activity will be 150 counts per minute
after 40 hours the activity will be 75 counts per minute.]

Answer **B**

Example 9

The element X has an atomic mass number of 238 and an atomic number of 92. It emits an α particle forming an element Y. Y can be represented by
A $^{234}_{90}Y$
B $^{236}_{90}Y$
C $^{235}_{91}Y$
D $^{238}_{92}Y$
E $^{238}_{93}Y$

Solution

[See section 3, 'Equations for radioactive decay'. An α particle is $^4_2$He, so the nucleon (mass) number goes down by 4 and the proton (atomic) number goes down by 2.]

Answer **A**

Example 10

An element P has an atomic mass of 239 and an atomic number of 92. It emits a β particle forming an element Q. Q can be represented by
A $^{239}_{91}Q$ B $^{239}_{92}Q$ C $^{239}_{93}Q$ D $^{238}_{92}Q$ E $^{235}_{90}Q$

Solution

[See section 3, 'Equations for radioactive decay'. A β particle is $^0_{-1}$e. So the nucleon (mass) number stays the same but the proton (atomic) number goes up by 1.]

Answer **C**

Example 11

(a) What is meant by radioactivity? (3 marks)
(b) Which type of radiation is
(i) the most penetrating?
(ii) the most ionising?
(iii) used to sterilise medical instruments?
(iv) absorbed by a piece of paper? (4 marks)
(c) Give one example of the use of β radiation. (4 marks)

Solution

(a) Radioactivity is the emission of α, β or γ radiation from the nucleus of unstable nuclei.
(b) (i) γ radiation; (ii) α radiation; (iii) γ radiation; (iv) α radiation.
(c) β radiation is used to monitor the thickness of the paper being produced in a paper mill. The source is

placed on one side of the paper and a detector (for example, a G–M tube connected to a ratemeter) on the other side of the paper. An increase in the count rate would mean that the thickness of the paper had decreased. The reading from the G–M tube is used to adjust the rollers which determine the thickness of the paper.

Example 12

Table 15.1 lists three radioactive sources which are used in medicine and/or industry.

Table 15.1

Radioactive source	Type of radiation emitted	Ionising ability of radiation	Penetration of radiation
Cobalt-60	Gamma	Poor	Halved by 1 cm of lead
Strontium-90	Beta	Moderate	Stopped by 3 mm of aluminium
Polonium-210	Alpha	Good	Stopped by a very thin sheet of paper

(a) Cancers deep inside the body can be treated by using radiation from a source outside the body. Which of these radioactive sources is best for this? Why? (2 marks)

(b) Which of these radioactive sources is best for making air conduct electricity? Why? (2 marks)

(c) Radiation from radioactive sources can be dangerous. Describe one way in which any one who passes by a source can be protected from it. (2 marks)

(d) Paper is made in very long rolls. The diagram shows apparatus for measuring the thickness of paper as it is being made.

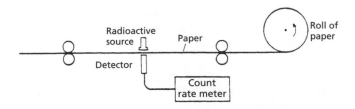

The source of radiation is above the paper and the detector is below.
What happens to the count rate recorded by the detector if the paper gets thinner as it is being made? (1 mark)

Which of the sources in the table is most suitable for this apparatus? (1 mark)
Give reasons why each of the other two sources are NOT suitable. (2 marks)
(MEG, Intermediate Tier)

Solution

(a) Gamma. Because it is the only radiation which would penetrate deep into the body.

(b) Alpha. Because it causes very much more ionisation than beta or gamma radiation.

(c) Wear special protective clothing [or keep well away from the source as you pass, or keep the source in a lead box].

(d) The count rate goes up.
Beta.
Alpha is not suitable because it would not penetrate the paper. Gamma is not suitable because the count rate would not vary if the paper thickness changed.

Example 13

This question is about radioactivity and its applications. When an alpha particle moves through a cloud chamber it leaves a straight, short, thick track.

(a) (i) What is an alpha particle? (1 mark)
(ii) What do the tracks consist of? (2 marks)
(iii) Why are the tracks short? (2 marks)
(iv) Why are the tracks thick? (2 marks)

(b) A radioactive tracer can be injected into the bloodstream to allow the flow of the blood through the blood vessels to be monitored. The illustration below shows the basic arrangement used. The graph on the next page shows the results of monitoring the blood in the leg of a patient.

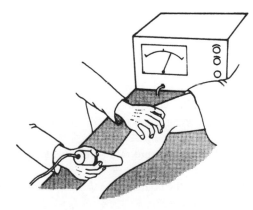

(i) Given that the particles from the tracer have to pass through the leg, what is the most likely type of particle to be emitted by the tracer? Explain your answer. (3 marks)
(ii) Why does the graph show a peak above one part of the leg? (2 marks)
(iii) Why should the half-life of the tracer be quite short? (1 mark)

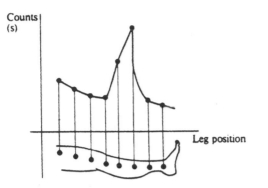

(iv) Why is it necessary for the radiographer operating the tracer detector to wear protective clothing? (1 mark)
(Total 14 marks)
(NEAB, Intermediate Tier)

Solution

(a) (i) A positively charged helium nucleus.
(ii) Liquid droplets which are formed when vapour condenses on the ions formed by the moving particle.
(iii) α particles only have a range of about 5 cm in air.
(iv) α particles cause thicker tracks than β particles because, being larger and having a greater charge, they cause more ionisation.
(b) (i) γ radiation. γ-rays are more penetrating than α or β particles. The γ-rays escape from the body and may be detected. No dangerous γ or β radiation is present which could cause internal ionisation damage.
(ii) This is where there is a maximum presence of the isotope.
(iii) So that the level of radiation in the body falls quickly once the flow of blood has been monitored.
(iv) Continual exposure to radiation is dangerous and can damage the body cells.

Example 14

The thickness of sheet metal produced at a steel mill is often monitored by means of radioactivity. A source is placed on one side of the sheet and a detector connected to a counter on the other side of the sheet. As the sheet passes along the production line, the readings on the counter over a period of 14 s are shown in Table 15.2.

Table 15.2

Time/s	0	2	4	6	8	10	12	14
Total count	0	60	120	180	250	320	380	440
Count in 2 s	–	60	60	60				

(a) Would an α, β or γ source be the best type of source? Explain why the other two types of source would not be so suitable. (3 marks)
(b) Explain the meaning of the word 'radioactivity'. (3 marks)
(c) Complete Table 15.2 shown above and explain what is happening to the thickness of the foil during the 14 s period. (6 marks)
(d) If the sheet metal got very thick, might the count rate fall to zero? Explain your answer. (2 marks)

Solution

(a) A β source would be best. α radiation would not penetrate the steel and the variation in γ radiation with small changes in thickness would be very small and not as easy to detect as the variation when a β source is used.
(b) The word 'radioactive' is used to describe the emission from the nucleus of an unstable atom of α, β or χ radiation.
(c)

Table 15.3

Time/s	0	2	4	6	8	10	12	14
Total count	0	60	120	180	250	320	380	440
Count in 2 s	–	60	60	60	70	70	60	60

After 6 s and before 8 s the thickness has started to decrease. By 12 s the decrease has been corrected, and the thickness from 12 s to 14 s is the same as in the first 6 s.
(d) The count rate would never fall to zero, because of the background count that would always be present.

Example 15

(a) Name *three* types of radiation emitted by radioactive sources. (1 mark)
(b) State, justifying your choice in each case, which of the radiations named in (a)
(i) carries a negative charge
(ii) is similar to X-rays
(iii) is most easily absorbed
(iv) travels with the greatest speed
(v) is not deflected by a magnetic field
(vi) is emitted when $^{238}_{9}U$ decays to $^{234}_{9}Th$
(vii) is similar in nature to cathode rays (7 marks)
(c) Carbon-14 ($^{14}_{6}C$) is an isotope of carbon. It is radioactive, decaying to nitrogen-14 ($^{14}_{7}N$).
(i) What is the meaning of the term isotope?
(ii) Write an equation for the decay of carbon-14. (3 marks)
(d) Carbon-14 has a half-life of 5600 years.
(i) What is the meaning of the term half-life?

(ii) Draw a graph to show the decay of carbon-14 from an initial activity of 64 counts per minute.
(7 marks)

(e) While trees and plants live they absorb and emit carbon-14 (in the form of carbon dioxide) so that the amount of the isotope remains constant.
  (i) What happens to the amount of carbon-14 after a tree dies?
  (ii) A sample of wood from an ancient dwelling gives 36 counts per minute. A similar sample of living wood gives 64 counts per minute. From your graph deduce the approximate age of the dwelling.
(3 marks)

Solution

(a) Alpha, beta and gamma radiation.
(b) (i) β radiation is normally negatively charged (although positive β particles are sometimes emitted).
  (ii) γ radiation and X-rays are both electromagnetic radiation.
  (iii) α radiation will not pass through a sheet of paper. The others easily penetrate a piece of paper.
  (iv) γ radiation like all electromagnetic radiation travels at the speed of light.
  (v) γ radiation. Both α and β radiation are charged and deflected by a magnetic field.
  (vi) α radiation. It is a helium nucleus $^{4}_{2}$He.
  (vii) β radiation. Cathode rays and β radiation are both fast moving electrons.
(c) (i) Isotopes are atoms of a given element which differ only in the number of neutrons in the

nucleus. Isotopes therefore have the same proton number but a different nucleon number.
  (ii) $^{14}_{6}C \rightarrow {}^{14}_{7}N + {}^{0}_{-1}e$
(d) (i) The half-life is the time for the activity to halve. Thus in 5600 years the activity will halve. In 11 200 years it will have become a quarter of its original value.
(e) (i) The carbon-14 decays and every 5600 years the activity will halve.
  (ii) 4500 years.

Example 16

In an experiment to determine the half-life of radon-220 ($^{220}_{86}$Rn) the results in Table 15.4 were obtained, after allowing for the background count:

Table 15.4

Time/s	0	10	20	30	40	50	60	70
Count rate/s^{-1}	30	26	23	21	18	16	14	12

(a) By plotting the count rate (vertically) against the time (horizontally), determine the half-life of $^{220}_{86}$Rn. Show clearly on your graph how you obtain your answer.
(6 marks)
(b) (i) What is the origin of the background count?
(2 marks)
  (ii) How is the background count determined?
(3 marks)
(c) $^{220}_{86}$Rn emits α particles.
  (i) What is an α particle? (2 marks)
  (ii) When $^{220}_{86}$Rn emits an α particle it becomes an isotope of the element polonium (Po). Write an equation to represent this change. (2 marks)
(d) When carrying out experiments with radioactive sources, students are instructed that
  (i) the source should never be held close to the human body.
  (ii) no eating or drinking is allowed in the laboratory.
Why is it important to follow these instructions?
(5 marks)

Solution

(a) [The half-life is the time for the count rate to halve no matter where it starts. You should always take a number of different starting points and take the average value for the half-life. In this case only two were taken because the readings stopped soon after the count rate halved. Normally at least three values would be calculated.]
(b) (i) The background count is always present in the atmosphere. It arises from cosmic rays entering the atmosphere, radioactive materials on the

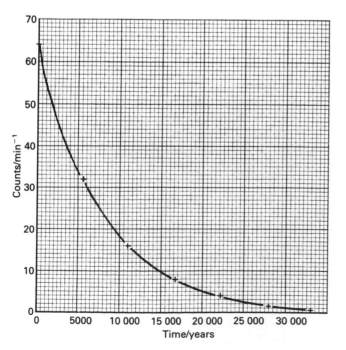

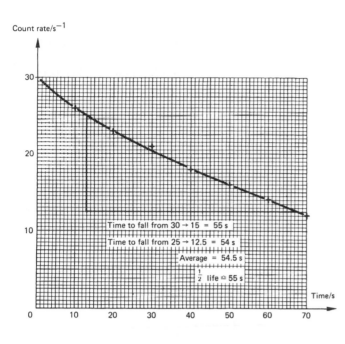

Count rate/s^{-1}

Time to fall from 30 → 15 = 55 s
Time to fall from 25 → 12.5 = 54 s
Average = 54.5 s
$\frac{1}{2}$ life ≈ 55 s

Time/s

surface of the Earth, radon in the atmosphere and X-rays from television screens.

(ii) In the above experiment the detector is removed from the radon source and left in the atmosphere. The count is taken over a period of say a minute and the count rate per second determined by dividing by 60. The average of about 10 counts would be taken.

(c) (i) An α particle is a helium nucleus, $^4_2$He. It has a charge of +2 and a nucleon number of 4.

(ii) $^{220}_{86}$Rn = $^{216}_{84}$Po + $^4_2$He

(d) Radiation is dangerous because it can harm living cells.

(i) The intensity of radiation falls off rapidly as the distance increases, so the further the source is from the human body the less is the risk.

(ii) α radiation cannot penetrate the skin so the greatest risk comes from swallowing minute traces. Indeed any radiation is more dangerous if it is inside the body; hence, eating and drinking should never take place in the vicinity of radioactive materials.

Harm to the body could result if these instructions were not followed.

Example 17

A student suggests that a stable isotope of an element X should be represented by the form $^7_3$X for another stable isotope.

(a) State why these two forms represent isotopes of the element X.

(b) How many neutrons are there in an atom of
(i) $^4_3$X (ii) $^7_3$X?

(c) Which is likely to be the form of the stable isotope? Give a reason for your answer. (6 marks)

Solution

(a) They have the same proton number, 3, and hence the same number of protons, the same number of electrons and the same chemical properties. They have different numbers of neutrons in the nucleus.

(b) (i) 1, (ii) 4 [(4 − 3) and (7 − 3)].

(c) Light elements are stable if the numbers of protons and neutrons are approximately equal. X is likely to be more stable because it has 3 protons and 4 neutrons.

Example 18

In a nuclear reactor the following reaction takes place
$^{235}_{92}$U + $^1_0$n → $^{236}_{92}$U
The $^{236}_{92}$U formed is unstable and disintegrates, with the release of two or three neutrons, together with the release of a considerable amount of energy.

(a) What do the numbers 236 and 92 represent?
(2 marks)

(b) Explain the meaning of the term 'isotope' with reference to $^{235}_{92}$U and $^{236}_{92}$U. (3 marks)

(c) Use the information given above to explain what is meant by a chain reaction. (3 marks)

(d) What are moderators? Why are they necessary in a nuclear reactor? (3 marks)

(e) What is the source of energy in the above reaction?
(2 marks)

(f) Explain how the heat is removed from the core of a reactor and how this heat is used to generate electricity. (3 marks)

(g) Many nuclear waste products have long half-lives. Explain the term 'half-life' and discuss why the long half-life of waste products presents a health hazard.
(4 marks)

Solution

(a) 236 is the nucleon number and represents the number of neutrons plus the number of protons in the nucleus. 92 is the proton number and is the number of protons in the nucleus.

(b) Isotopes are atoms of a given element which differ only in the number of neutrons in the nucleus. Isotopes have the same number of protons in the nucleus (92 in $^{235}_{92}$U and $^{236}_{92}$U, but a different number of neutrons: (235 − 92) = 143 in $^{235}_{92}$U and (236 − 92) = 144 in $^{236}_{92}$U.

(c) If the two or three neutrons released when $^{236}_{92}$U disintegrates are captured by other $^{235}_{92}$U atoms, then more neutrons are released. As the process continues, more and more neutrons are released. A chain reaction has started.

(d) Moderators are used to reduce the speed of released neutrons. Slow-moving neutrons are more easily captured by $^{235}_{92}$U than fast-moving ones.

(e) The source of energy is the reduction of the mass of the atom. Some of the mass is turned into energy.

(f) A heat extractor fluid flows around the reactor and the fluid's temperature rises as it absorbs heat from the reactor. A heat extractor converts this heat into steam; the steam is used to drive a turbine generator.

(g) The half-life is the time for the activity of a radioactive substance to fall to half its value. If the half-life is 1 h, after 1 h the activity is halved, after 2 h it is a quarter and after 3 h it is an eighth, so after a very short time the activity is reduced to a very small amount. On the other hand, if the half-life is 1000 years, then the activity remains at a very high level for a very long time and this can result in a health hazard.

Example 19

(a) A hospital patient is believed to have a blood circulation problem. The doctors investigate this problem by introducing a radionuclide (radioisotope) into the patient's bloodstream to act as a tracer.
   (i) What type of radiation should the radionuclide give out?
   (ii) Explain your answer to part (i).
   (iii) What three features should the radionuclide have? (7 marks)

(b) The diagram shows an ultrasound scan being carried out on a pregnant woman. The scanner uses ultrasonic waves to check on the progress of the foetus.
   (i) What are ultrasonic waves?
   (ii) Explain how ultrasonic waves produce an image of the foetus. (You may use the space below to draw a diagram to help your answer.)
   (5 marks)
   (NEAB, Jun 95, Higher Tier, Q3)

Solution

(a) (i) Gamma radiation.
   (ii) Only gamma radiation would be able to penetrate through the body to be picked up by the detector. The other types of radiation would be absorbed by the body and would ionise the cells in the process and could cause damage and even cancer.
   (iii) A short half-life so that the activity falls quickly to safe levels.
      Half-life should not be too short otherwise there would not be enough time available to perform the test.
      A short half-life means that the initial activity will be high and therefore easy to detect.

(b) (i) Beams of very high frequency sound waves which have a frequency higher than humans can hear.
   (ii) The ultrasound detector is placed on the surface of the body using a gel to make sure that the contact is good. The sound travels into the body where it is reflected at the interfaces between different types of tissue. These reflected pulses reach the detector and a visual image of the information is shown on a screen.

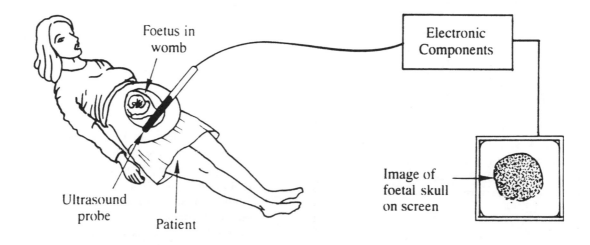

Examination questions

(Numerical answers and hints on solutions will be found at the end of the chapter.)

Question 1

(a) An isotope of uranium has a mass number of 238 and a proton number of 92.
State
   (i) the number of neutrons in the nucleus,
                                                        (1 mark)
   (ii) the number of electrons orbiting the nucleus.
                                                        (1 mark)

(b) Another isotope of uranium has a mass number of 235.
State
   (i) the number of protons in the nucleus, (1 mark)
   (ii) the number of neutrons in the nucleus.
                                                        (1 mark)

(c) A radioactive isotope which emits alpha particles can be dangerous if swallowed. Explain why this is so.
                  (WJEC, Jun 95, Intermediate Tier, Q19)

Question 2

In 1986, when the nuclear power station at Chernobyl exploded, radioactive material blew across Western Europe. Scientists detected radioactivity in the winds on the north-east coast of England.

(a) Give one way in which radioactive materials are different from non-radioactive materials.   (1 mark)

(b)   (i) What could the scientists use to detect the radioactivity?   (1 mark)
       (ii) Using this apparatus, how could the scientists know that there was radioactive material in the winds?   (1 mark)

Rains brought some of the radioactive material down and contaminated some area of grassland. Sheep eating the grass had to be destroyed so that the radioactive materials were not eaten by people.

(c) Why are ionising materials hazardous to living things?   (1 mark)

(d) Why were scientists more concerned about radioactive materials in food than in the clouds?
                                                        (1 mark)

(e) Give one difference between the alpha particles and gamma rays given out by the radioactive material.
                                                        (1 mark)
                  (MEG Nuffield, Jun 95, Intermediate Tier, Q9)

Question 3

The diagram shows part of a smoke detector.

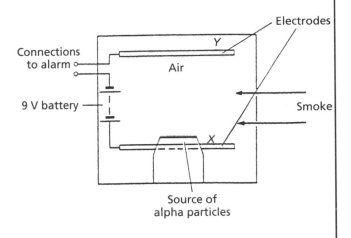

An alpha particle source is used to produce *ions* in the air between plates X and Y.

(a) What are *ions*?   (2 marks)
(b) Why is an alpha particle source more suitable than a gamma ray source?   (1 mark)
(c) Plates X and Y are connected to a battery. What happens to the positive ions and to the negative ions?
                                                        (2 marks)
(d) When smoke enters the detector, the number of ions reaching the plates X and Y falls. The current falls and the alarm sounds. What else might cause the current to fall and make the alarm sound?   (1 mark)
(e) Why should the alpha particle source have a long half-life?   (1 mark)
(f) The smoke detector is usually placed on the ceiling. Why is this?   (1 mark)
(g) Explain whether or not this source of alpha particles causes a safety problem.   (1 mark)
                  (MEG, Jun 94, Intermediate Tier)

Question 4

The diagram shows a Magnox nuclear reactor.

(a) What is the purpose of each of the following?
   (i) the graphite moderator;   (1 mark)
   (ii) the concrete shield;   (1 mark)
   (iii) the control rods.   (1 mark)

The fuel rods contain uranium-235. The reaction taking place is:

(b)   (i) What is the name of that kind of nuclear reaction shown in the diagram?   (1 mark)

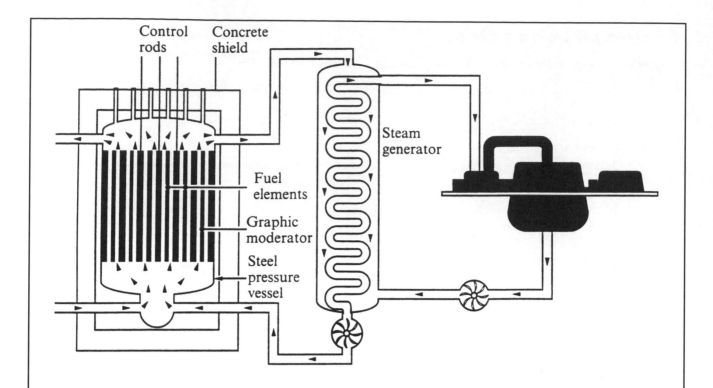

(ii) The particles labelled *A* in the diagram are fragments from the original U-235 nucleus. What are the names of the particles labelled *B*? (1 mark)

(iii) This nuclear reaction can be the start of what is called a *chain reaction*. Explain what is a chain reaction, what advantages there can be from it and what precautions must be taken? (3 marks)

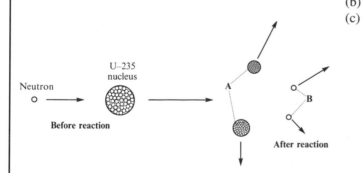

(c) Fill in the table below to show the charges and masses associated with the named type of radiation. (2 marks)

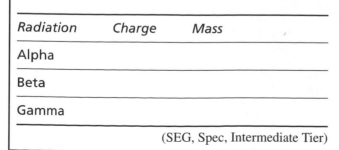

Radiation	Charge	Mass
Alpha		
Beta		
Gamma		

(SEG, Spec, Intermediate Tier)

Question 5

Trees contain carbon. Most of the carbon atoms are of the isotope known as carbon-12, which contains 6 protons and 6 neutrons in its nucleus. A few of the carbon atoms are carbon-14. Carbon-12 has a stable nucleus but carbon-14 is radioactive.

(a) What does the nucleus of a carbon-14 atom contain? (2 marks)

(b) What does the word *radioactive* mean? (3 marks)

(c) A tree is cut down and used to make a boat. Many centuries later the boat is found by an archaeologist who wishes to know how old it is. He asks a scientist to find out what proportion of the carbon atoms in the wood are carbon-14 atoms. This information is used to find out how long ago the tree was cut down.

On the axes on the next page, plot points and sketch a graph to show how the number of carbon-14 atoms in a sample of wood from the boat will change with time. Assume that the sample contained 6000 carbon-14 atoms when the tree was cut down. The half-life of carbon-14 is 5700 years.

(d) Explain why this method of dating a piece of wood is not suitable for finding the age of a boat made from wood cut down in the last 50 years. (2 marks)

(MEG, Jun 95, Higher Tier)

Question 6

(a) A cloud chamber can be used to detect alpha, beta or gamma radiation. Write down the name of one other type of radiation **detector**. (1 mark)

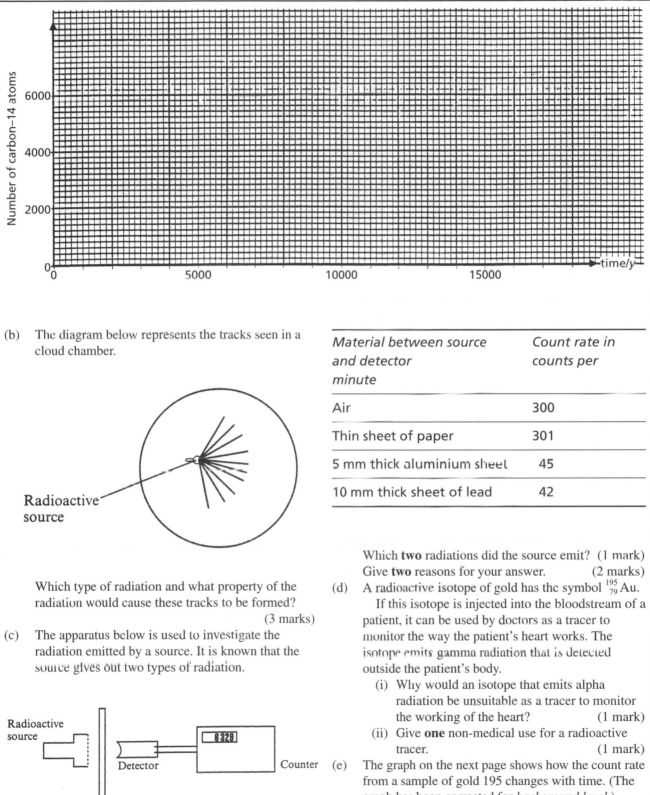

(b) The diagram below represents the tracks seen in a cloud chamber.

Radioactive source

Which type of radiation and what property of the radiation would cause these tracks to be formed?
(3 marks)

(c) The apparatus below is used to investigate the radiation emitted by a source. It is known that the source gives out two types of radiation.

Radioactive source

Detector

Counter

Absorbing material

→ 2 cm ←

Various materials were placed between the source and the detector. For each material the count rate, corrected for background radiation, was recorded as in the next table:

Material between source and detector	Count rate in counts per minute
Air	300
Thin sheet of paper	301
5 mm thick aluminium sheet	45
10 mm thick sheet of lead	42

Which **two** radiations did the source emit? (1 mark)
Give **two** reasons for your answer. (2 marks)

(d) A radioactive isotope of gold has the symbol $^{195}_{79}$Au. If this isotope is injected into the bloodstream of a patient, it can be used by doctors as a tracer to monitor the way the patient's heart works. The isotope emits gamma radiation that is detected outside the patient's body.

(i) Why would an isotope that emits alpha radiation be unsuitable as a tracer to monitor the working of the heart? (1 mark)

(ii) Give **one** non-medical use for a radioactive tracer. (1 mark)

(e) The graph on the next page shows how the count rate from a sample of gold 195 changes with time. (The graph has been corrected for *background level*.)

(i) What is meant by *background level*? (1 mark)

(ii) Use the graph to find the half-life of gold 195. You should show clearly on your graph how you obtain your answer. (2 marks)

(iii) Give **two** benefits of using an isotope with a short half-life as a tracer to monitor the heart. (2 marks)

(SEG, Jun 95, Higher Tier)

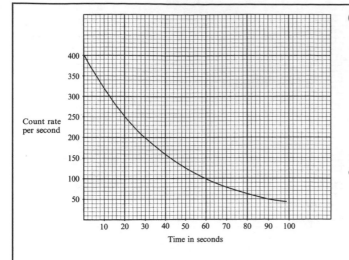

Question 7

Radon is a naturally occurring gas which comes from the uranium in all rocks and soils. Radon-222 ($^{222}_{88}$Ra) undergoes radioactive decay.

(a) (i) Write down the number of protons and the number of neutrons in the nucleus of an atom of $^{222}_{88}$Ra. (2 marks)

    (ii) An atom of $^{222}_{88}$Ra decays by the mission of an α particle. Write down the atomic number (proton number) and the mass number (nucleon number) of the resulting isotope. (2 marks)

    (iii) The half-life of $^{222}_{88}$Ra is 38 s. State what this means. (2 marks)

    (iv) Explain why the breathing in of radon increases the risk of contracting lung cancer. (1 mark)

(b) Radon gas can enter homes through any gaps in the floor or walls. The graph below shows how the radon gas concentration varies in a home during a period of 24 hours.

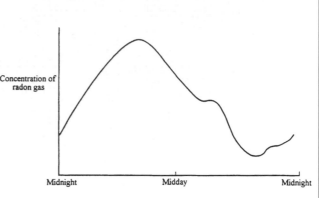

    (i) Suggest ONE reason why the concentration of radon gas increases during the night. (1 mark)

    (ii) Suggest ONE way you might prevent the concentration from increasing so much during the night. (1 mark)

(c) (i) In a particular sealed room the count rate produced by the radon gas was 200 Bq. What will be the count rate produced by the radon gas after 114 s if the half-life of radon is 38 s? You may assume no air was allowed to enter or leave the room. (3 marks)

    (ii) Why might the measured count rate be higher than your calculated value? (1 mark)

(ULEAC, Syll A, Jun 95, Higher Tier, Q1)

Answers to examination questions

1. (a) (i) Neutrons = 238 − 92 = 146
    (ii) Electrons = number of protons = 92
  (b) (i) Since it is the same element, proton number = 92
    (ii) Neutrons = 235 − 92 = 143
  (c) Because of their large size, alpha particles have the greatest ionising ability. Cells near the isotope will be ionised and this will damage the tissue and could even lead to cancer.

2. (a) (i) They contain unstable nuclei which become stable by giving out radioactive particles or rays.
  (b) (i) A Geiger–Müller tube and counter.
    (ii) The G–M tube would be attached to the counter and a reading of the counts in a certain time period would be taken.

  (c) They can cause cancer.
  (d) When people eat food, some of it is kept in the body and the ionisation it produces over a period of time can lead to cancer.
  (e) Gamma rays are electromagnetic waves (like light) but alpha particles are the nuclei of helium atoms.

3. (a) Ions are charged atoms which have either lost or gained one or more electrons giving them a positive or negative charge respectively.
  (b) Alpha particles are much better at producing ionisation than gamma rays.
  (c) Positive ions: attracted towards the negative plate *Y*. Negative ions: attracted towards the positive plate *X*.
  (d) The battery running down.
  (e) So that it does not need constant replacing.

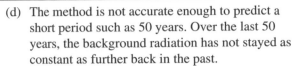

(f) Smoke is usually warm and hot air rises by convection so the smoke will reach the ceiling first.

(g) The source is small and of low activity so there is no safety problem.

4. (a) (i) The neutrons produced during fission are not easily absorbed because they have too much energy and this could prevent further fission. The purpose of the graphite moderator is to slow the neutrons down so they may be more easily captured.

(ii) This stops the radiation escaping from inside the reactor.

(iii) These absorb the neutrons. By lowering them into the reactor the number of neutrons may be reduced and the reaction may be slowed down. Raising them will speed up the reaction.

(b) (i) Fission reaction.

(ii) Neutrons.

(iii) Each of the neutrons labelled *B* can react with two further U-235 nuclei with each nuclei producing 2 more neutrons. The process is repeated. Some mass is lost during this fission due to it being converted into a huge amount of heat energy. To stop the reaction going out of control, control rods are used to remove some of the neutrons preventing them from reacting with further U-235 nuclei.

(c)

Radiation	Charge	Mass
Alpha	Positive (+2)	4
Beta	Negative	1/1860
Gamma	None	None

5. (a) 6 protons and (14 − 6) = 8 neutrons.

(b) The nucleus is unstable and gives out radiation (alpha particles, beta particles or gamma rays) until the nucleus becomes stable again.

(c)

(d) The method is not accurate enough to predict a short period such as 50 years. Over the last 50 years, the background radiation has not stayed as constant as further back in the past.

6. (a) A Geiger–Müller tube or photographic plate (these are the commonest).

(b) Alpha particles. Because they are large they are more likely to knock electrons out of their orbits and therefore cause ionisation.

(c) Beta and gamma radiation.
There is no significant drop in the count rate with the paper, so alpha can't be present. There is a drop with the aluminium owing to the beta being absorbed. Gamma is still getting through since the lead will have absorbed all the beta radiation.

(d) (i) Alpha would be absorbed and would not be detected outside the body.

(ii) Finding leaks in gas pipes.

(e) (i) The level of radiation had there been no source present due to naturally occurring radiation in the environment.

(ii) 30 seconds.

(iii) Only causes a small amount of damage to the body.
Procedure can be repeated after a short time (e.g. before and after exercise).

7. (a) (i) Protons = 88
Neutrons = 134

(ii) Atomic number = 86
Mass number = 218

(iii) This is the time taken for the activity (or count rate) of the sample to fall to half of its original value.

(iv) Alpha particles are absorbed by the tissue in the lung.

(b) (i) Windows and doors are usually shut at night so there are no currents of air to ventilate and dilute the concentration of radon gas.

(ii) Leave windows open at night.

(c) (i) 114 s = 3 × 38 s = 3 half-lives. After 3 half-lives the count rate will be $\frac{1}{8}$ of 200 = 25 Bq.

(ii) When it decays, a new isotope is produced and this may also be radioactive.

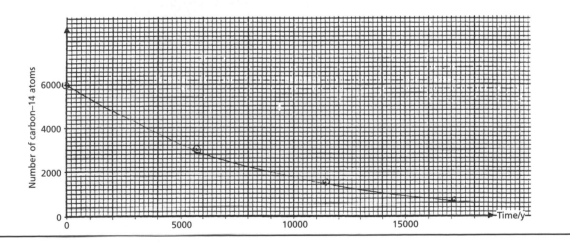

16

Spectra and Photons

Topic

WJEC & NEAB	ULEAC Syll A	ULEAC Syll B	MEG	Topic	MEG Salters'	MEG Nuffield	SEG	NICCEA
✓	✓	✓	✓	**The photoelectric effect**	✓	✓	✓	✓
✓	✓	✓	✓	**Spectra (continuous and line)**	✓	✓	✓	✓
✓	✓	✓	✓	**Energy levels**	✓	✓	✓	✓
	✓	✓	✓	**Emission and absorption of photons**		✓	✓	
✓	✓	✓	✓	**X-rays: their production and properties**	✓	✓	✓	✓

1 The photoelectric effect

If electromagnetic radiation above a certain frequency is shone onto a clean metal surface, then electrons, called photoelectrons, are emitted from the surface and this effect is called the photoelectric effect.

By performing an experiment it can be shown that light above a certain frequency is needed for the photoelectrons to be produced. The minimum frequency needed is called the threshold frequency. It is also found that the intensity of the radiation does not influence whether the photoelectrons are produced or not as might be expected. The intensity only determines the rate of emission of the photoelectrons (provided the frequency of the radiation is high enough to produce photoelectrons in the first place).

The photoelectric effect can be demonstrated using the following apparatus.

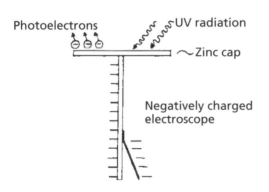

An electroscope which has a cap made of zinc is illuminated with ultra-violet (u.v.) radiation. Before the illumination the electroscope is charged negatively but when the u.v. radiation is shone onto the cap, the leaves start to move closer together which indicates that some of the electrons are being lost. If the u.v. light is replaced by visible light, then this does not happen since visible light has too small a frequency for the photoelectric effect.

The relevance of the photoelectric effect

The photoelectric effect was hard to explain but eventually a theory was put forward called the quantum theory. According to this, light was emitted in packets of energy called *photons*. Each photon could be considered to be a particle of light energy and the amount of energy the photon has is determined by the frequency of the light. Red light has a lower frequency than blue light and therefore a photon of red light has a smaller amount of energy than a quanta of blue light. There is a certain quantity of energy, called the work function, which needs to be supplied before an electron is released so this explains why if the frequency of the radiation is too low then no photoelectrons are released. If the energy of the photons have a greater energy than the work function, then the excess energy is given to the electrons as kinetic energy.

The next illustration shows this process.

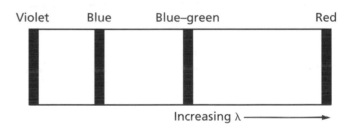

2 Spectra (continuous and line)

Ionisation energy

The ionisation energy of an atom is the energy which an atom must absorb to remove it completely from an atom.

Line spectra

If the temperature of a gaseous element such as hydrogen, helium, sodium or mercury is raised, the atoms collide with each other at high speed. These collisions cause electrons in their normal orbits (called the ground state) to be promoted to higher orbitals which have a higher energy. The electron in the higher orbit is unstable and soon falls back to its original energy level with the emission of light which has an energy equal to the energy difference between the energy levels of the orbits. Many different transitions between energy levels are possible and each one will give rise to a particular line in the line spectrum. Line spectra can be used to identify a particular element.

Violet	Blue	Blue–green	Red

Increasing λ ⟶

Continuous spectra

Continuous spectra are produced when liquids or gases are heated. Since there are so many interactions between the atoms owing to their closeness, the electrons have a continuous range of energies.

Violet Blue Yellow Red
Ultra-violet   Infra-red
Indigo Green Orange

Increasing λ ⟶

3 Energy levels

Excitation is the name of the process whereby electrons absorb energy without ionisation occurring. The energy is usually supplied in the form of heat, light or electrical energy.

It is found by examining spectra that atoms absorb and emit energy in fixed amounts so this means that electrons inside the atom have definite values of energy. The lowest energy an atom can have is called its ground state and there are a variety of different energy levels that an atom can have before ionisation occurs.

4 Emission and absorption of photons

Emission spectra are produced when the excited atoms are the source of the emitted photons. Photons are emitted when the electrons in the atom are excited after gaining energy by heat. The electrons move to a higher energy level making the atom unstable. The stability is regained by the electrons dropping down to lower energy levels with the emission of the excess energy as photons of electromagnetic radiation. The larger the energy level gap, the higher the frequency of the emitted radiation.

When white light is passed through mercury vapour, the mercury atoms absorb those frequencies of light which correspond to the lines produced in the emission spectrum for mercury. Electrons are promoted to higher energy levels but soon drop back to their original levels with the emission of light at the same frequency at which it was absorbed. The original light was absorbed in the same direction as the incident beam, but the re-emitted light is emitted in all directions and as a result the light in the straight through direction appears to have those wavelengths removed which have been absorbed. If the spectrum is examined using a device called a spectrometer, it can be seen that the light consists of a continuous spectrum with black lines superimposed at the positions where mercury lines would have appeared in the emission spectrum. This spectrum is called an absorption spectrum. This type of spectrum are produced when light passes through hot gaseous atoms and such a spectrum can be seen in light from the Sun.

5 X-rays: their production and properties

Thermionic emission

When a metal filament is heated to a high temperature by current, electrons are released from the surface of the filament and this process is called thermionic emission.

If the electrons are released into a vacuum (air would slow the electrons down) and accelerated by an electric field, they can be made to travel at a very high speed. When the electrons hit a metal target (usually tungsten), they rapidly lose kinetic energy as they enter the metal turning

much of their energy into heat with the emission of some of it as X-rays.

Controlling the intensity of the X-rays

The filament current determines the intensity with a high current, producing more electrons by thermionic emission and therefore a greater rate of production of X-rays.

Controlling the penetrating power

This is determined by the voltage between the cathode (filament) and the anode (target). The greater the voltage, the faster the electrons travel when they reach the target and the shorter the wavelength of the X-rays produced.

Properties of X-rays

1 They are electromagnetic and have a much smaller wavelength and a much higher frequency than light.
2 They are able to pass through soft tissue but their intensity is considerably reduced by hard materials such as bone (hence their use for the detection of broken bones in medicine).
3 They can blacken a photographic plate (hence their use in radiography).
4 They are ionising radiation and therefore dangerous in large doses.
5 X-rays are best absorbed by materials of high atomic number because there are large numbers of orbital electrons to interact with.

Worked examples
Example 1

The diagram shows four possible movements between the energy levels of hydrogen. The emissions all fall in the visible region of the spectrum.

Which of **A**, **B**, **C** or **D** is most likely to lead to a red line?

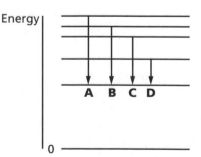

Solution

[In the visible spectrum, red light has the greatest

wavelength. Hence the frequency and the energy of the red photons are the lowest so the energy gap will be small.]

Answer **D**

Example 2

The diagram shows apparatus which may be used to produce a beam of X-rays.

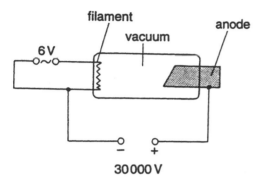

(a) The electron current between the filament and the anode is 20 mA.
Calculate the power input by the 30 000 V supply.
(3 marks)

(b) A small fraction of the energy input by the power supply turns to X-rays when the electrons hit the anode. What happens to the rest of the energy?
(1 mark)

(c) X-rays of high frequency are more penetrating than X-rays of low frequency. State and explain the change that is made to produce this more penetrating radiation.
(3 marks)

(d) When a beam of X-rays of a certain frequency passes directly through matter, the *intensity* of the beam falls but the *frequency* of the radiation is unchanged.
Explain this, by writing about photons. (3 marks)
(MEG, Jun 95, Further Tier)

Solution

(a) Power = Current × Voltage = $(20 \times 10^{-3}\,\text{A})$ × $(30\,000\,\text{V}) = 600\,\text{W}$

(b) It is changed into heat as the electrons are retarded in the anode.

(c) Increasing the voltage across the X-ray tube increases the kinetic energy of the electrons and increases the energy of the X-ray photons produced. X-ray photons which have a higher energy have a higher frequency.

(d) X-rays are electromagnetic radiation and consists of photons of high energy owing to them having a high frequency. The photons all have the same speed, frequency and energy and the intensity depends on the number arriving per m^2 per second. When they travel through a material some of them will be absorbed and used to promote electrons to higher energy levels while the others are able to pass through unimpeded without any change in energy, frequency and speed.

Example 3

Ben's science teacher showed his class several demonstrations.
The arrangement he used is shown below.

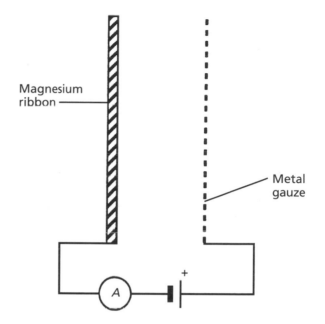

Light is able to pass through the openings in the metal gauze on to the magnesium ribbon. The results of the different demonstrations are shown in the diagrams below.

(a) The teacher asked the class to suggest what was happening.
  (i) Simon suggested that as the ultra-violet light travelled through the air it created ion-pairs between the magnesium ribbon and the gauze and so caused a current to flow. Which demonstration showed that this is not true? Explain your answer. (3 marks)
  (ii) Joanne thought that perhaps the ultra-violet light carried a positive charge and so was attracted to the magnesium ribbon. As the charged ultra-violet light moved between the gauze and the magnesium ribbon a current was produced.
     Explain why this idea could not be correct. Support your answer by reference to the demonstrations. (2 marks)
  (iii) Claire correctly suggested that when the ultra-violet light landed on the magnesium ribbon electrons were ejected from it.
     How could Claire's theory be used to explain the observations in Demonstrations 1 and 2? (3 marks)

(b) The current measured in demonstration one was 5×10^{-9} A. How many electrons every second must have been ejected from the magnesium ribbon by the ultra-violet light?
(Charge of an electron = 1.6×10^{-19} C.) (2 marks)

(c) Maria knew that white light was an electromagnetic

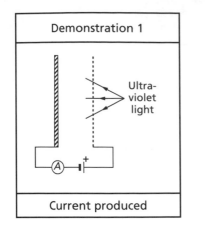

Demonstration 1

Ultra-violet light

Current produced

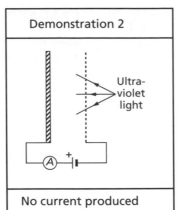

Demonstration 2

Ultra-violet light

No current produced

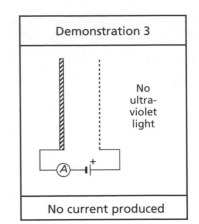

Demonstration 3

No ultra-violet light

No current produced

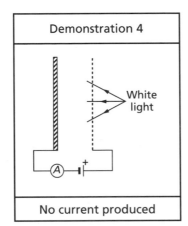

Demonstration 4

White light

No current produced

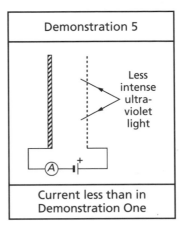

Demonstration 5

Less intense ultra-violet light

Current less than in Demonstration One

wave so it should knock out electrons from the magnesium ribbon.

Jonathan suggested that perhaps the white light was not intense enough. Which demonstration could he have given in support of his argument? Give reasons for your choice. (3 marks)

(NEAB, Syll B, May 93, Higher Tier)

Solution

(a) (i) Demonstration 2. If ionisation took place, reversal of the battery would reverse the polarity of the gauze and the magnesium ribbon but there would still be movement to the positive ions to the negative plate and vice versa. Since there is no current registered on the ammeter this process does not occur.

(ii) Reversal of the battery in Demonstrations 1 and 2 would, if the u.v. was charged, cause the current to move around the circuit in different directions. Since in Demonstration 2 there is no current, the u.v. cannot be charged in this way.

(iii) In Demonstration 1, photoelectrons being negatively charged are attracted across the gap to the metal gauze where they pass through the wire to the battery. The flow of charge in this way produces a current.

In Demonstration 2, the magnesium is made positive so the photoelectrons will no longer be attracted to the gauze and no current is produced.

(b) For a current of 5×10^{-9} A, 5×10^{-9} coulombs of charge must pass per second and since each charge is 1.6×10^{-19} C, the number of electrons per second is given by

$$\frac{5 \times 10^{-9}}{1.6 \times 10^{-19}} = 3.125 \times 10^{10}$$

(c) Demonstration 5 because in Demonstration 1 a current was produced but with the less intense u.v. of Demonstration 5 a current, albeit smaller, was produced. If the emission depended on the intensity then Demonstration 5 would not produce a current.

Examination questions

(Numerical answers and hints on solutions will be found at the end of the chapter.)

Question 1

For airport security, X-rays are used to examine the contents of bags.

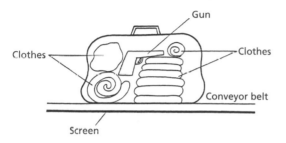

The X-rays can pass through bags and clothes but are stopped by metal objects.

(a) (i) On the diagram above, draw some paths of X-rays to show how a shadow of the gun is formed on the screen.   (3 marks)

(ii) Why must the conveyor belt not be made of metal?   (1 mark)

(b) (i) Why must exposure of people to X-rays be as little as possible?   (1 mark)

(ii) How can people near to this X-ray machine be protected?   (2 marks)

(MEG, Jun 95, Intermediate Tier)

Question 2

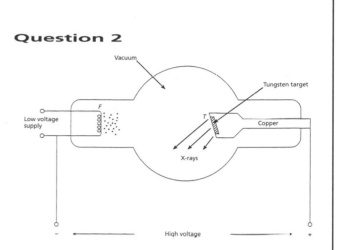

The diagram shows a simple X-ray tube.

(a) Name and describe the process by which electrons are emitted by the filament F.   (2 marks)

(b) State the purpose of the high positive voltage applied to the target.   (1 mark)

(c) Write down the energy transfer which occurs at the target T.   (2 marks)

(d) Why is the target made of tungsten which has a high melting point?   (1 mark)

(e) State and explain whether the tube would produce X-rays if the high voltage were an alternating voltage.

(2 marks)

(f) Briefly describe one medical use of X-rays.

(2 marks)

(WJEC, Jun 95, Higher Tier, Q14)

Answers to examination questions

1. (a) (i) Rays should be drawn from the source which pass through everything except the gun. Those rays which have passed through the case will also pass through the conveyor belt and show up on the screen.

(ii) Metal will not allow the X-rays through and will show up as a dark area on the screen. The shape of the gun would show up as a shadow.

(b) (i) X-rays produce ionisation and this can lead to cancer.

(ii) By making sure that they do not place their hands inside the box.
Use a box lined with lead to shield the user from the radiation.

2. (a) The filament is heated by the current passing

through it and the electrons are emitted from the filament. This process is called thermionic emission.

(b) It attracts the electrons and accelerates them towards it.

(c) The kinetic energy of the electrons is turned mainly into heat but with some energy converted into X-rays.

(d) The heat generated at the target could cause it to melt.

(e) Yes, because the tube would still work when the current was flowing such that F was the cathode and T the anode. When the other way round it would not operate since thermionic emission would be unable to occur at the target.

(f) Using them in medicine to detect broken bones. The X-rays pass through the soft tissue to the photographic plate which is blackened. The bones absorb most of the rays so the plate will appear light.

17

Communications

WJEC & NEAB	ULEAC Syll A	ULEAC Syll B	MEG	Topic	MEG Salters'	MEG Nuffield	SEG	NICCEA
✓				**Development of communication systems**	✓	✓	✓	✓
✓	✓	✓		**Communication using light and radio waves**	✓	✓	✓	✓
	✓	✓		**Analogue and digital signals**	✓	✓		✓
	✓			**Operation of an amplitude modulated radio system**		✓		

1 Development of communication systems

Most communication systems consist of the following building blocks:

- Encoding
- Transmission
- Reception
- Decoding
- Amplification
- Storage
- Retrieval
- Input and output transducers.

In ship-to-ship signalling using a flashing lamp to send a message, the following blocks from the list are used:

- Encoding – performed by the person sending the message
- Transmission – this is the flashing signal lamp
- Reception – the human eye receives the light from the signal
- Decoding – the brain interprets the series of flashes as a message
- Storage – the message is written down.

Ways of communicating over a distance have been developed over the years, starting with small distances and graduating to distances from one side of the world to the other. Starting from the smallest distances and working up the methods include:

Line of sight

Here a visual signal is used such as smoke signals, hoisting flags, semaphore, flashing lights, etc. The range of the line of sight systems could be increased by giving the observer a telescope.

Telegraph

The telegraph uses an electric current to carry messages by using one of the properties of the electric current. Several properties were used such as the electrolysis of water to give hydrogen and oxygen but eventually they settled on the magnetic effect. This used the current to work a buzzer each time a switch was closed and led to the development of Morse code. Long wires, having a high resistance, meant that the current was too small to operate the buzzer so a relay was used to switch a higher current circuit on and the process was repeated every few kilometres.

Telephone

Telephones overcame the problem of having to remember lots of codes. In a telephone the sound signals are converted to electrical signals which vary thousands of times per second. Again there were problems over the resistance of the wires used, so repeaters were still used to boost the signal after it had travelled a certain distance.

Newer techniques were then discovered which extended the distance over which communication could take place.

2 Communication using light and radio waves

Radio and television

Both the telegraph and telephone needed wires to connect the sender to the receiver and this restricted their use. Marconi combined the techniques of the telegraph with the discovery that electromagnetic waves could be used to send signals without wires. Radio waves are able to be reflected, diffracted and can pass through walls of buildings. An explanation of the workings of a radio are covered in Example 5.

In a radio based system, a transmitter is used to send the signal over the airways at the speed of light. A receiver is used at the other end to recover the signal before being sent to the transducer to convert it back into its original energy form.

Television involves splitting up a picture into a set of lines or dots which are sent as a radio signal having an extremely high frequency. The TV set reassembles the picture and adds the colours.

Satellites

Communication satellites are used to pass TV and radio signals from one part of the world to another. Signals are sent from the transmitting station up to the satellites where they are picked up and re-transmitted to the receiving dish at a different place on the Earth. Because communication satellites are placed in geostationary orbits, there is no need for the disk to have to move to track them. Also, the satellite is always overhead and therefore in range of the transmitter and receiver on the ground. For further information on satellites, see Example 3 or chapter 19 on Space Physics.

Optical fibre systems

Optical communication systems are now replacing the old wire based communication systems. Telephone lines, cable TV, etc. all use pulses of light travelling along glass fibres.

The main advantage in using glass fibre is that with conventional wire based systems, a current is passed through a wire and if the wire is long, the current needs to be amplified every 2 km. Because the wires have to be kept thick to keep the resistance of the wire low, the wires tend to be bulky. Optical fibres do not have this problem and may be made very thin.

3 Analogue and digital signals

Analogue signals

Analogue signals have values which vary continuously with time. For instance the electrical signal passed along a telephone line will have a waveform which is a replica of the speech waveform. Analogue signals have an infinite number of inbetween positions whereas the other type of signal jumps from one value to the next. A watch with a dial and hands can show an infinite number of times whereas a digital one will jump from one minute or one second to the next.

Digital signals

Digital signals consist of pulses of voltage or current which represent the information to be processed. The digital voltages can either be high or low. A high voltage is used to denote a 1 and the low voltage a 0. In this way, information can be sent as a series of binary pulses (1s and 0s). It is cheap to produce circuitry to deal with digital signals and because of the binary nature of the signals, noise is not so much of a problem since if the signal becomes distorted, then the signal may be regenerated and made like new again. With analogue signals this is not possible because of the complex nature of the waveform so the effect of noise is to continually degrade the signal.

4 Operation of an amplitude modulated radio system

Modulation is the periodic variation of one or more properties of a signal or wave and the technique is used in communication where information needs to be sent from one place to another. When information is transmitted using an electromagnetic wave such as a radio wave, its amplitude fluctuates above and below a zero level at regular intervals. Amplitude modulation varies the height or amplitude of the wave in order to carry information. The frequency of some radio waves is measured in millions of

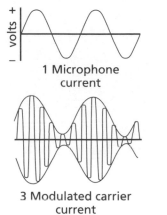

1 Microphone current

2 Radio-wave carrier

3 Modulated carrier current

4 De-modulated current in receiver

Hertz but speech or music have frequencies of less than 20 000 Hz. Low frequencies cannot travel large distances so the low frequency wave is superimposed on a high frequency carrier wave. This process is shown in the diagram.

When the amplitude modulated (AM) wave reaches a radio receiver tuned to the correct frequency, it is de-modulated (i.e. it has the carrier wave removed) to reveal the original wave that was carried.

Worked examples

Example 1

The diagram shows a motorway sign advertising a local radio station

TRAVEL NEWS	
1548 kHz	
194 m	MW
103.4 VHF	FM

The wavelength of the medium wave transmitter advertised on this sign is
A   1548 kHz
B   194 m
C   103.4 VHF
D   103.4 MHz

(ULEAC, Syll B, Jun 95, Intermediate Tier)

Solution

[VHF stands for very high frequency. Any unit with Hz in it is a frequency. Wavelengths are measured in m.]

Answer **B**

Example 2

Television sets receive radio waves from a transmitter.

Sometimes you get a type of interference called 'ghosting' on your TV set. 'Ghosting' is a second faint picture that overlaps and so tends to blur the main picture. It may occur as a result of reflection by a nearby building – see diagram on the next page.

The ghost picture occurs because of the reflected waves
A   interfere with waves from a local radio station
B   cause destructive interference
C   cause constructive interference
D   arrive a small time after the main waves

Solution

[The waves which are reflected off the office block travel a greater distance than those reaching the receiver direct. They therefore take longer and arrive slightly after the

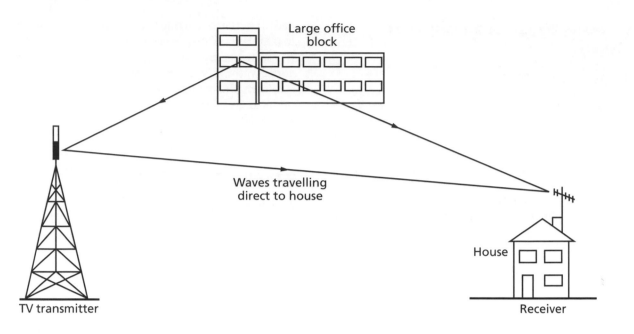

signals taking the direct route. Hence, there are two signals a short time apart reaching the receiver.]

Answer **D**

Example 3

INTELSAT IVA is a communications satellite in a geostationary orbit around the Earth. Its orbit is at a height of 35 800 km above the Earth's surface.

(a)   (i) Explain what is meant by a geostationary orbit.   (2 marks)

    (ii) State an advantage of having a geostationary orbit for a communications satellite.   (1 mark)

(b)   (i) If the radius of the Earth is 6400 km, show that the speed of INTELSAT IVA is 3100 m/s when it is in geostationary orbit.   (4 marks)

    (ii) Explain the condition required for a satellite to remain in circular orbit. Hence calculate the size if the Earth's gravitational field strength at a height of 35 800 km above the Earth's surface.   (4 marks)

(c)   (i) INTELSAT IVA transmits to Earth on a frequency of 4.0 GHz (4.0×10^9 Hz). If the speed of the carrier wave is 3.0×10^8 m/s, calculate the wavelength used for transmission.   (3 marks)

    (ii) Draw a labelled diagram to explain why the transmitting station would be unable to communicate with the satellite if it used a carrier wave of frequency 3.0 MHz (3.0×10^6 Hz).   (2 marks)

    (iii) Explain how the size of transmitting dish and the wavelength used for transmissions determine the maximum number of communication satellites that can be in geostationary orbit.   (3 marks)

(ULEAC, Syll A, Jun 95, Higher Tier)

Solution

(a)   (i) A geostationary orbit is one where the satellite is directly over the same point of the Earth's surface, all of the time. This means it has to have a period of 24 hrs which is the same as that of the Earth.

    (ii) Since the satellite is always in the same position relative to the Earth, communication using it can be made 24 hours per day. [You could also have mentioned that you do not need to track the satellite by moving the dish.]

(b)   (i) $\text{Speed} = \dfrac{\text{distance travelled}}{\text{time taken}}$

[The distance travelled will be a circle of radius $(6400 + 35\,800) = 42\,200$ km and centre the same as that of the Earth. The distance travelled by the satellite will be the circumference of this circle.]

$$\text{Speed} = \frac{2 \times \Pi \times r}{T}$$

[To obtain the speed in m/s using this formula, it is necessary to change the radius from kilometres to metres and the time from hours to seconds.]

$$\text{So, speed} = \frac{2 \times \Pi \times (42\,200 \times 1000 \text{ m})}{24 \times 60 \times 60 \text{ s}}$$

$$= 3069 \text{ m/s}$$

    (ii) A centripetal force is needed, acting towards the centre of the circle and this is provided by the gravitational force.

If g is the gravitational field strength at a radius r from the Earth's centre:

$$mg = \frac{mv^2}{r}$$

Dividing both sides by r gives

$$g = \frac{v^2}{r} = \frac{(3069 \text{ m/s})^2}{42\ 200\ 000 \text{ m}} = 0.22 \text{ N/kg}$$
$$\text{(or m/s}^2)$$

(c)   (i)   Wavelength $= \dfrac{\text{velocity}}{\text{frequency}}$

$$= \frac{3.0 \times 10^8 \text{ m/s}}{4.0 \times 10^9 \text{ Hz}}$$

$$= 7.5 \times 10^{-2} \text{ m} = 0.075 \text{ m}$$

(ii)   The carrier has a smaller frequency and therefore a higher wavelength.
[Remember, as the frequency goes down the wavelength goes up.]
A larger wavelength would be reflected by the ionosphere and would not reach the satellite.

(iii)   Signals sent from the transmitting dish are refracted and so spread out. If there are too many satellites sharing the same geostationary orbit, the signals from each dish would overlap and interfere with each other. Diffraction may be reduced by using a smaller wavelength for the transmission signal.

Example 4

The illustration below shows the passage of three rays of light along an optical fibre.

(a)   (i)   Which of the rays of light shown travels the greatest distance?
Explain your answer.   (2 marks)

(ii)   Which ray of light will reach the end of the optical fibre first?   (1 mark)

(b)   In an experiment to measure the speed of light the arrangement shown in the diagram at the top of the next column was used.
The subsequent illustration shows the traces produced in the CRO for one experiment.

(i)   If the time taken for the pulse to travel along the 20 m of cable was 9×10^{-8} s, calculate the speed of light in the cable.   (3 marks)

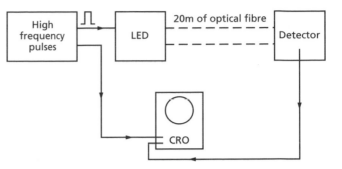

(ii)   Give a reason why the height of the pulse is reduced as it travels along the cable.   (2 marks)

(iii)   Explain why the pulse becomes spread out as it travels along the optical fibre.   (2 marks)

(iv)   One way of reducing the spreading of the pulse is to arrange for the speed of light to change as a ray moves out from the central axis towards the outer cladding.
Explain how and why the speed would have to change so that all the rays reach the end of the optical fibre at the same time.   (4 marks)
(NEAB, Syll B, Jun 92, Higher Tier)

Solution

(a)   (i)   Ray 1 because it will make a greater number of reflections with the side of the cable compared to the other rays.

(ii)   Ray 3 [It travels the least distance.]

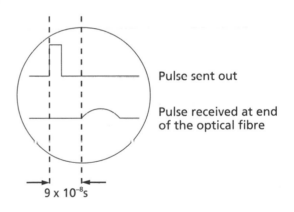

Pulse sent out

Pulse received at end of the optical fibre

9×10^{-8}s

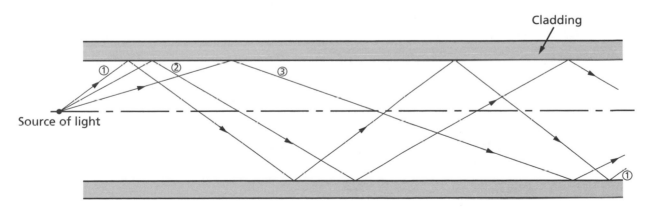

Cladding

Source of light

.(b)   (i) $\text{Speed} = \dfrac{\text{distance}}{\text{time}} = \dfrac{20\text{ m}}{9 \times 10^{-8}\text{ s}}$

$$= 2.2 \times 10^8\text{ m/s}$$

(ii) Some of the light is absorbed by the material so the amplitude of the pulse decreases with increasing distance through the fibre.

(iii) The light is diffracted as it passes through the fibre and this spreads the pulse out and rounds off the corners of its original shape.

(iv) Rays making more reflections with the side of the cable travel further and take more time than the other rays. The speed of light in the fibre depends on the refractive index of the material. By making the refractive index progressively less either side of the central line the speed of the pulse can be made to travel faster the further it is from the central line.

Example 5

The diagram below shows the essential parts of a simple radio.

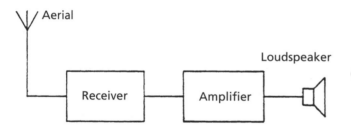

The aerial consists of a vertical length of aluminium and is used to pick up the radio waves which are carrying the signal. The radio waves are electro-magnetic waves.

(a)   (i) Which part of the electro-magnetic wave is able to create small a.c. currents in the aerial?

(ii) Explain how this part of the wave causes the currents to be produced.

(iii) It is noticed that when the aerial is turned through 90°, so that it is horizontal, the signal received by the radio falls substantially. What does this tell you about the electro-magnetic waves being transmitted?

(iv) For the diagram shown outline the energy transfers taking place.   (8 marks)

(b) Some radio receivers do not use an aerial but instead rely on a coil in the receiver to pick up the changing magnetic field of the electro-magnetic field of an electro-magnetic wave. This is shown in the diagram below.

(i) What will be produced across the ends of the coil by the changing magnetic field in the coil.

(ii) Inside the coil there is normally something called a ferrite rod.
Explain the function of this ferrite rod.

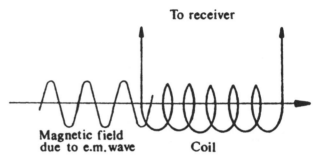

(iii) When a coil is arranged as shown below the signal detected is considerably reduced. Give a reason for this.   (4 marks)

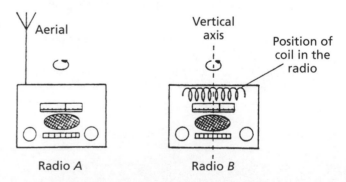

(c) The diagram below shows two radios A and B. Radio A makes use of a vertical aerial for detecting the signal, radio B makes use of a coil. Explain why if the radios are rotated through 360° about a vertical axis, there is no variation in the amplitude of the signal received for radio A whereas for radio B the amplitude does vary.

Radio A   Radio B

(d) A particular radio station transmits electro-magnetic waves with a wavelength of 1500 m. The speed of the electromagnetic wave is 3×10^8 m/s. Calculate the frequency of transmission.   (2 marks)

     (ULEAC, Syll A, Sample Questions, Higher Tier)

Solution

(a)   (i) The electric field.

(ii) The electric field produces a force on the free

electrons in the aerial and makes them start to move and the moving electrons constitutes a current.

(iii) They have been polarised in the vertical direction.

(iv) Electromagnetic wave energy to electrical energy input to the receiver to sound energy at the loudspeaker.

(b) (i) An induced e.m.f.

(ii) It increases the strength of the magnetic field inside and around the coil.

(iii) There aren't as many field lines being cut because the field is parallel to the coil.

(c) For *B* to receive the maximum signal, the coil must be situated with the field along the axis of the coil. This will only occur twice in every revolution of the radio. Radio *A* however will always have its aerial parallel to the electric field.

(d) Frequency (Hz) = $\dfrac{\text{Speed (m/s)}}{\text{Wavelength (m)}}$

$= \dfrac{3 \times 10^8 \text{ m/s}}{1500 \text{ m}}$

$= 200\,000 \text{ Hz} = 200 \text{ kHz}$

Examination questions

(Numerical answers and hints on solutions will be found at the end of the chapter.)

Question 1

(a) The diagram below shows an early method used to send line-of-sight messages over short distances. The message was transmitted by moving the side-arms into various positions.

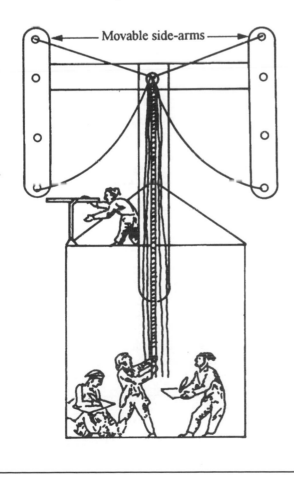

Movable side-arms

(i) On the diagram label the following:
- the transmitter
- the receiver
- the decoder
- the storage

(4 marks)

(ii) Which feature, shown on the diagram, indicates that the inventor had thought about one of the limitations of transmitting messages this way? Explain your answer.

(2 marks)

(b) The diagram below shows the basic building blocks of a more modern method of communication.

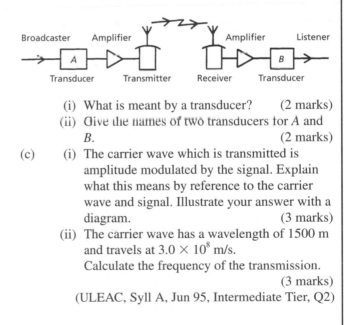

(i) What is meant by a transducer? (2 marks)

(ii) Give the names of two transducers for *A* and *B*. (2 marks)

(c) (i) The carrier wave which is transmitted is amplitude modulated by the signal. Explain what this means by reference to the carrier wave and signal. Illustrate your answer with a diagram. (3 marks)

(ii) The carrier wave has a wavelength of 1500 m and travels at 3.0×10^8 m/s. Calculate the frequency of the transmission. (3 marks)

(ULEAC, Syll A, Jun 95, Intermediate Tier, Q2)

Question 2

(a) In a large office block telephones are linked by cables.
These cables are usually made of copper.

(i) What are the advantages of using cable in such a system? (2 marks)

(ii) Why is copper preferred to other metals? (1 mark)

(b) Long distance intercontinental telephone calls are often transmitted by using microwaves.

(i) What advantages does microwave transmission have over transmission by cable? (2 marks)

(ii) State the disadvantage of using microwaves. (1 mark)

(c) Radio waves are to be transmitted between two Earth stations, *A* and *B*.
Radio waves of frequency well below 30 MHz are reflected back to Earth by the ionosphere.
Radio waves well above this frequency pass through the ionosphere.

Ionosphere

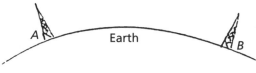

A   Earth   *B*

(c) (i) Mark on the diagram three possible paths by which the radiowaves can pass from *A* to *B*.

(ii) Station *A* is able to transmit and receive signals of the following frequencies:
300 kHz
3 MHz
30 MHz
300 MHz
State with reasons which frequency would be most suitable for communication by satellite.

(iii) The satellite is 3600 km above the Earth. Calculate the minimum time for signals to pass from *A* to *B* via satellite.
(Speed of radio waves = 300 000 000 m/s.)

(iv) Calculate the wavelength of the waves you have selected.

(v) If *A* and *B* were two local radio stations less than 100 km apart, which frequencies could be used?
State with reasons which you think would be best.

(ULEAC, Syll A, Sample Questions)

Answers to examination questions

1. (a) (i)
   - transmitter is either the movable arms or the person operating the ropes
   - receiver is the person with the telescope
   - decoder is the person standing with the pen
   - storage is the person sitting down with the pen.

   (ii) The receiver has a telescope so the towers can be further away than the naked eye can see. [Could also have had the fact that the side-arms are adjustable.]

   (b) (i) It changes one form of energy to another. For example, in the diagram *A* changes sound energy to electrical energy.

   (ii) *A* = microphone   *B* = loudspeaker.

   (c) (i) [See section 4 for diagram and explanation.]

   (ii) frequency = $\dfrac{\text{velocity}}{\text{wavelength}}$

   $= \dfrac{3 \times 10^8 \text{ m/s}}{1500 \text{ m}}$

   $= 200\,000 \text{ Hz} = 2 \text{ kHz}$

2. (a) (i) Information may be kept secure. Wires are cheaper than using radio communication.

   (ii) Copper is a very good conductor of electricity.

   (b) (i) Cheaper than having to lay long cables which sometimes have to pass along the ocean bed.
   You can make use of a satellite.

   (ii) They may be intercepted thus breaching security.

   (c) (i) [The diagram should show the following:]
   A beam from *A* to the satellite and then to *B*.
   A beam from *A* to the ionosphere and then to *B*.
   A beam directly from *A* to *B*.

   (ii) 300 MHz, since this will be able to pass through the ionosphere without being reflected.

   (iii) Time = $\dfrac{36\,000\,000 \text{ m} \times 2}{300\,000\,000 \text{ m/s}}$

   $= 0.42 \text{ s}$ [Signal has to travel two distances.]

(iv) Wavelength $= \dfrac{\text{Speed}}{\text{Frequency}}$

$= \dfrac{300\ 000\ 000 \text{ m/s}}{300 \times 1\ 000\ 000 \text{ Hz}} = 1 \text{ m}$

(v) The lowest frequency, 300 kHz, because there is less energy lost at lower frequencies and the wavelength will be longer so the waves will travel further.

18

The Earth, atmosphere and weather

Topic

WJEC & NEAB	ULEAC Syll A	ULEAC Syll B	MEG		MEG Salters'	MEG Nuffield	SEG	NICCEA
✓	✓			**Weathering and soil formation**				
✓	✓			**Types of rock and their formation**				
✓	✓	✓		**The Earth's structure**			✓	✓
✓		✓		**Plate tectonics, earthquakes and volcanoes**				✓
✓	✓	✓	✓	**Climate and what determines it**	✓	✓	✓	
✓		✓	✓	**Tides and phases of the moon**	✓	✓	✓	✓

1 Weathering and soil formation

Weathering is the name given to the breaking down of larger rocks by the water and air in the environment and is brought about in a variety of ways. Water can get into cracks which then freezes and forces the rock apart. Mountain streams break up pieces of rock. Pollution, such as acid rain, causes rocks such as limestone to dissolve.

Soil is finely divided rock particles, humus which is formed from the decay of plant and animal remains, and air. Soil can be divided into two: topsoil where most of the plants and animals live and subsoil which is mainly devoid of living material.

2 Types of rock and their formation

The solid rock which makes up the Earth's crust may be of the following types:

- **Igneous rocks:** These are rocks which are formed when the molten rock (called magma) is forced into the upper layers of the Earth's crust producing a rock such as granite or formed from erupting volcanoes producing a rock called basalt. Both types of rock cool slowly which gives them a crystalline appearance and the slower the cooling the larger the crystals.
- **Sedimentary rocks:** Weathering breaks rocks into smaller pieces and can end up being washed into rivers and eventually the sea. The resulting pebbles, sand and mud are collectively called sediments and these start to build up in layers. The weight of the sediments above squeeze the layers together and over millions of years they become cemented together to form sedimentary rock. Sandstone, limestone and shale are examples of sedimentary rocks.
- **Metamorphic rocks:** These are other types of rock which have changed owing to the fact that movement of the Earth has meant that the rocks are now buried deep underground and are now subjected to high temperatures and pressures. This causes a change in the texture and structure of the rock without melting it. Examples of metamorphic rocks include marble which is formed from limestone and slate which is formed from mudstone.

3 The Earth's structure

Information about the structure of the Earth has been obtained mainly from the study of how the shock waves travel inside the Earth during an earthquake. A machine called a seismometer is used to pick up the shock waves after travelling through the Earth and it is from this that we can find out about the structure of the Earth.

When there is an earthquake there are three types of shock waves called P, S and L waves which travel into the Earth. The P waves are the fastest and are able to pass through liquids as well as solids. In contrast, S waves can only travel through solids. L waves are only able to travel along the surface of the crust.

P and S waves travel in bent paths inside the mantle owing to refraction. From the direction that the waves are refracted we can tell that the waves travel faster in the mantle than in the crust and that the deeper they go into the mantle, the faster they travel.

Only the P waves pass through the core and this suggests that part of the core is made up of liquid.

L waves are restricted to the crust and tell us little about the structure other than the crust has a different density to the next layer.

From information regarding the refraction of these P waves we find that there is a large loss in speed as the waves travel from the mantle to the core. P waves also go faster as they get deeper inside the core and there are small increases in speed until about two-thirds of the way into the core after which the speed stays almost constant.

From this seismic data we find:

- That the Earth has a layered structure.
- That the Earth has a thin crust which is thicker under the continents (typically 5–10 km) than under the oceans (typically 25–90 km).
- That the Earth has a core and that the core has two parts with the outer one being liquid and the inner being solid. The core is about half the diameter of the Earth.
- That the mantle consists of rocks which are denser than those in the crust. The mantle consists of a viscous material (like bitumen which is used for surfacing roads).

The diagram shows the layered structure of the Earth.

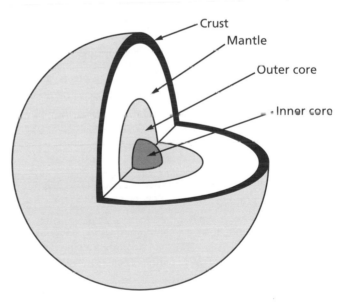

The magnetic properties of the Earth are due to electric currents inside the liquid outer core.

4 Plate tectonics, earthquakes and volcanoes

At one time it was thought that as the Earth cooled down it shrank and its surface became rippled and this accounted for the geological features such as valleys and mountains. This theory has now been discounted and the following alternative theory has been put forward called the Plate Tectonic Theory.

It is thought that at one time all the continents and land masses were joined together and have eventually broken apart and moved away from each other. The main evidence for this theory comes from the following facts:

1 The land masses have shapes which fit together fairly closely (compare the west coast of Africa with the east coast of South America).
2 The fossils, types of rock and structures of the rocks are similar where they would have moved away.

The map below shows how the land masses fit together.

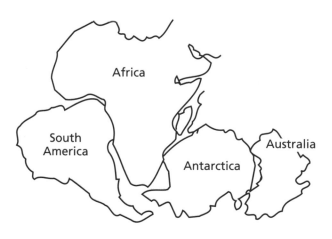

The theory is that the Earth's crust is cracked into a few large pieces and these pieces, called tectonic plates, are moving at a rate of a couple of centimetres per year owing to convection currents being set up inside the Earth due to the large amount of heat produced during the process of radioactive decay occurring inside the Earth's crust.

These tectonic plates may do any of the following:

- They may slide past each other causing earthquakes along the point of contact. This occurs along the coast of California.
- They may move towards each other. If this happens the thinner oceanic plate is driven down below the denser continental plate. This is called subducting and causes the thinner plate to be pushed down nearer the core where it melts. The continental plate is forced upwards as a result. In the areas where this occurs both volcanoes and earthquakes are common. This process is occurring along the western edge of South America.
- They may move further away from each other leaving a gap in the crust. Magma rises to fill the gap and this produces new basaltic oceanic crust. This process forms the ocean ridges such as the mid-Atlantic ridge.

5 Climate and what determines it

Recording the weather

The following measurements are taken to record details about the current weather which serves as an aid for predicting the future weather.

- Temperature (current temperature, maximum and minimum temperature)
- Air pressure
- Windspeed and direction
- Rainfall (usually in millimetres)
- The number of hours of sunshine per day
- Humidity (a measure of the amount of moisture in the air)

The water cycle

Water is continually evaporating from the surface of seas, oceans, rivers and lakes. Water also evaporates from the soil and plants and trees lose a considerable amount of water from their leaves in the process called transpiration.

Water vapour as a result of this evaporation is carried upward into the atmosphere. However, as it rises it cools down and condenses into small droplets which gather together and form clouds. As these droplets get bigger there comes a point when they are released from the cloud as either rain or snow, depending on the temperature. In this way the water is returned to the ground and the cycle, called the water cycle, carries on.

The Sun heats the Earth up by radiation and this radiation causes the land, sea and atmosphere to warm up during the day. Some of this heat is lost during the night with clear nights being colder due to the heat being rapidly lost. Clouds tend to insulate against this heat loss.

The Sun does not heat the whole Earth up evenly and this means there are temperature differences between places which causes convection currents to be set up. In the atmosphere, the air starts to move with the warm air rising causing a reduction in the pressure. The denser, colder air moves to replace it and the air pressure in this region will be higher. The movement of air in this way we call wind so the Sun is responsible for winds.

High and low pressure areas and their effect on the weather

High and low pressures are indicated on a weather map as a series of lines called isobars. Isobars are lines joining places which have equal pressures. In the diagram on the next page the high and low pressure regions are indicated by the words high or low or alternatively by using the letters *H* or *L*.

Isobars close together means that the pressure difference for a small change in distance is great and this indicates that strong winds are likely. Consequently widely spaced isobars indicate only slight winds.

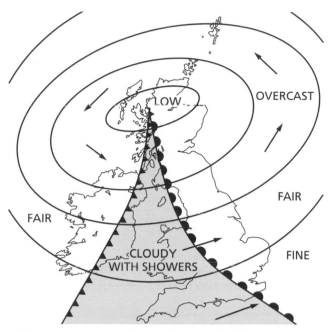

Low pressure areas

Winds around these blow parallel to the isobars in an anticlockwise direction. The weather in these areas is usually unsettled with wet and windy weather. The typical weather associated with a low pressure region is shown in the illustration.

High pressure areas

Winds around these blow parallel to the isobars in a clockwise direction. The weather in these areas is usually settled with clear, calm and fine weather.

Note: The above wind directions refer only to the Northern Hemisphere and in the Southern Hemisphere the situation would be reversed.

You may think that the winds would be likely to blow from high to low pressure regions. Things are not quite this simple owing to the influence of the Earth's rotation, so instead we have the following:

Air masses and the weather they bring

The weather in the British Isles is influenced by airstreams and where they have come from. The illustration below shows the main airstreams.

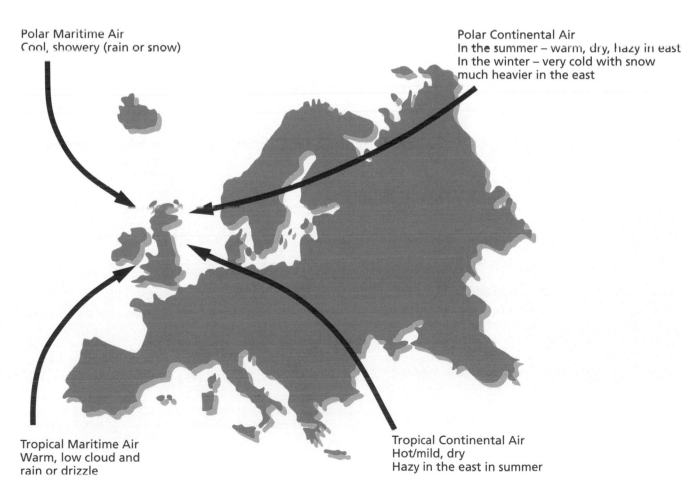

Polar Maritime Air
Cool, showery (rain or snow)

Polar Continental Air
In the summer – warm, dry, hazy in east
In the winter – very cold with snow much heavier in the east

Tropical Maritime Air
Warm, low cloud and rain or drizzle

Tropical Continental Air
Hot/mild, dry
Hazy in the east in summer

Air masses and the weather they bring

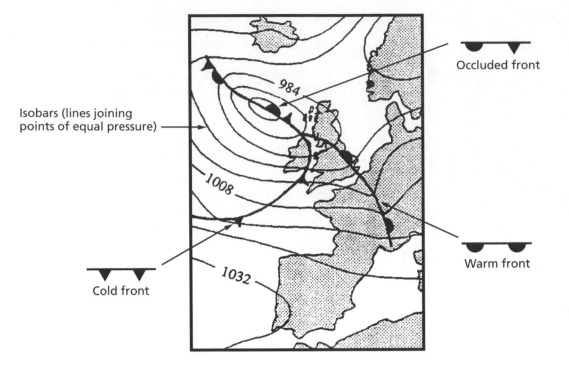

Isobars (lines joining points of equal pressure)

Occluded front

Cold front

Warm front

The main points from the diagram are:

- If the air mass has recently travelled over a large area of water it will bring wet weather.
- Drier air, after passing over large areas of land brings dry air.
- Polar air is much cooler than tropical air.

Fronts

When a warm and a cold air mass meet, they do not mix. Instead one of them displaces the other and takes over the situation and the point where they meet is called a front and is shown on a weather map as a thick black line.

A *warm front* is where a warm mass of air meets a cold mass and takes over it. When the warm front meets cold air, it rises above the cold air and tries to push it back. If a cold mass of air meets a warm mass and takes control of it by forcing the warm air over the top of the cold air mass, then this is called a *cold front*.

The illustration above shows a weather map showing fronts. As the fronts move over a particular point, the weather changes depending on the type of front, with cold fronts carrying cold dry air and warm fronts carrying warm moist air. The area in between the two fronts will experience changeable weather.

Sometimes the cold front catches up with the warm front and then overtakes it. When this happens, the warm front rises, cools and the water vapour it contains condenses and forms rain. This is called an *occluded front*.

6 Tides and phases of the Moon

Tides

Tides are due mainly to the gravitational pull that the Moon exerts on the water in the World's oceans and seas in the region nearest the Moon and also in a similar region on the opposite side of the Earth. The next illustration shows this. The Sun also exerts a pull but it is much smaller owing to its greater distance from the Earth.

The Earth makes one complete revolution per day so there are two high tides per day (one for the side nearest the Moon and once when the Earth has rotated and the position is now on the opposite side of the Earth. The water to make the high tides means that the other areas experience low tides and each position will have two such tides per day.

Phases of the Moon

The Moon is only seen because of the sunlight reflected off its surface. The Moon takes 28 days to orbit the Earth and during this time the various phases can be observed depending on the positions of the Moon and Earth in relation to the Sun. Different proportions of the Moon can be seen, from the new Moon where the Moon cannot be seen owing to the illuminated surface being positioned away from the Moon to the full Moon where the whole of one side of the Moon may be observed.

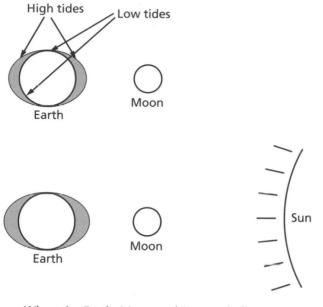

When the Earth, Moon and Sun are in line, a larger tide, called a spring tide, is produced

When the Sun is at right angles to the Earth and the Moon, the Moon's gravitational field cancels out slightly with that of the Sun, resulting in a smaller tide called a neap tide

The production of tides.

Worked examples

Example 1

The diagram shows some mountains which are near the sea.

(a) Choose the right words from the list below to complete these sentences which follow.   (3 marks)

**conduction   convection   diffusion
evaporation   expansion   radiation**

The air above the sea becomes moist due to the
.... of water.
The moist air is warmed by .... from the Sun.
The moist air rises because of ....

(b) Explain why are clouds more likely to form at X rather than Y.   (3 marks)
(SEG, Spec, Intermediate Tier)

Solution

(a) **evaporation**, **radiation**, **convection**.
(b) The moist air having just passed over the sea, starts to rise and as it rises it starts to cool down. As the water vapour in the air cools, it starts to condense and form small water droplets which are the clouds. As the clouds are pushed upwards the water droplets get larger and fall as rain on the side of the mountain nearest the sea. Y is in the shadow of the mountain.

Example 2

Use the weather map below to answer the questions which follow.

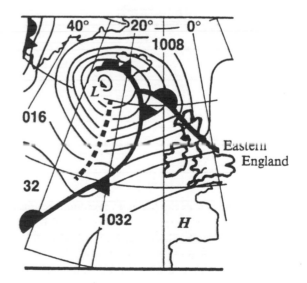

(a) What does the letter L on the map stand for?   (1 mark)
(b) Mark with a X on the map where gale force winds are forecast.   (1 mark)
(c) What type of frontal system is shown over the British Isles?   (1 mark)
(d) Explain why rain is forecast for Eastern England. You will be awarded up to 3 marks if you write your ideas clearly.   (6 marks)
(SEG, Spec, Intermediate Tier)

Solution

(a) *L* stands for the centre of the low pressure region.
(b) *X* should be marked anywhere inside the low pressure region. [In such a region the winds move around in an anticlockwise direction.]
(c) A warm front.
(d) There is a warm front over Eastern England and as the warm moist air rises over the cooler air it cools as it gets higher and the water vapour starts to condense and form droplets. As the water droplets get bigger they start to fall as rain.

Example 3

(a) Unlike the Sun, the Moon is not a source of light. Explain why you can see the Moon at night.

(2 marks)

The diagram shows the orbit of the Moon around the Earth. Not to scale.

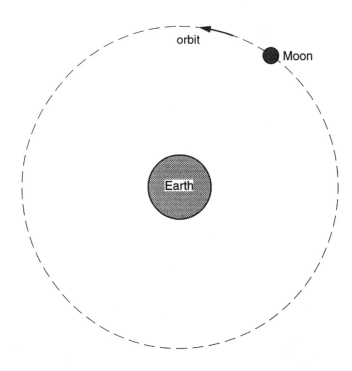

(b) What force causes the Moon to go round the Earth?
(c) (i) Mark the diagram with letters *X* and *Y* to show where high tides occur on Earth when the Moon is in the position shown. (1 mark)
    (ii) For either *X* or *Y*, explain why there is a high tide there.

(MEG Nuffield, Jun 95, Intermediate Tier)

Solution

(a) You are seeing the sunlight from the Sun reflected off the surface of the Moon.
(b) The centripetal force [anything moving in a circle has this force acting on it].
(c) (i) [The side of the Earth facing the Moon and the

opposite side both experience a high tide and either one should be labelled.]
  (ii) For the high tide nearest the Moon; the gravitational force of attraction due to the Moon attracts the water in the oceans and causes it to build up.

Example 4

During an earthquake the sudden movement of the rocks in the Earth's crust creates three types of shock wave:

P waves   speed 5 km/s; can travel through solids and liquids;
S waves   speed 3 km/s; can travel through solids only;
L waves   speed very slow; can travel only within the Earth's crust.

The diagram shows a point *E* which is the epicentre of an earthquake and the positions of three seismological stations *F*, *G* and *H*.

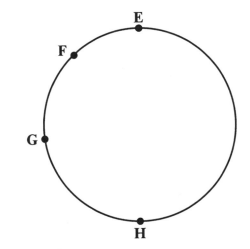

The information was recorded in table 18.1.

Table 18.1

	Time when P waves arrived	Time when S waves arrived
Station *F*	13.00	13.20
Station *G*	13.50	none detected
Station *H*	14.20	none detected

(a) Use the information given to calculate the distance between station *F* and the epicentre, *E*, of the earthquake. (3 marks)
(b) How does the information recorded at the stations help provide evidence for the Earth's internal structure? (3 marks)

(SEG, Spec, Higher Tier)

Solution

(a) Let *t* be the time in seconds taken by the **P** waves. To travel the same distance, the **S** waves take

$(20 \times 60 \text{ s}) = 1200$ s longer. Hence the time for the **S** waves is $(t + 1200)$ s. Since the distances travelled are the same we can say:

$3(t + 1200) = 5t$

Solving, gives $t = 1800$ s

The **P** waves take 1800 s.

Distance = speed $\times$ time = 5 km/s $\times$ 1800 s
= 9000 km

(b) See section 3.

Example 5

In an earthquake the sudden movement of the rocks produces 3 types of seismic waves, known as the **P**, **S** and **L** waves.

The diagram shows how these waves travel away from the epicentre of the earthquake.

(a) Explain in detail how the movement of the **P** and **S** waves through the Earth provides evidence to support the layered structure of the Earth shown in the diagram.

(b) Why do the **L** waves provide no evidence to support the layered structure of the Earth?

(c) What evidence is there to support the idea that the liquid in the outer core is moving?

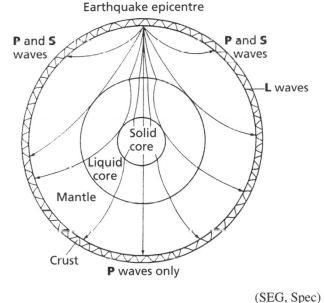

Earthquake epicentre

P and **S** waves

P and **S** waves

L waves

Solid core

Liquid core

Mantle

Crust

P waves only

(SEG, Spec)

Solution

(a) See section 3.

(b) **L** waves are only able to travel in the crust and therefore do not travel in the layers and therefore provide no evidence about them.

Example 6

Using physical principles, explain why or how

(a) Sea breezes blow towards the land during the day and away from it at night.

(b) Fog tends to hang around the ground.

(c) Smog is formed in polluted areas.

(d) Ground frost is formed.

Solution

(a) During the day, radiation from the Sun heats the land and sea up. The land heats up quicker than the sea and to a higher temperature. Air above the land rises and convection currents move this warm air upwards which brings in the cooler air over the sea to replace it. This flow air causes the wind to blow towards the land.

(b) Fog mainly occurs in the autumn when the air is saturated with moisture. During the night, especially during clear nights, the temperature drops owing to heat being radiated from the Earth. The air nearer the ground is cooler than that higher up and the water in this air starts to form droplets which we call fog.

(c) Smoke and dust in the air around cities and other airborne pollution causes something on which the droplets of water can condense. This is now called smog since it is a combination of smoke and fog.

(d) On a clear night the land loses heat quickly and the water vapour in the nearby air starts to condense. The resulting dew freezes and this gives rise to a ground frost.

Example 7

The diagram shows a section through the crust and upper mantle of the Earth.

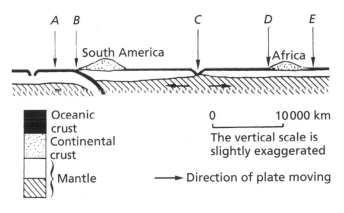

A B   C   D   E

South America

Africa

Oceanic crust

Continental crust

Mantle

0   10 000 km

The vertical scale is slightly exaggerated

→ Direction of plate moving

(a) (i) At which **one** of the places labelled *A* to *E* is oceanic crust being subducted (driven down)?

(ii) At which **one** of the places labelled *A* to *E* would you expect to find the youngest part of the oceanic crust? (2 marks)

(b) Oceanic crust and continental crust have different properties.

(i) Which type of crust has the higher density?

(ii) Which type of crust is granitic in composition? (2 marks)

(c) The rocks of the ocean floor are affected by the Earth's magnetic field.
  (i) Where is this magnetic field formed?
  (ii) Describe the pattern of magnetisation shown by the rocks on the ocean floor.
  (iii) Explain how this pattern supports the theory of plate tectonics. (8 marks)
      (NEAB, Jun 95, Higher Tier, Q10)

Solution

(a) (i) *B* (ii) *C*
(b) (i) Oceanic (ii) Continental.
(c) (i) In the liquid of the outer core.

  (ii) There is an alternate pattern of normally and reversely magnetised rocks with the bands running parallel to the ocean ridge. The pattern is symmetrical about the ocean ridge.
  (iii) The ocean ridge is formed from magma rising up from below and this causes sea floor spreading as this new material is added. This movement supports the theory of plate tectonics. When this new material cools, it takes on the magnetic field pattern of the Earth. However, since the magnetic field has reversed direction many times in the past, this gives rise to the alternating bands. The alternating bands prove that the sea floor has spread over time.

Examination questions

(Numerical answers and hints on solutions will be found at the end of the chapter.)

Question 1

(a) The diagram below shows what happens to the energy from the Sun.

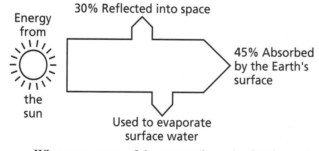

What percentage of the energy from the Sun is used to evaporate surface water? (1 mark)

(b) The following diagram is a cross-section of a mountain range showing rainfall. There are two mistakes. Cross out each of the mistakes and put in the correct information. (4 marks)

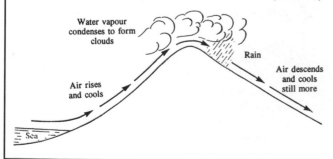

(c) Explain how the energy absorbed by the Earth's surface is transferred to the air around us. You may wish to draw a diagram. (3 marks)

(d) The diagram shows a town in a valley. In very hot weather it is possible for a layer of air in the atmosphere to be warmer than the air near to the ground.

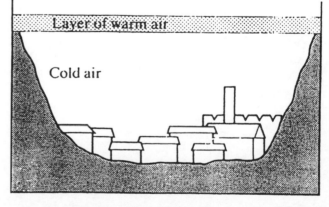

Show on the diagram how you would expect the smoke from the chimney to behave. (1 mark)
(SEG, Spec, Intermediate Tier)

Question 2

(a) On the diagram on the next page add arrows to show the water cycle. (1 mark)

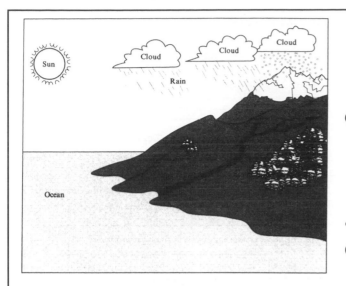

(b) The Sun is a very important part of the water cycle. Why? (2 marks)

(c) Give **two** ways the rain shown in the diagram can get back to the ocean. (2 marks)

(SEG, Jun 95, Intermediate Tier)

Question 3

The diagram below shows the different air streams (air masses) that can affect the British weather. Use the diagram to help you answer the questions that follow it.

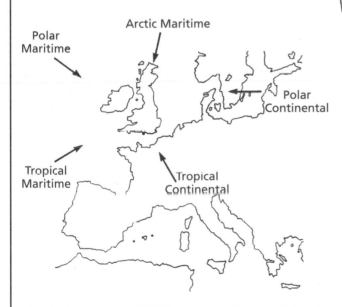

(a) In the summer of 1989, there were long periods of hot, dry weather. People noticed that their cars were often covered with a fine, dry red dust.

   (i) Which **one** of the five airstreams dominated the weather during the summer of 1989? (1 mark)

   (ii) Mark on the map with a **X** where the red dust is likely to have come from. (1 mark)

   (iii) How did the red dust get to Britain? (1 mark)

(b) The weather in the south-west of Britain is usually warm and humid (damp).

   (i) Which **two** airstreams are most likely to bring damp weather to Britain? Give a reason for your answer. (2 marks)

   (ii) Which **one** of the airstreams gives the south-west of Britain its usual weather conditions? Give a reason for your answer. (3 marks)

(c) Which **one** of the airstreams would cause the weather conditions described below?
Very dry, cool in summer, very cold in winter.
  (1 mark)

(SEG, Spec, Intermediate Tier)

Question 4

(a) Explain what is meant by the following terms:
   (i) igneous rocks
   (ii) sedimentary rock
   (iii) metamorphic rock (6 marks)

(b) A student examined two samples of the same type of igneous rock using a microscope at the same magnification.
She made the drawings below.

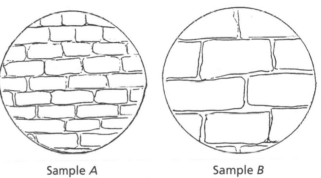

Sample *A*   Sample *B*

What difference in the origin of the two samples is likely to have caused this difference in appearance?
  (2 marks)

(c) The diagram represents a volcano erupting

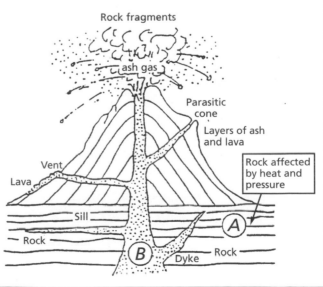

(i) What type of rock would you expect to be found at *A*?

(ii) In time another type of rock will form at *B*. Name the type of rock and explain how it is formed. (12 marks)

(ULEAC, Syll A, Spec, Intermediate Tier)

Question 5

The map shows a number of tectonic plates on the surface of the Earth. The arrows show the direction in which they are moving.

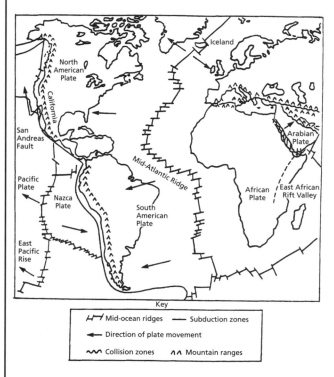

(a) What evidence suggests that South America and Africa may have been joined a long time ago? (2 marks)

(b) What is the evidence from the Mid-Atlantic Ridge that the American and African continents are moving apart? (2 marks)

(c) How does the movement of the tectonic plates produce earthquakes in California (West coast of North America)? (2 marks)

(d) What effects are produced when plates move towards each other, for example, on the West coast of South America? (2 marks)

(WJEC, Jun 95, Higher Tier, Q7)

Question 6

The map below shows the weather over the British Isles.

(a) Name the following weather features:
   (i) line *A* .... (1 mark)
   (ii) line *B* .... (1 mark)
   (iii) line *C* .... (1 mark)
   (iv) region *D* .... (1 mark)

(b) What quantity is being measured by the numbers 980, 984, 988? (1 mark)

(c) By means of arrows on the diagram, indicate wind directions at *E* and *F*. (1 mark)

(d) (i) How does the strength of the wind at *E* compare with that at *F*? (1 mark)
   (ii) Explain your answer. (1 mark)

(e) Rain will fall near feature *B* or *C*. Choose **one** of these and explain why rain is produced. (4 marks)

(WJEC, Jun 95, Higher Tier, Q5)

Answers to examination questions

1. (a) 25% [All the percentages have to add up to 100.]
   (b) The rain is drawn on the wrong side of the mountain and needs redrawing on the other side.
       'Air descends and cools still more' is wrong. Needs replacing with 'Air descends and warms up'.
   (c) Radiation from the Sun heats the Earth's surface which then heats up the nearest layer of air by conduction. The air expands, becomes less dense and rises. The cooler air moves in to replace this air.
   (d)

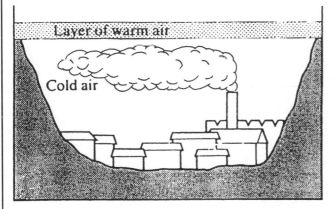

2. (a) [Arrows need to be drawn in an anticlockwise direction rising up from the sea and falling over the mountains and moving down the river and back to the sea.]
   (b) The Sun provides the heat energy needed to evaporate the water.
   (c) Water is carried by the rivers back to the oceans. Snow melts and is carried into oceans.

3. (a) (i) Tropical Continental.
       (ii) **X** should be drawn on North Africa (sand is from the Sahara desert).
       (iii) It is picked up and carried by the Tropical Continental air mass.
    (b) (i) [Any air mass which has travelled over water] Polar Maritime, Arctic Maritime and Tropical Maritime. They have all recently travelled over the sea and will be rich in water vapour.

   (ii) Tropical Marine because the air is warm and moist.
   (c) Polar Continental.

4. (a) (i), (ii) and (iii) see section 2.
   (b) Different rates of cooling. The larger crystals in sample *B* are caused by slower cooling compared to sample *A*.
   (c) (i) Metamorphic.
       (ii) Igneous rock is formed due to the cooling and subsequent crystallisation of the lava.

5. (a) They have shapes which fit together quite closely. The rocks and the fossils found in them are similar for the two land masses.
   (b) The Mid-Atlantic Ridge is getting larger due to volcanic activity. This means that the ocean floor in this region is spreading out.
   (c) It is in this region that the two plates are sliding past each other. Every now and again they stick and then move and this gives a jolt which we call an earthquake.
   (d) This is the process called subduction where the edge of the denser oceanic plate sinks under the less dense continental plate. On moving down, the oceanic plate starts to melt as it meets the mantle.

6. (a) (i) Isobars (ii) Cold front (iii) Warm front
       (iv) Low pressure (or depression).
   (b) Atmospheric pressure.
   (c) Arrows should point anticlockwise around the low on the diagram.
   (d) (i) Wind is stronger at *E* than at *F*.
       (ii) The isobars are closer together at *E* which means that the pressure change with distance is greater so there will be strong winds.
   (e) The cold air from the west at *B* will push underneath the warm air causing the warm air to rise. The rising warm air cools down as it gains height and the water it contains starts to condense thus forming rain. Once the cold front has passed over, the rain will stop.

19

Space Physics

WJEC & NEAB	ULEAC Syll A	ULEAC Syll B	MEG	Topic	MEG Salters'	MEG Nuffield	SEG	NICCEA
✓	✓	✓	✓	**Gravity and gravitational field strength**	✓	✓	✓	✓
✓	✓	✓	✓	**Satellites**	✓	✓	✓	✓
✓	✓	✓	✓	**The universe and its expansion**	✓	✓	✓	✓
✓	✓	✓	✓	**Days and seasons**	✓	✓	✓	✓
✓	✓	✓	✓	**Our solar system**	✓	✓	✓	✓
✓	✓	✓	✓	**Solar and lunar eclipses**	✓	✓	✓	✓
✓	✓	✓	✓	**The lifecycle of stars**	✓	✓	✓	✓

1 Gravity and gravitational field strength

Gravity is the force that holds the whole of the universe together and is responsible for the planets moving in their orbits around the Sun. Gravitational force is the force between two masses and it depends on the size of the masses with larger masses exerting a stronger pull on each other. It also depends on the distance between the centres of the masses with masses closer together exerting a larger gravitational force. The variation of force with distance follows an inverse square law.

Our Moon travels through space and if there was no force acting on it, it would move in a straight line. The Earth's gravitational force causes the Moon to move in an almost circular orbit.

Gravitational field strength is the number of newtons of force a 1 kg mass would experience in a certain place in the gravitational field. On the Earth's surface, for instance, the gravitational field strength has a value of 10 N/kg. The Moon is much smaller and the field strength on the surface of the Moon is only one sixth of that on the Earth.

2 Satellites

To overcome the pull of gravity it is necessary to give a satellite a speed of at least 28 000 km/h in a direction parallel to the surface of the Earth. Once it is above the Earth's atmosphere it will circle endlessly around the Earth in its own orbit.

The main uses of satellites are as follows:

- **Communication:** For sending telephone calls, faxes, computer data, TV pictures around the world.
- **To monitor conditions on the Earth:** You can see weather patterns developing and give accurate weather forecasts.
- **For astronomical purposes:** Distant galaxies can be looked at without the Earth's atmosphere getting in the way.

Polar orbiting satellites

Weather satellites are often placed in low Polar orbits and

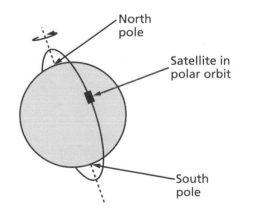

A polar orbiting satellite.

this means that they pass over both the North and South Poles. Since the Earth spins on its axis it means that in this type of orbit a satellite will be able to scan the whole Earth.

Geo-stationary orbiting satellites

Most communication satellites need to stay in the same position above the Earth so they need to be in an orbit that rotates at the same rate as the Earth. This means that they need to have a period of rotation of 24 hours. When viewed from the Earth, these satellites appear in the same position. Such an orbit is called a geo-stationary orbit.

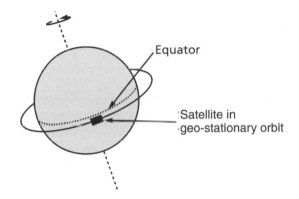

A geo-stationary orbiting satellite.

3 The universe and its expansion

The universe is mainly space (nothing) but which also has matter dotted in it in the form of galaxies. It is reckoned that there are billions of galaxies in the universe. Each galaxy consists of billions of stars very similar to our Sun and looks like a giant spinning Catherine wheel. It is almost certain that around these Suns could be planets and some could be similar to our Earth. By observing distant galaxies we can see that the universe is expanding.

The Sun

Originally it was thought that the Sun was hot because of violent chemical reactions occurring in its interior. Now we know that these reactions are nuclear reactions. The nuclear reaction which occurs in the Sun is called a fusion reaction because smaller nuclei (the nuclei of hydrogen atoms) fuse (stick) together to form larger helium nuclei. The illustration shows this happening.

4 Hydrogen nuclei   Helium nucleus
                                     + large release of energy

Nuclear fusion.

The mass of the larger helium nuclei is less than the total mass of the two hydrogen nuclei used to fuse together and this difference in mass occurs because some of the matter has been changed into energy. In fusion reactions, a small loss of mass releases a huge amount of energy compared to the energy released from a similar mass during a chemical reaction like that of burning a chemical such as oil in air.

4 Days and seasons
Days

The Sun appears to an observer on the Earth to rise in the East and slowly move across the sky reaching its highest point at mid-day and then falling till it sets in the west, but it does not actually do this since it is really the Earth which is moving from west to east.

The Earth spins around on its axis once every 24 hours from west to east and to an observer on the Earth it makes the Sun appear to move in the opposite direction.

As well as this rotation, the Earth travels in an orbit around the Sun and the time taken to complete one orbit is 365.25 days. The period for the orbit we call a year. Every four years the extra quarter of a day is taken into account by calling the year a leap year and making the extra day into the 29 February.

The seasons

The seasons are due to the angle of tilt of the Earth to the plane of the Sun (23.5°).

When the Earth is on position A the top half of the Earth (i.e. the Northern Hemisphere) is further away from the Sun than the bottom half (the Southern Hemisphere) and this means that at this time of year there will be Winter in the Northern Hemisphere and Summer in the Southern. The situation is reversed in position B.

For places tilted towards the Sun

(a) the period of daylight is longer than that of the night, and

(b) the Sun rises higher in the sky.

The net effect of (a) and (b) is that more energy is received from the Sun and the weather is therefore warmer.

5 Our solar system

Our solar system consists of the Sun and its orbiting planets with their moons.

Planets

There are nine planets, including the Earth, which circle around the Sun. Putting them in order with the one nearest the Sun first we have:

Mercury, Venus, Earth, Mars, Jupiter, Saturn, Uranus, Neptune and Pluto.

These nine planets, along with our Sun, make up our solar system. Although the planets can appear bright in the sky, they give out no light of their own and only appear bright because of the Sun's rays being reflected off their surface.

Terrestrial planets (sometimes called rocky planets)

These are the Earth-like planets because they are small and dense and are made of rock. Earth, Mercury, Venus and Mars are classed as terrestrial planets and they are the four nearest planets to the Sun. Like the Earth, these planets have layered structures and show evidence of volcanic activity.

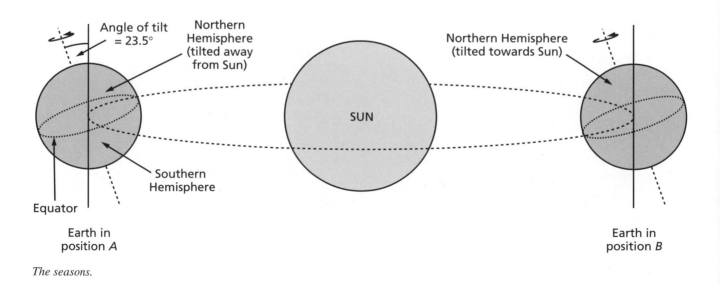

The seasons.

Icy planets

These are the larger planets which are quite unlike the Earth since they have deep atmospheres above deep oceans of liquid gas (mainly hydrogen). The atmospheres of these planets show intensely violent weather patterns. Icy planets are further from the Sun than the terrestrial planets and because of this they are much colder so many of the elements which would be gases on the Earth are now liquid on these planets.

Saturn, Uranus, Neptune and Pluto are classed as icy planets, although Pluto could be considered as a rocky planet as well.

The Moon

The Moon is not a planet but instead a satellite of the Earth. Like the planets, the Moon gives out no light of its own so the light we see is due to sunlight being reflected off its surface.

If you look at the Moon over a period of a month, each night it will look slightly different owing to the fact that the Moon takes 4 weeks to rotate around the Earth. Some of the appearances or phases of the Moon are shown in the illustration.

Many of the other planets in our solar system also have moons. Mars has 2 moons and Saturn has at least 22.

The Moon is much smaller than the Earth and does not therefore exert as strong a gravitational pull on nearby objects. On the Moon's surface the gravitational field strength is only one-sixth of what it is on the Earth's surface.

The Moon's gravitational pull is responsible for the tides. Chapter 18 explains the production of tides.

6 Solar and lunar eclipses

Solar eclipse (eclipse of the Sun)

This occurs when the Moon passes between the Earth and the Sun and casts a shadow on the Earth's surface. A small area (called the umbra) experiences total darkness and the eclipse here is said to be total. Around this area is an area where some light still gets through (called the penumbra) and the eclipse is partial. The illustration below shows this.

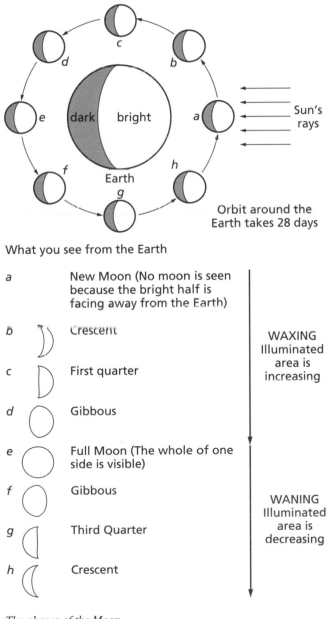

What you see from the Earth

a		New Moon (No moon is seen because the bright half is facing away from the Earth)
b	Crescent	WAXING Illuminated area is increasing
c	First quarter	
d	Gibbous	
e	Full Moon (The whole of one side is visible)	
f	Gibbous	WANING Illuminated area is decreasing
g	Third Quarter	
h	Crescent	

The phases of the Moon.

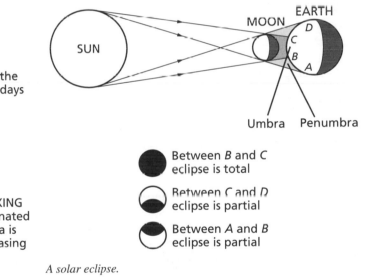

A solar eclipse.

Lunar eclipse (eclipse of the Moon)

This occurs when the Earth moves between the Sun and the Moon. At position A the whole of the Moon is in total darkness (i.e. inside the umbra) whereas at B and C the Moon is in the penumbra. The next illustration shows the arrangement.

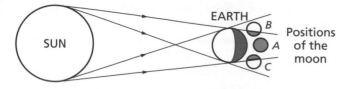

A lunar eclipse.

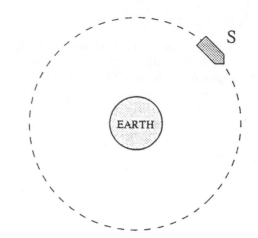

7 The lifecycle of stars

Our Sun is a star and is the nearest star to the Earth. The stars are millions of times further apart than the planets are.

Stars are formed when gases and dust are pulled together by gravity. The much smaller planets are produced in a similar way. From their initial 'birth' stars go through a whole lifecycle. When formed, the stars are so massive that the force of gravity that they have pulls the matter together very tightly and the resulting high temperatures produce a force in the opposite direction. The forces are therefore balanced and the star is in its stable period. Our Sun is in this stable period at the moment.

In the next part of its lifecycle it expands to become a red giant and later on it starts to contract under its own gravity to become a white dwarf with a density millions of times greater than any material on the Earth. Some very massive stars may then explode to become a supernova which throws dust and gases into space leaving a neutron star behind.

The stars stay in almost fixed positions with regards to each other and the arrangements are called *constellations*. The planets give the appearance of stars from the light reflected off them but they do not stay in fixed positions when compared to the stars. The positions of the planets against the background of the stars depends on where they and the Earth are in their orbits around the Sun.

Theories regarding the origin of the universe

It is found that the light from other galaxies is shifted towards the red end of the spectrum (i.e. towards the longer wavelengths). It is also found that the further away the galaxies are, then the greater this red shift. Any theories as to the origins of the universe has to be able to explain these observations.

The most popular theory is that the universe is expanding so the galaxies are all moving further apart and that as they move further away from us, they move faster. If the universe is expanding then this suggests that at one time it must have been in one place billions of years ago and has exploded outwards. The huge explosion which formed the universe is aptly called the 'Big Bang'.

Worked examples

Example 1

Satellites and our Solar System

The diagram represents the path of a satellite, *S*, travelling at constant speed in a circular orbit around the Earth.

(a)   (i) Draw an arrow labelled *v* to show the direction of the velocity of the satellite, *S*.   (1 mark)
       (ii) Draw an arrow labelled *a* to show the direction of the acceleration of the satellite, *S*.   (1 mark)

(b)   The satellite is in orbit over the North and South Poles. It takes 90 minutes to complete one orbit. A tracking station in Britain observes the satellite overhead at 9 a.m. It next observes the satellite overhead at 9 a.m. the following day. Explain the reason for this.   (2 marks)

(c)   A television company wishes to place a satellite in orbit so that it is always over the same point on the Earth's surface.
       (i) Where on the Earth's surface should this point be?   (1 mark)
       (ii) How long will the satellite orbit the Earth?   (1 mark)

(MEG Nuffield, Jun 94, Intermediate Tier)

Solution

(a)   (i) Arrow should be straight out from *S* (i.e. at a tangent to the circle).
       (ii) Arrow should point towards the centre of the circle since this is direction of the centripetal force.

(b)   In the 24 hours between 9 a.m. and 9 a.m. the next day the Earth will have made one complete revolution. The Polar orbiting satellite will have

made $\frac{24}{1.5} = 16$ complete orbits. After 16 orbits

of the satellite and one complete revolution of the Earth, the paths coincide again.

(c)   (i) Anywhere above the Equator.
       (ii) 24 hours (1 day).

Example 2

(a)   The next illustration shows the phase of the Moon and the times of sun-rise and sun-set in London on 9 November 1990.

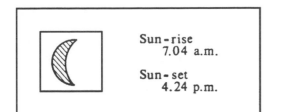

Sun-rise
7.04 a.m.

Sun-set
4.24 p.m.

(i) What phase of the Moon is shown in the figure?
(ii) If you looked at the Moon over the next few weeks its phase would change.
Using the letters shown below write down the correct order of the phases you would see over the next few weeks.

Full moon

· A   B   C

Correct order would be ....
(iii) About how many weeks does it take the Moon to go through all of its phases?
(iv) Using the words *earlier* or *later* complete the following two sentences.
The time of sun-rise on 16 November will be .... than 7.04 a.m.
The time of sun-set on 1 November was .... than 4.24 p.m.
(v) How do you know that on the surface of the Moon there are mountains and craters?

(6 marks)

(b) The constellation known as The Plough appears one evening as shown below.

* Pole star

(i) Add to the diagram the position of The Plough 6 hours later.
(ii) Explain why the constellation appears to move during the course of the night.
(iii) Explain why the positions of the stars relative to each other do not appear to change.

(4 marks)
(ULEAC, Syll A, Spec, Intermediate Tier)

Solution

(a)　(i) Third quarter
　　(ii) *C A B*
　　(iii) 4 weeks
　　(iv) Later.
　　　　Later.

(v) Can observe shadows.

(b)　(i) [The Earth will have rotated by one-quarter of a turn from west to east]
90° in a clockwise direction.
　(ii) Because the Earth is rotating about its axis.
　(ii) They are very distant and even if they were moving quite fast, the motion would not be noticed on the Earth.

Example 3

(a) Choose the correct words from the box below to complete the sentences which follow:　　(3 marks)

energy	galaxy	gravity	Jupiter
Mercury	Moon	planet	Pluto
Saturn	solar system	Sun	universe

Every .... goes around the Sun. .... is nearest the Sun, and .... is furthest away from it.

(4 marks)

(b) Rewrite the following list in order of their sizes.
* galaxy
* Moon
* planet
* solar system
* universe

smallest:　.............
　　　　　.............
　　　　　.............
　　　　　.............
largest:　.............

(c) Read through the following article and then answer the questions which follow.

People have been interested in Mars for many years because it is the planet which comes nearest to Earth. We now know that Mars has days which are 24 hours 37 minutes long and that it takes Mars 687 days to orbit the Sun. Mars has seasons, though they are longer than those on Earth. Its daytime temperature rarely reaches 16°C, even on the Martian equator. Gravity on the planet is about 40% of that of Earth. This suggests that the planet's atmosphere is thin and so the nights get very cold. The atmosphere on Mars is mostly carbon dioxide gas. There is no liquid water on the surface of the planet, though it has polar ice caps made of ice and solid carbon dioxide.

(i) In the space below draw a table to show:
two features which make Mars similar to the Earth, **and**
two features which make Mars different from the Earth.
You will be awarded up to two marks for the clarity of your table.　　(6 marks)
(ii) Suggest and explain **one** reason why the gravity on Mars is less than that on Earth.　　(3 marks)
(iii) Suggest and explain **one** reason why the nights on Mars are very cold.　　(3 marks)

(SEG, Spec, Intermediate Tier)

Solution

(a)   planet, Mercury and Pluto.

(b)   Moon
       planet
       solar system
       galaxy
       universe.

(c)   (i)   *Similarities.* Any **two** from: both have seasons, both have polar ice caps, their days are of similar lengths and they both orbit the Sun.
              Differences. Any **two** from: Longer year/seasons/days, less gravity, colder days, no liquid water, more CO_2 in the atmosphere.
              The above should be displayed in table form with sensible headings.
       (ii)   Mars has less mass than the Earth, since gravity is proportional to mass.
       (iii) Mars has a thin atmosphere with not much air and there are no clouds and because of this the insulation around the planet is not as effective so the heat is much more readily radiated into space.

Example 4

(a)   Scientists think that the Sun and other stars were created from dust particles and gas molecules which were attracted together.
       What force caused this attraction?   (1 mark)

(b)   What happened to the kinetic energy of the particles as result of this attraction?   (1 mark)

(c)   The centre of a dust cloud which is turning into a star becomes so hot that the process of nuclear fusion can take place.
       Describe the process of *nuclear fusion*.   (3 marks)

(d)   The nucleus of any atom is positively charged.
       Explain why this makes it very difficult for nuclear fusion to occur unless the temperature is very high.
       (3 marks)

(e)   The nuclear fusion reaction in the centre of our Sun will eventually run out of fuel. What is expected to happen to the Sun after that?   (3 marks)
       (MEG, Jun 95, Higher Tier)

Solution

(a)   Gravitational force.

(b)   The kinetic energy starts to increase and is eventually turned into heat.

(c)   The nuclei are moving so fast and are at such a high temperature that they can be fused together to form larger nuclei with the release of a huge amount of energy. This process is called fusion and is responsible for giving the star the huge amount of power it needs to shine.

(d)   There will be a large force of electrostatic repulsion as the nuclei move closer together since they are both positively charged. If they have enough kinetic energy due to their high temperature they may be able to overcome this and fuse together.

(e)   It will grow larger and go cooler and become a red giant. Some of the matter will be lost from the red giant and produce a smaller body called a white dwarf. It will then run out of energy and become a black dwarf.

Example 5

(a)   Every object attracts every other object with a force.
       (i)   What type of force is acting between the Earth and the Moon?   (1 mark)
       (ii)   What **two** factors does this force depend on?   (2 marks)
       (iii) What evidence is there on Earth to show the force of the Moon?   (1 mark)

(b)   The asteroids can be considered as bits of a planet that fell apart. They orbit between Mars and Jupiter. Use the data in table 19.1 to suggest similar data for these asteroids. Give a brief explanation for your suggestions.   (6 marks)

Solution

(a)   (i)   Gravitational.
       (ii)   The mass of the objects and their distance apart.
       (iii) Tides.

(b)   Distance from the Sun can be anywhere between that for Jupiter and Mars.
       Surface temperature again has to be somewhere between the values for Jupiter and Mars.
       The total relative mass would be between the masses of Jupiter and Mars.

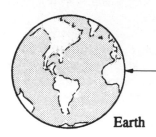

Earth   **Moon**

Table 19.1

Planet	Relative mass (Earth = 1)	Surface gravity field N/kg	Distance from Sun/Mm	Surface temp °C
Mercury	0.05	3.6	58.0	350
Venus	0.81	0.87	107.5	460
Earth	1.0	9.8	149.6	20
Mars	0.11	3.7	228	−23
Jupiter	318.0	25.9	778	−120
Saturn	95.0	11.3	1427	−180
Uranus	14.0	10.4	2870	−210
Neptune	17.5	14.0	4497	−220
Pluto	0.003	–	5900	−230

(SEG, Spec, Higher Tier)

The surface gravity depends on the sizes of the asteroids and not on the original size of the planet which has broken up so it is impossible to give a value for this.

Example 6

The diagram shows sound waves from
A a stationary source
B a source that is moving to the left.

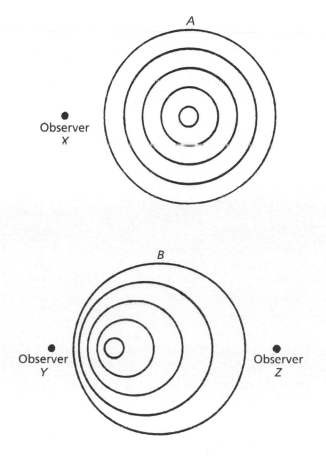

(a) (i) Describe the sound that observer X will hear from source A.
  (ii) Compare the frequencies of the sounds heard by observers Y and Z as the sound waves reach them from source B.   (3 marks)

(b) The diagram below shows the line emission spectra from
A a nearby galaxy
B a distant galaxy.

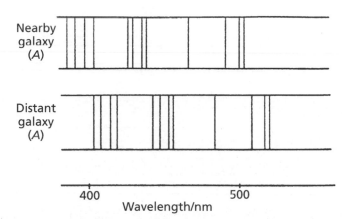

(i) Outline how a line emission spectrum can be produced in a school laboratory.
(ii) What does a line emission spectrum tell you?
(iii) What evidence shown in the diagram supports the view that the universe is expanding? Explain how it supports this view.   (5 marks)

(c) In 1929 Hubble proposed that the speed of a galaxy moving away from us was given by $v = H \times D$ where D was its distance and H the Hubble constant.
  (i) Suppose the galaxy has moved with a velocity v for a time t, where t is the age of the universe. How far would it have travelled in this time?
  (ii) Use the expression in (c)(i) and $v = H \times D$ to find an equation for t the age of the universe.

(iii) If $H = 2 \times 10^{-18}$/s how old is the age of the universe in years?
(1 year $= 3 \times 10^7$ s) (5 marks)

(d) The graph below shows two possible ways in which the distance of a galaxy might vary with time from the beginning of the universe.

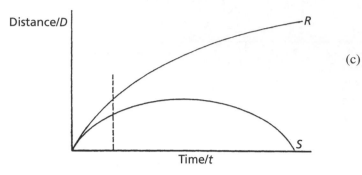

(i) Both curves show that at present the expansion of the universe is slowing down. State how the graphs show this and explain what is causing the rate of expansion to decrease.

(ii) What does graph (S) suggest what will happen to the universe in the future?

(iii) State what condition needs to be satisfied so that the universe will continue to expand as shown in graph (R). (7 marks)
(ULEAC, Syll A, Spec, Higher Tier)

Solution

(a) (i) A note of constant frequency will be heard.
(ii) A higher frequency will be heard at B
A lower frequency will be heard at C.

(b) (i) A sodium lamp excites the gaseous sodium atoms using electrical energy. The light is viewed through a diffraction grating using a spectrometer.
(ii) The line spectrum may be used to identify the gaseous element/s in the gas.
(iii) The line spectrum has lines at a longer wavelength than would be expected. [The lines are shifted towards the red end of the spectrum.]

(c) (i) $D = vt$
(ii) $D = H \times D \times t$ and dividing both sides by D gives $1 = H \times t$

Hence, $t = \dfrac{1}{H}$

(iii) $t = \dfrac{1}{(2 \times 10^{-18})} = 5 \times 10^{17}$ s

$\qquad = \dfrac{(5 \times 10^{17})}{(3 \times 10^7)}$ years $= 1.67 \times 10^{10}$ years

(d) (i) The gradient of the graph indicates the speed. The gradient decreases with time so the expansion is slowing down.
Other galaxies are attracting it.
(ii) It will expand up to a maximum distance and then start to contract.
(iii) The universe can only be an open one [an open one will continue to expand forever] if the initial KE produced during the Big Bang is less than the total gravitational potential energy caused by the total mass contained in the universe.

Examination questions

(Numerical answers and hints on solutions will be found at the end of the chapter.)

Question 1

The diagram below is not to scale.

(a) (i) Add to the diagram **TWO** arrows to show the direction of:
(A) the pull of the Earth on the Moon (label this E);
(B) the pull of the Moon on the Earth (label this M). (2 marks)

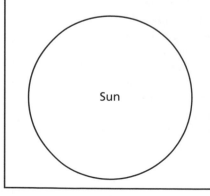

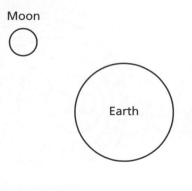

(ii) Explain why the Moon orbits the Earth.
(4 marks)

(b) Both the Sun and the Moon can be observed from the Earth's surface.
  (i) What process occurs in the Sun which produces light? (1 mark)
  (ii) The Moon does not produce its own light. How are we able to 'see' the Moon? (2 marks)

(c) Table 2 gives the surface temperatures of Mercury and Venus and their distance from the Sun.

Table 2

	Maximum surface temperature (°C)	Distance from the Sun (millions km)
Mercury	350	58
Venus	460	108

  (i) Explain why the maximum surface temperature of Mercury ought to be higher than that of Venus. (2 marks)
  (ii) It is thought that the atmosphere of Venus consists of a dense layer of carbon dioxide and sulphur dioxide. Explain in detail why this might cause the surface temperature of Venus to be higher than that of Mercury. (3 marks)

(d) Place the following in increasing order of size, starting with the smallest.
  galaxy, Sun, universe, solar system, planet

(e) Why would it be difficult for humans to travel to a distant galaxy? (3 marks)
  (ULEAC, Syll A, Jun 95, Intermediate Tier)

Question 2

A television satellite is in a circular orbit around the Earth

(a) Explain as fully as you can, why the satellite stays in orbit. (4 marks)

(b) A second satellite, which has exactly the same mass as the first, is put into a higher orbit. The second satellite moves at half the speed of the first.
  Compare the kinetic energy of the satellites as they move around their orbits.
  Explain your answer. (3 marks)
  (NEAB, Jun 94, Modular Double Award, Higher Tier)

Question 3

At the very high temperatures in the Sun, hydrogen is converted into helium. It takes four hydrogen nuclei to produce one helium nucleus.

Table 3 shows the relative masses of hydrogen and helium nuclei.

Table 3

Nucleus	Relative Mass
Hydrogen	1.007825
Helium	4.0037

Hydrogen nucleus   Helium nucleus

(a) Use these figures to calculate what happens to the mass of the Sun as hydrogen is converted to helium. (3 marks)

(b) Use your answer to part (a) to explain how the Sun has been able to radiate huge amounts of energy for billions of years. (2 marks)
  (NEAB, Jun 94, Modular Double Award, Higher Tier)

Question 4

This question is about the solar system and the universe.

(a) Name and describe the process by which the Sun releases energy. (4 marks)

(b) Currently the Sun is stable. The gravitational forces holding it together are balancing the internal forces trying to make it expand. Describe the processes which occur when a massive star passes through the stages of being
  (i) a red giant, (1 mark)
  (ii) a white dwarf, (1 mark)
  (iii) a supernova. (2 marks)

(c) What is the final state of a star after it has passed through the supernova stage? (1 mark)

(d) (i) What is the evidence which suggests that
    (I) galaxies are moving away from us (1 mark)
    (II) the farther away a galaxy is, the faster it is moving? (1 mark)
  (ii) Name the theory of the universe which this supports. (1 mark)
  (WJEC, Jun 95, Higher Tier, Q16)

Question 5

There have been several manned space flights to the Moon but there have been none so far to any of the planets.
  Use your knowledge of space science to explain why.
  (NEAB, Jun 95, Higher Tier)

Answers to examination questions

1. (a) (i) (A) Arrow should point from the Moon towards the Earth
(B) Arrow should point from the Earth towards the Moon.
(ii) The Moon would travel in a straight line without the presence of the Earth. The Earth's gravitational force supplies the centripetal force necessary for the Moon to move in its circular orbit.
(b) (i) Nuclear fusion.
(ii) Light from the Sun is reflected off the surface of the Moon.
(c) (i) Mercury is much closer to the Sun so the amount of radiation per m^2 per second will be higher.
(ii) The greenhouse effect will occur to a greater extent. This means heat radiation (infra-red) will enter the planet's atmosphere but due to a change in the wavelength it will not be re-radiated from the planet's surface causing the planet to warm up.
(d) planet, Sun, solar system, galaxy, universe.
(e) [Any of the following points should be mentioned] The provisions needed for the journey would be too great/distances are so large.
A large amount of fuel would be needed.
Provisions such as water, food and air would be too heavy.

2. (a) A satellite stays in orbit because gravity provides the centripetal force which is needed for the circular motion. The size of the centripetal force, F, is given by

$$F = \frac{mv^2}{r}$$

Where m is the mass of the satellite, v its velocity and r, the radius of its orbit. This force acts inwards towards the centre of the Earth and is provided by the gravitational force mg, where g is the gravitational field strength at the radius of the satellite from the Earth's centre.
(b) Kinetic energy $= \frac{1}{2}mv^2$.
Since the mass stays the same but the velocity doubles and the KE depends on the square of the velocity, doubling the velocity quadruples the kinetic energy.

3. (a) Relative mass of four hydrogen
nuclei $= 4 \times 1.007825 = 4.0313$
Difference in the mass when four hydrogen nuclei form a single helium
nuclei $= 4.0313 - 4.0037 = 0.0276$. This figure is the mass which is lost during the conversion of hydrogen to helium.
This is the mass which is converted into energy during the fusion. This energy is released as heat energy.
(b) A very small amount of mass is released as a huge amount of energy during nuclear fusion. Since the mass of the Sun is large there is an almost unlimitless supply of fuel for the above process.

4. (a) The Sun releases energy by fusion where the nuclei of two smaller atoms are fused together to form a heavier nuclei. The energy comes from the conversion of a small mass difference during the fusion into a huge amount of energy. This energy is given out as heat (infra-red) and visible light.
(b) (i) The star expands because the outward internal forces are greater than the inward gravitational force.
(ii) Here the red giant ejects matter and leaves it core behind and the core is called a white dwarf.
(iii) Here a large mass star has produced a red giant which has then exploded to produce a supernova.
(c) A very small star, called a neutron star is produced.
(d) (i) The light from the distant galaxies is shifted to the red end of the spectrum.
(ii) The more distant the galaxy, the greater the red shift.
(e) The Big Bang Theory.

5. The planets are much further from the Earth than the Moon so the flight would be much longer and more fuel, air, water, food etc. would be needed. The rocket to power such a heavy craft would be large. Costs of such a flight would be high.
There might be medical problems owing to the long time spent in weightlessness.
Most of the planets have either too high or too low temperatures and some have dangerous atmospheres.

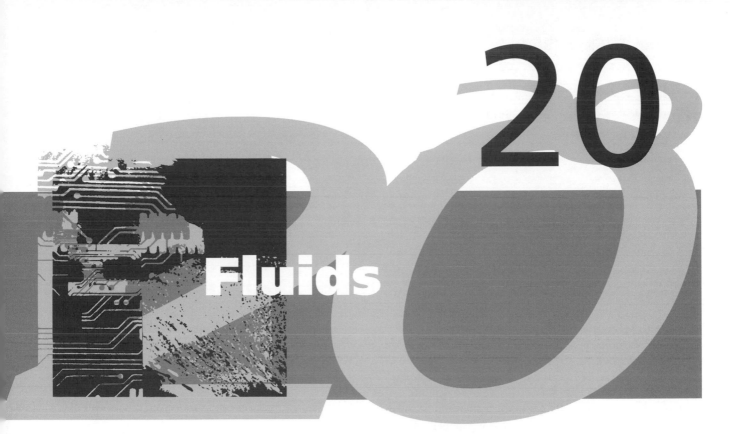

Fluids

20

				Topic	MEG Salters'	MEG Nuffield	SEG	NICCEA
WJEC & NEAB	ULEAC Syll A	ULEAC Syll B	MEG					
				Surface tension and its effects		✓		
				Streamline and turbulent flow		✓		
				Viscosity and its effects		✓		
				Newtonian and thixotropic fluids		✓		
				The Bernoulli principle		✓		
				Principles of flight		✓		

1 Surface tension and its effects

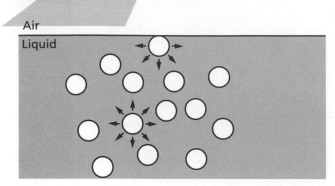

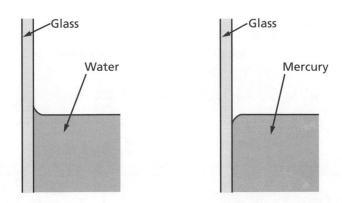

A molecule well inside the liquid is attracted in all directions by the surrounding molecules and the forces tend to cancel each other out so the net force on the liquid molecule is zero. A molecule at the surface is only attracted by those molecules below and to the side of it. Although there may be molecules which have formed a vapour above these they will be fewer so the net force will tend to pull the surface molecule inwards. This means that the surface molecules are more strongly held and the surface behaves as though it is covered by a thin invisible skin.

Surface tension effects

A needle may be floated on water even though the metal is denser than water. The attractive forces between the surface water molecules and the weight of the needle cause the surface to distort.

In the absence of external forces, liquids form spherical droplets owing to the forces acting on the surface molecules acting inwards. The surface tension binds the drop together and keeps it spherical.

Factors which affect surface tension

- The addition of soap, detergents and other contaminants to a pure liquid reduces the surface tension.
- An increase in temperature gives molecules more kinetic energy and gives them energy to escape the attractive forces thus reducing the surface tension.

Adhesive and cohesive forces

Cohesive forces are those forces between like molecules whereas adhesive forces are the forces between unlike molecules.

In the diagrams the water rises upward where it is in contact with the glass whereas the mercury moves downward. The reason for this is that with water the attraction between the water molecules (the cohesive

forces) is smaller than the attractive forces between the water and the glass (the adhesive forces). In the case of mercury it is the other way around.

It is for a similar reason that when split, the water spreads out over a sheet of glass whereas the mercury will gather itself into small globules and does not 'wet' the glass.

Capillary rise

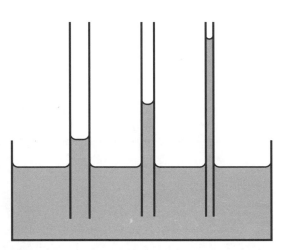

The adhesive forces between the water and the glass act upward on the surface of the water causing it to rise up the tube. As it rises the weight of the water being supported increases until it becomes equal to the upward cohesive force when the rising stops.

Practical effects of surface tension

Blotting paper

The spaces between the fibres in the paper cause capillary action causing the water to be soaked up by the paper.

Washing

Detergents and soap destroy surface tension and this makes it easier for the water to soak into fabric and allow the dirt to pass into the water. Heating also reduces surface tension so a hot wash will clean clothes better than a cold one.

Damp proof courses

Bricks, like the blotting paper, have small air spaces which cause the water to rise up by capillary action and this can produce damp in houses. To prevent this a layer of a waterproof material is included low down in the brick course so that the water cannot rise in this way. This layer is called a damp proof course.

2 Streamline and turbulent flow

Streamlines are curves which follow the direction of the velocity of the fluid particles. When the fluid flow is steady, all the fluid particles passing a certain point will follow identical paths.

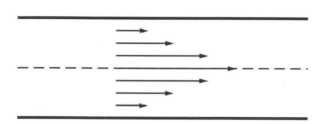

When water flows through a pipe like the one shown, the streamlines are parallel provided that the speed of the fluid is not too great. The velocity of the fluid varies along the streamlines with the velocity greatest at the centre and then decreasing to almost zero next to the walls of the pipe.

If the velocity of the fluid increases there comes a point when the speed is so great that the streamlines are no longer parallel and the velocities and streamlines are variable. When this happens, the flow is called *turbulent flow*. When a solid moves through a fluid turbulence can occur at any sharp edges and it is for this reason that fast sports cars have smooth lines which enables them to cut through the air with very little turbulence. By carefully designing cars the fuel consumption can be reduced significantly.

3 Viscosity and its effects

Viscous fluids (liquids or gases) offer a resistance to any solid trying to move in them. Viscous liquids tend to flow slowly and include liquids like treacle. When solids move through viscous fluids friction between the solid and the liquid converts some of the kinetic energy into heat.

Because the particles are more spaced out in gases and they have much smaller forces of attraction between them, gases have less viscosity than solids.

When a viscous fluid moves, the layers of the fluid slide over each other and the friction between the layers creates the viscosity.

Viscosity can be useful in the case of lubricants which reduce wear between moving parts.

4 Newtonian and thixotropic fluids

Newtonian fluids are fluids whose viscosity stays constant. In contrast, thixotropic fluids are fluids whose viscosity decreases with time when they have a force applied to them. Non-drip paint is an example of a thixotropic fluid. When a force is applied to the paint by the paintbrush being used to apply it, the viscosity decreases and this makes it easy for the paint to be spread evenly over the surface. Once the paint is on the surface there is no longer a force applied so the viscosity increases and this prevents the paint from forming drips.

5 The Bernoulli principle

The Bernoulli principle is a relationship between the pressure and the velocity of a moving fluid. When the velocity of a moving fluid increases, the pressure inside the fluid or the pressure exerted by the fluid decreases. This may be observed by using the apparatus shown in the diagram.

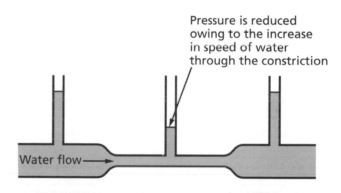

Pressure is reduced owing to the increase in speed of water through the constriction

Water flow

As the water flows through the pipe and meets the constriction, its speed increases and this causes the pressure in this region to decrease, as can be seen by the lower level of water in the manometer.

6 Principles of flight

The Bernoulli principle can be used to explain the force, or lift as it is called, experienced by an aircraft's wing. Examination of the cross-section of the wing reveals that the upper surface of the wing is more curved than the lower surface. This means that air passing over the upper surface moves faster and hence experiences a lower pressure than the bottom surface. It is the greater pressure on the underside of the wing that supplies the lift which is necessary for flight.

Worked examples

Example 1

One event at a summer fair is a plastic duck race along the local river. The ducks, all the same, are let go at the same

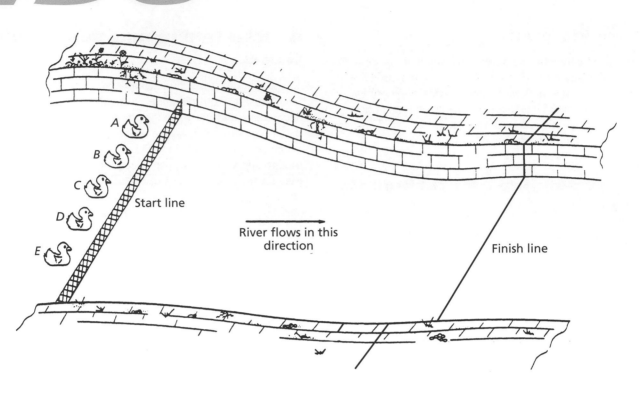

time from different places along a start line. The first plastic duck to cross the finish line is the winner.

(a) (i) Which one of the five ducks should win the race? (1 mark)
     (ii) Why do you think this duck will win? (2 marks)

(b) Someone watching the race notices that when each duck gets close to the finish line it travels faster. Why does this happen? (2 marks)
(SEG, Spec, Intermediate Tier)

Solution

(a) (i) Duck *C* should win.
     (ii) The speed of the river is greatest in the centre since the speed decreases from the centre towards the banks. Also the river is likely to be deeper in the middle which means there will be less friction due to contact with the bottom on the upper layer of water.

(b) The river gets narrower near the finish line. The same volume of water per second has to go through a smaller cross-sectional area and will therefore need to speed up.

Example 2

(a) Describe and explain, in terms of molecules, how the viscosity of syrup changes as its temperature increases. (3 marks)

(b) The diagram below shows some simple apparatus used to investigate viscosity.

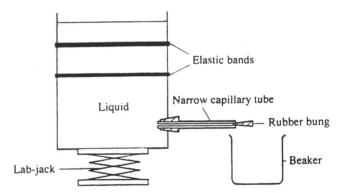

Describe how you would use the apparatus to show experimentally that engine oil and gearbox oil have different viscosities. (4 marks)

Your description should include:

• the quantities you would keep constant
• the measurements you would make
• how would you decide which of the two liquids is the more viscous.

(c) The flow of a gas through a pipe is affected by the viscosity of the gas.

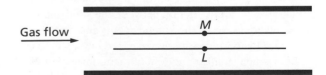

By considering the movement of the molecules in the two layers of gas marked *L* and *M*, explain why viscous forces exist in a gas. (6 marks)
(SEG, Spec, Higher Tier)

Solution

(a) As the temperature increases the molecules of a fluid move further apart so this makes it easier for them to move past each other so the viscosity decreases and the liquid flows more easily.

(b) The following were kept constant:
1 the temperature of the two liquids
2 the distance between the elastic bands
3 the height of the lab-jack
The bung is removed and a stop clock started. The liquid was timed and the time taken for the level to reach the lower elastic band was recorded. The experiment was repeated for the other liquid and the liquid which takes the longest time to travel the distance between the two elastic bands is the most viscous.

(c) L moves faster than M. Molecules in L may cross over to the slower layer M and slower molecules in M may move to the faster layer L but the total number of molecules in each layer remains constant. However, since the molecules swapped between layers have different velocities L will slow down slightly and M will speed up slightly. This means there is a change in momentum in each layer so a viscous force has acted on the layers.

Example 3

(a) Water flows through the pipe shown in the diagram.

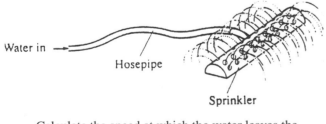

Describe and explain why the speed of the water entering the pipe is different to the speed of the water leaving the pipe. (3 marks)

(b) A garden hosepipe with a water sprinkler attached is connected to a tap which supplies water at the rate of 50 cm³/s. The sprinkler has a total of 25 holes, each 0.01 cm² in area.

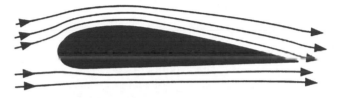

Calculate the speed at which the water leaves the sprinkler. (4 marks)
(SEG, Spec, Higher Tier)

Solution

(a) Since the volume flow per second into the pipe must equal the volume flow per second out of the pipe, and the area of the outlet is smaller than the area of the inlet, the water must speed up.

(b) Area of the outlet = $25 \times (0.01 \text{ cm}^2) = 0.25 \text{ cm}^2$
If v is the speed of the water in cm/s leaving the sprinkler, the volume output per second = $0.25 \times v \text{ cm}^3/\text{s}$.

Volume per second in = volume per second out
$$50 \text{ cm}^3/\text{s} = 0.25 \times v \text{ cm}^3/\text{s}$$
Giving v = 200 cm/s

Example 4

(a) The four main forces acting on an aircraft in flight are thrust, drag, lift and weight.
The diagram below shows the flow of air past a moving aircraft wing.

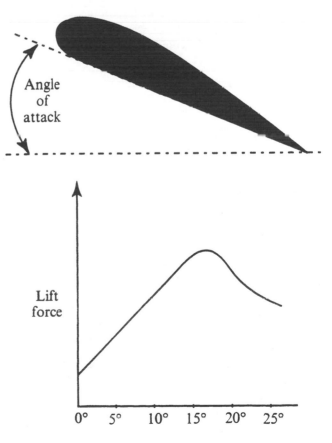

Explain how the shape of the wing results in a lift force acting on the aircraft.
You will be awarded up to two marks if you write your ideas clearly. (6 marks)

(b) When an aircraft flies, its wings are pushed into the air at an angle called the angle of attack. The graph shows how the angle of attack could affect the lift force on an aircraft wing when flying at constant speed.

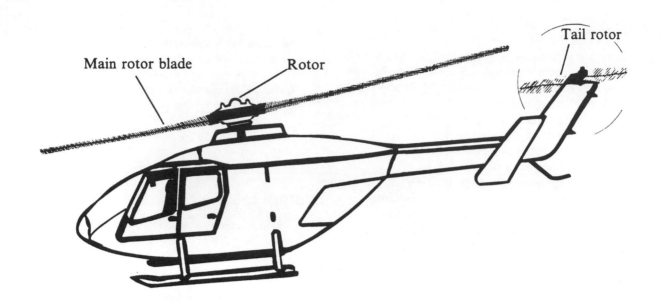

Main rotor blade   Rotor   Tail rotor

The pilot of an aircraft wants the aircraft to go faster but stay at the same height above the ground. Explain with reasons what the pilot must do. (4 marks)

(c) The helicopter below is shown in forward flight. Describe briefly the function of each of the following parts: (3 marks)
   (i)   main rotor blades
   (ii) rotor
   (iii) tail rotor

(SEG, Jun 95, Higher Tier)

Solution

(a) The curve on the upper part of the wing causes the air to go further over the top surface of the wing. Since it has to meet up with the air travelling the smaller distance below the wing it has to move faster over the top surface. The pressure above the wing is reduced compared with the bottom. A force acts from the high to low pressure and this gives the lift force.

(b) They need to increase the power supplied by the engines by opening the throttle which will increase the forward force and cause the aircraft to accelerate. At the same time, since increasing the speed will increase the lift force, to stay at the same height it will be necessary to lower the angle of attack by lowering the nose. This will decrease the lift force and ensure that the aircraft stays at the same height.

(c) (i)   The main rotor blades supply the lift force.
     (ii) The rotor tilts the blades so that a forward thrust can be supplied to the blades to drive the helicopter forward.
     (iii) It prevents the body of the helicopter from rotating.

Examination questions

(Numerical answers and hints on solutions will be found at the end of the chapter.)

Question 1

1 Look at the figure opposite showing two tubes A and B.
   (a) The diagrams show capillary action in two liquids; mercury and water.
       Which liquid is being used with tube A?
   (b) Explain why one of the liquids rises whilst the other one falls.
   (c) When houses are built a waterproof membrane is placed low down in the brick course called a damp proof course. Explain the reason why this is necessary.

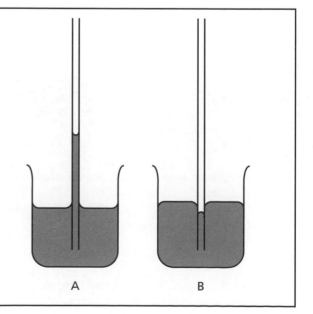

A   B

Question 2

(a) New cars are tested in a wind tunnel to see if the air flow is streamlined. Instead of the car moving through the air, the air is moved past the car.

The diagram shows the air flow produced around the two different cars.

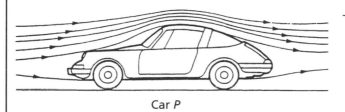

Car P

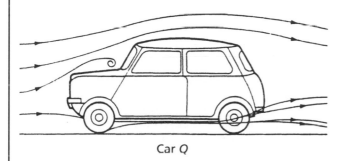

Car Q

(i) Which one of the two cars P or Q is the more streamlined? (1 mark)

(ii) Mark on one of the diagrams a X to show the position of an eddy. (1 mark)

(iii) Why is it important for the designers of cars to try and make their cars streamlined? (1 mark)

(b) The diagrams below are of different animals labelled A to I.

(i) Which of the animals above have a streamlined shape? (1 mark)

(ii) Choose **one** of the animals and describe how the streamlined shape is important to its survival. (2 marks)

(c) A 'hydrofoil' is a boat in which the main section is lifted out of the water, in a similar way to an aircraft rising in the air.

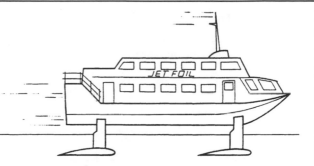

The diagram shows that as the hydrofoil moves forward, a streamlined flow of water passes the underwater aerofoils.

(i) Explain how the lift force on the hydrofoil is produced. (4 marks)

(ii) What happens to the size of the lift force as the speed of the hydrofoil increases? Explain your answer. (2 marks)

(SEG, Spec, Intermediate Tier)

Question 3

(a) In terms of the molecules in water, explain why the molecules at the surface of a liquid such as water are more tightly held than those in the main body of the liquid.

(b) The figure below shows water rising in two capillary tubes of different diameter.

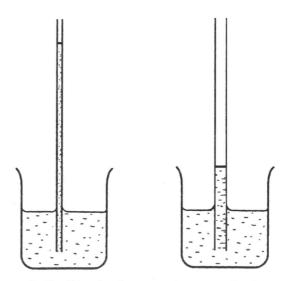

(i) Explain why the water rises up each tube.

(ii) Explain why the water rises further up the tube with the smaller diameter.

(c) A force between molecules may be either *adhesive* or *cohesive*. Explain the two words in italics.

Question 4

(a) The graph below shows the relationship between the volume of gas flowing through a pipe every second and the pressure difference between the ends of the pipe. The pipe has an internal diameter of 8 mm.

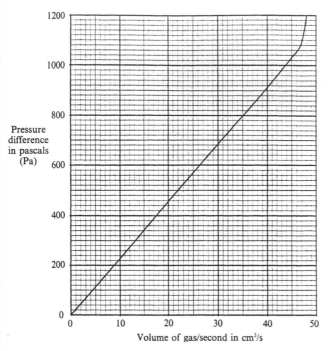

(i) Calculate the average speed of the gas flowing through the pipe when the pressure difference between the ends of the pipe equals 800 Pa. Show clearly how you obtain your answer.
(2 marks)

(ii) Even with a constant pressure difference between the ends of the pipe the gas within the pipe will flow with a range of speeds. Explain why. (2 marks)

(b) The diagram below shows part of a weather map. The isobars join points of equal atmospheric pressure. The arrows show the direction of the wind.

(i) The wind does not follow a straight path from high to low pressure. Why? (1 mark)

(ii) How and why will the windspeeds at point X and point Y on the map be different?
(2 marks)

(iii) The air pressure at X will increase as the cold front passes over it. Explain in terms of the air molecules why this happens. (2 marks)
(SEG, Jun 95, Higher Tier)

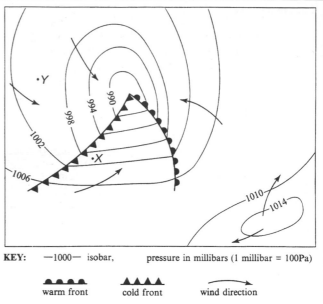

KEY: —1000— isobar, pressure in millibars (1 millibar = 100Pa)

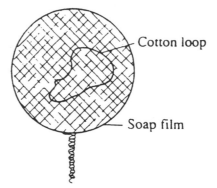

warm front   cold front   wind direction

Question 5

(a) A light cotton loop can rest on a soap film which has been formed on a metal ring.

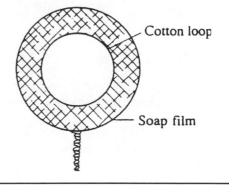

(i) Draw arrows on the diagram to show the forces acting on the thread. (1 mark)

(ii) What name is given to the forces acting on the cotton thread? (1 mark)

(iii) Why do these forces seem to have no effect on the surface of the soap film? (1 mark)

(b) The diagram below shows what happens when the soap film inside the thread is punctured, and the thread is drawn out into a circle. Explain why this should happen. (1 mark)

(c) Explain in terms of the forces between molecules why the surface of a liquid behaves like a stretched elastic skin that is always trying to shrink. Your answer should include a diagram. (6 marks)
(SEG, Spec, Higher Tier)

Question 6

Fluid flow can be described as *Newtonian* or *thixotropic*.

(a) Explain the meaning of each word when applied to fluid flow. (4 marks)
(b) Many paints are thixotropic in their behaviour. Explain how this helps the painter to apply the paint evenly. (2 marks)
(SEG, Spec, Higher Tier)

Answers to examination questions

1 (a) Water
(b) In the case of water, the forces between the water molecules (i.e. the cohesive forces) are attractive and the forces between the water and the glass molecules (i.e. the adhesive forces) are also attractive but larger so the net effect is for the water to move up the tube.
In the case of mercury, the cohesive forces are larger than the adhesive forces and this means that mercury molecules are attracted into the body of the liquid.
(c) It stops the moisture in the ground rising through small gaps in the brick by capilliary action thus preventing the damp rising and causing mould inside.

2. (a) (i) *P*
(ii) You need to mark on the diagram where the flow lines start to curl near the windscreen of the car.
(iii) Less drag means more efficient (i.e. more miles per gallon).
(b) (i) *I*, *C* and *A* (dolphin, fish and bird) all have a streamlined shape.
(ii) *I* (dolphin). Enables them to swim faster in water to catch food. They are also able to use less energy.
(c) (i) The water travels further over the top surface and therefore needs to travel faster to be able to meet up with the water travelling under the hydrofoil. The pressure is therefore less on the top than on the bottom of the

hydrofoil. The pressure difference causes the lift force.
(ii) The pressure difference increases so the lift force will increase.

3 (a) See section 1. The diagram in this section should be drawn to aid your explanation.
(b) (i) See the first part of the answer to question 1(b).
(ii) There is a net force on the molecules which cause the water to move upward but as it does so, the downward force of the weight of the water rises until it equals the upward force and equilibrium is achieved. The water now remains at a certain height in the tube.
(c) Adhesive forces are those forces of attraction between like molecules whereas cohesive forces are those attractive forces between unlike molecules.

4. (a) (i) From the graph, volume per second
= 35 cm³/s
Cross-sectional area of pipe
$= \Pi \times (0.4)^2 = 0.503$ cm²

$$\text{Speed} = \frac{\text{Volume per second}}{\text{Area}}$$

$$= \frac{35}{0.503} = 69.6 \text{ cm/s}$$

(ii) There are viscous forces between the layers of gas owing to the fact that the speed of the gas is zero touching the pipe to a maximum speed at the centre of the pipe.

(b)
  (i) Because of the rotation of the Earth.

  (ii) Speed at X is greater owing to the fact there is a larger pressure gradient shown by the isobars being more closely packed.

  (iii) Cold air has a greater density than warm air and this means the molecules are more closely packed. This results in there being a greater number of collisions between molecules per unit time thus making the pressure greater.

5. (a)
  (i) All the forces are balanced so for every arrow drawn there should be another drawn in the opposite direction.

  (ii) Surface tension.

  (iii) Because they act equally on both sides of any line in the surface.

(b) The forces pulling the thread outwards are no longer counterbalanced.

(c) See section 1.

6. (a) Newtonian fluids are fluids whose viscosity stays constant when a force is applied to them. Thixotropic fluids have a viscosity which decreases with time when they have a force applied.

(b) See section 4.

SOME BASIC UNITS

Unit and symbol	Quantity measured and usual symbol	Comments
second (s)	time (t)	The unit of time. 60 s in 1 minute.
metre (m)	length, distance (l, s)	Approximately the length of a good-sized stride.
kilometre (km)	length, distance (l, s)	1 km = 1000 m (a bit more than half a mile).
kilogram (kg)	mass (m)	The mass of the average bag of sugar is about 1 kg.
newton (N)	force (F)	The pull of the Earth (weight) on an apple of average size is about 1 N.
pascal (Pa)	pressure (P)	1 Pa = 1 N/m^2. The pressure exerted when you push hard on a table with your thumb is about 1 million pascals.
joule (J)	energy (E)	1 J = 1 Nm. About the energy needed to place an apple of average size on a shelf 1 metre high. 4200 J (specific heat capacity) is needed to raise 1 kg of water through 1 K. About 2 million joules (specific latent heat) is needed to boil away 1 kg of water.
watt (W)	power (p)	1 W is a rate of working of 1 J/s. It is also the energy produced every second when 1 V causes a current of 1 A. Household mains lamps are usually between 40 W and 100 W. Watts = volts × amperes.
degree Celsius (°C)	temperature (t, θ)	The temperature of water changes by 100 °C when going from melting point to the boiling point.
kelvin (K)	temperature (T)	A temperature change of 1 kelvin is the same as a temperature change of 1 degree C.
hertz (Hz)	frequency (f)	1 Hz is one cycle per second. BBC radio broadcasts are about 1 MHz; VHF about 90 MHz.
ampere (A)	current (I)	The current in most torch and household bulbs is between 0.1 A and 0.4 A.
volt (V)	potential difference (V)	Many cassette players and torches use batteries which are 1.5 V. The mains voltage is 240 V.
ohm (Ω)	resistance (R)	A p.d. of 1 V across 1 Ω produces a current of 1 A. A torch bulb has a resistance of about 10 Ω. $V = IR$.
m/s	speed (v)	45 miles per hour is about 20 m/s. The speed of light is 3×10^8 m/s.
m/s^2	acceleration (a)	Objects falling on Earth accelerate at about 10 m/s^2. Family cars accelerate at about 2 m/s^2.

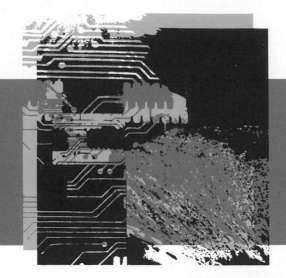

Index

A
Acceleration 14
Air masses 213
Amplitude modulation 203
Analogue signals 203
Atomic structure 179
Atoms and molecules 49

B
Bernoulli principle 235
Biomass 78
Bistables 148
Boiling and melting point 60
Bridge rectifier 150
Brownian motion 49

C
Camera 92
Capacitors 148
Capillary rise 234
Cathode ray oscilloscope 125
Charge 125
Circular motion 15
Climate 212
Communication 202
Conduction 67
Conductors and insulators 125
Convection 67
Critical angle 90

D
Days 224
Density 2
Diffraction 106
Digital signals 203
Dynamo 165

E
Earthing 135
Earthquakes 216–17
Earth's structure 211
Eclipses 225
Efficiency 33,42
Electrical circuits 133
Electrical energy 135
Electrical power 135
Electrical safety 135
Electromagnetic spectrum 105–6
Electromagnetism 164
Emission and absorption of photons 196
Energy changes 31
Energy, forms of 30
Energy levels 196
Equations of motion 15
Equilibrium 41
Eye 92

F
Faraday's law 165
Fission 180
Flight, principles of 235
Floating and sinking 3
Flow of fluids 235
Fluids 234–5
Fossil fuels 76
Fronts 214

Fusion 182

G
Gas laws 50
Gates 148
Geothermal energy 77
Graphs of motion 14
Gravity 223
Greenhouse effect 76

H
Half-life 180
Heat 60
Heat transfer 67
High pressure regions 212–13
Hooke's Law 3
Hydraulics 5
Hydroelectric power 78

I
Insulation in the home 67
Interference 106
Isotopes 179

K
Kinetic energy 31
Kinetic theory 49

L
Left-hand-rule 165
Lenses 91
Longitudinal waves 105
Low-pressure regions 212–13

M
Machines 42
Magnetism 164
Mass 2
Mirrors, curved 92
Moments 41
Momentum 14
Motor 165

N
Newton's laws of motion 14
Noise 105
Non-renewable resources 76
Nuclear fuel 76
Nucelon number 179

O
Ohms law 133
Optical fibres 91, 202

P
Parallel circuits 133–4
Phases of the Moon 214, 225–6
Photoelectric effect 195
Planets 224
Plate tectonics 212
Polarisation 107
Potential divider 134
Potential energy 31
Power 33
Pressure 2
Projectiles 15

Projector 92
Proton number 179
Pulleys 42
Pumped storage schemes 79

R
Radiation, heat transfer by 67
Radioactive decay 180
Radioactivity, safety 180
Radioactivity, uses of 180
Red shift 229
Reflection 90
Refraction 90
Refrigerator 60
Relays 148
Renewable resources 77
Resonance 105
Right-hand-rule 165
Ring main 134
Rock types 211

S
Satellites 202, 233
Scalar 14
Seasons 224
Series circuits 133
Soil formation 211
Solar panel 77
Sound 105
Specific heat capacity 60
Specific latent heat 60
Spectra, line and continuous 195
Stability 41
Stars 226
Subduction 217–18
Surface tension 234

T
Telescope 92
Temperature 60
Terminal velocity 15
Thermos flask 67
Tidal power 78
Transducers 148
Transformers 165
Transistors 148
Transverse waves 105

U
Units 2
Universe 223, 226

V
Vector 14
Velocity 14
Viscosity 235

W
Water cycle 212
Wave power 78
Weight 2
Wind power 78
Work 32

X
X rays 185